The Humanistic Background
of Science

SUNY series in American Philosophy and Cultural Thought
―――――
Randall E. Auxier and John R. Shook, editors

> March 27, 1963
>
> Dr. Phillip Frank
> 1558 Massachusetts Avenue
> Cambridge, Massachusetts
>
> Dear Dr. Frank:
>
> Tucked away in compartment 521, which you were using for a period, I have discovered some papers which I suspect are yours. There are a few letters addressed to you, but the majority of the papers consist of a manuscript having to do with the philosophy of science.
>
> If you are interested, I will be pleased to have these documents wrapped up and mailed to you.
>
> Sincerely yours,
>
> Philipp
>
> RHS:nd

A letter, found in Philipp Frank's papers, about the manuscript for *The Humanistic Background of Science*. Image courtesy of Harvard University Archives.

The Humanistic Background of Science

Philipp Frank

Edited by George A. Reisch and Adam Tamas Tuboly

over, *The Unicorn is in Captivity and No Longer Dead*, one of the series of seven tapestries *The Hunt of the Unicorn* circa 1495 and circa 1505. Gift of John Davison Rockefeller, Jr. (January 29, 1874–May 11, 1960), 1937.

Published by State University of New York Press, Albany

© 2021 State University of New York

All rights reserved

Printed in the United States of America

No part of this book may be used or reproduced in any manner whatsoever without written permission. No part of this book may be stored in a retrieval system or transmitted in any form or by any means including electronic, electrostatic, magnetic tape, mechanical, photocopying, recording, or otherwise without the prior permission in writing of the publisher.

For information, contact State University of New York Press, Albany, NY
www.sunypress.edu

Library of Congress Cataloging-in-Publication Data

Names: Frank, Philipp, author. | Reisch, George A., editor. | Tuboly, Adam Tamas, editor.
Title: The humanistic background of science / by Philipp Frank ; edited by George A. Reisch and Adam Tamas Tuboly.
Description: Albany : State University of New York Press, [2021] | Series: SUNY series in American philosophy and cultural thought | Includes bibliographical references and index.
Identifiers: LCCN 2021030164 (print) | LCCN 2021030165 (ebook) | ISBN 9781438485515 (hardcover : alk. paper) | ISBN 9781438485522 (pbk. : alk. paper | ISBN 9781438485539 (ebook)
Subjects: LCSH: Science—Philosophy. | Science and the humanities. | Frank, Philipp, 1884–1966.
Classification: LCC Q175 .F779 2021 (print) | LCC Q175 (ebook) | DDC 501—dc23
LC record available at https://lccn.loc.gov/2021030164
LC ebook record available at https://lccn.loc.gov/2021030165

10 9 8 7 6 5 4 3 2 1

Contents

List of Illustrations — xi

Chronology of Philipp Frank's Life — xiii

Philipp Frank: A Crusader for Scientific Philosophy — 1

Part I

Chapter 1. Introduction: Science, Facts, and Values — 71
1. Science and Poetry — 71
2. Charges against the Monopoly of Science — 73
3. Twentieth-Century Science and Philosophy — 76
4. The "Real World" Is Not Describable — 78
5. The Humanities Are Trailing behind the Natural Sciences — 80
6. The "Special Sciences" Don't Exhaust "Science" — 82
7. Semantic and Pragmatic Components of Science — 83
8. Philosophical Schools Woo the Support of Science — 86
9. Principles of Science and Human "Values" — 88

Chapter 2. The Longing for a Humanization of Science — 93
1. Dissatisfaction with Nineteenth-Century Science — 93
2. Emerson on the Changing Role of Science — 94
3. Lord Herbert Samuel for Modern Science — 96
4. Dehumanization of Science — 98
5. Soviet Philosophy and Modern Science — 99
6. The Birth of Modern Science Was the Birth of Dissatisfaction — 100

7. Bacon on the Copernican System 102
8. How Science Has Been "Humanized" 105
9. Analogies as Humanizing Elements 107
10. "Humanization," "Metaphysics" and the "Inner Eye" 108
11. Metaphysics, Common Sense, and the Inner Eye 110
12. The Nature of Metaphysical Statements 112
13. The Inner Eye and Intuition 113

Chapter 3. Metaphysical Interpretations of Science 115
1. The Founder of Pragmatism on Science and Philosophy 115
2. Peirce's Conception of Philosophy 116
3. Metaphysics Nearer to Common Sense than Science 117
4. The Purpose of Metaphysical Interpretation 121
5. Metaphysics as Science 123
6. The Laws of Physics and Their Metaphysical Interpretation 125
7. How Scientists Have Interpreted Their Own Theories 128

Chapter 4. The Sociology of Metaphysical Interpretations 131
1. Can Science Be "Purged" of Philosophy? 131
2. Science and Chance Philosophies 133
3. The Attitudes of Scientists and Authorities 135
4. The Battle of Worldviews 138
5. Purging Physics and Metaphysics 141
6. Science and Reality 143
7. Max Planck and the Real World 145
8. Meanings and Examples of "Real" 146
9. Sociological Role of "Reality" 148
10. "Reality" in Soviet Philosophy 151

Chapter 5. Philosophy of Science and Political Ideology 155
1. Sociology of Knowledge 155
2. The General Sense of Ideology 157
3. Mannheim, Ideology, and Sociology of Knowledge 158
4. Forms of Social Influence 160
5. Facts and Interpretation 161
6. Sociology of Science 163
7. Social Class and Social Situation 166
8. The Solution to the Puzzle 168

Chapter 6. Sociology of Science and the Search for a
Democratic Metaphysics   171
 1. Validation and Theory Building   171
 2. Science as a Compromise between Technology and
    Political Philosophy   172
 3. The Scientific Conscience   176
 4. Philosophical Interpretations and Democracy   178
 5. The Physical and the Socio-cosmic Universe   184

Part II

Chapter 7. Scholastic Philosophy and Thomism   189
 1. The Meanings of Rational and Intelligible   189
 2. The Role of Philosophical Schools   191
 3. Science and "Thomism"   192
 4. The Thomistic Theory of Matter   194
 5. The Social Significance of Thomistic Philosophy   197
 6. On Angels and Genuine Laws   199
 7. Thomism and Physical Laws   201
 8. Analogical and Scientific Thinking   204

Chapter 8. The Physical Universe as a Symbol   209
 1. The Moral Universe   209
 2. Physical Science in the Bible   211
 3. The Physical Universe and Human Behavior   213
 4. Scholastic "Scientism" and Modern "Positivism"   215
 5. Shifting the Problem to Revelation   217
 6. Realism and Nominalism   221
 7. The Situation in the Nineteenth and Twentieth Centuries   222

Chapter 9. Union, Divorce, and Reunion between Science
and Philosophy   227
 1. Science and Philosophy in the British and Soviet
    Encyclopedias   227
 2. "Truce" through a Naturalization of Science   229
 3. Attempts at a Reunion by a Positive Philosophy   231
 4. The Role of "Sociology" in Positive Philosophy   234

5. The "Truth" of General Principles in Positive Philosophy   236
6. The Relative Truth of Theories   238
7. Positive Philosophy and Marginal Metaphysics   240
8. Science and Philosophy after the Reunion   242
9. The Name "Philosophy" as a Challenge   244

Chapter 10. Science, Democracy, and the New Wave of Positivism   247
1. Science after the French Revolution   247
2. Positivism in the Second Half of the Nineteenth Century (Stallo)   248
3. Positivism in the Second Half of the Nineteenth Century (Mach)   251
4. The Reception of Mach and Stallo?   253
5. Conventionalism (Poincaré, Le Roy)   255
6. Abel Rey and the Bankruptcy of Science   256
7. Duhem's Accommodation of Positivism and Metaphysics   258

Chapter 11. The Vienna Circle: Moritz Schlick, Rudolf Carnap, and Otto Neurath   261
1. The Turning Point in Positivism   261
2. Logical Positivism and the Theory of Correspondence   263
3. Philosophy as Activity and the Unified Picture   265
4. Cross-connections among the Sciences   267
5. Changes in the Science of Meaning   268
6. The Vienna Circle and the Pragmatics of Metaphysics   270
7. Cognitive Significance and Scientific Value   272

Chapter 12. Pragmatism   275
1. Pragmatism (William James, Charles S. Peirce, and John Dewey)   275
2. Peirce's Pragmatism and Positivism   276
3. James's Pragmatism and Metaphysics   278
4. Dewey and Political Interpretations of Science   279
5. A New Development: Scientific Empiricism   281
6. The Meaning and Significance of Bridgman's Operationalism   283
7. Nagel's Contextualistic Naturalism   287

Contents ix

Chapter 13. Mechanistic and Dialectical Materialism 289
1. Mechanistic Materialism 289
2. La Mettrie's Materialism 290
3. Purposiveness in Nature 292
4. Materialism Refuted? 294
5. Materialism versus Positivism 296
6. Soviet Attacks against Positivism 299
7. The Conversion of Mass and "Star-Spangled" Operationalism 301

Chapter 14. The Laws and Politics of Dialectical Materialism 305
1. Dialectical versus Mechanistic Materialism 305
2. Diamat and Philosophy 307
3. Diamat and Realism 308
4. The Dialectical Laws 310
5. Quantitative and Qualitative Changes 311
6. Social Change and Natural Science 313

Conclusion: Einstein's Philosophy of Science 317
1. The Positivistic Basis 317
2. The Metaphysical Basis 319
3. The Analogical-Religious Basis 321

Notes 323

Bibliography 361

Index 375

Illustrations

Figure I.1	The young Philipp Frank, taken presumably in the 1910s, as depicted on his *carte de visite*.	5
Figure I.2	In the woods of Czechoslovakia, Philipp Frank sits second from the right.	10
Figure I.3	The Fifth Meeting of the German Physicists and Mathematicians, with Philipp Frank standing fourth from the left.	13
Figure I.4	Philipp Frank, Rudolf Carnap, Susan Stebbing, Heinrich Neider, Carl G. Hempel, Eva Hempel, Jørgen Jørgensen, and Uuno Saarnio (likely photographed in the Belgian Ardennes, 1935).	17
Figure I.5	Brochure advertising Frank's American lecture series, October–November 1938.	19
Figure I.6	Philipp Frank and Rudolf Carnap in America.	26
Figure I.7	Undated photograph of Philipp and Hania Frank presiding at a "Vienna Coffee House" at their home in Cambridge.	35
Figure I.8	Philipp Frank and P. W. Bridgman's joint retirement conference, 1954.	35
Figure I.9	A representative page from Philipp Frank's manuscript.	55

Chronology of Philipp Frank's Life

1884	Born on March 20, 1884 in Vienna, Austro-Hungary.
1902–1903	Enrolled at the University of Vienna (Winter Semester).
1906	Defended his PhD dissertation at the University of Vienna, "On the Criteria of Stability of the Motion of a Material Point and its Relation to the Principle of Least Action."
1907–1912	Participated in an informal discussion group, the "First Vienna Circle," with Hans Hahn and Otto Neurath in Viennese coffeehouses.
1909	Defended Habilitation thesis at the University of Vienna.
1910–1912	*Privatdozent* at the University of Vienna.
1912–1938	Professor of Theoretical Physics at the Department of Physics, German University of Prague; in time he became the head of the Department and developed a flourishing institute.
1928	Lectured in the Soviet Union
1929	Organized the first conference on *Erkenntnislehre der exakten Wissenschaften* (Epistemology of the Exact Sciences), Sept. 15–17, 1929.
1924–1936	Participated occasionally in meetings of the Vienna Circle under Moritz Schlick's leadership.
1939–1941	Visiting lecturer in mathematics and physics at Harvard's Department of Theoretical Physics.

1941–1954 Lecturer at Harvard's Department of Theoretical Physics. Teaches also at City College of New York and Radcliffe College (beginning in 1941), Brown University (1944), Purdue University (1947).

1940–1949 Conference on Science, Philosophy and Religion, New York City.

1941 Between Physics and Philosophy (Harvard University Press), a collection of his most important philosophical papers, translated into English.

1946 Contributes "Foundations of Physics" to Otto Neurath's *International Encyclopedia of Unified Science*.

1947–1966 The Institute for the Unity of Science, Boston.

1947 Einstein, His Life and Times (Knopf).

1949 Modern Science its Philosophy (Harvard University Press), an updated collection of papers.

1950 Relativity—A Richer Truth (Beacon), a short book of his talks presented at the *Conference on Science, Philosophy and Religion*.

1950 Lectured in Italy and England.

1954 Retired from Harvard University.

1954 Edited The Validation of Scientific Theories, containing his "The Variety of Reasons for the Acceptance of Scientific Theories."

1957 Philosophy of Science: The Link Between Philosophy and Science (Prentice-Hall).

1962 Interviewed by Thomas Kuhn about the history of physics.

1965 Moved to a nursing home with his wife, Hania Frank.

1965 Boston Studies in the Philosophy of Science, vol. 2, dedicated to Frank as a *Festschrift*.

1966 Dies July 21, 1966, in Cambridge, Massachusetts.

Philipp Frank

A Crusader for Scientific Philosophy

GEORGE REISCH AND ADAM TAMAS TUBOLY

Philipp Frank was an accomplished physicist and philosopher. He was a biographer of Einstein, Einstein's successor to the chair of the Department of Physics in Prague, a member of the Vienna Circle, a fixture in philosophical life at Harvard University, and—to some extent—in the intellectual life of the postwar United States. Yet, for various sociocultural and philosophical reasons, Frank and his writings did not enter the mainstream nor the canon of twentieth-century philosophy of science. He is known usually—and simply—as Einstein's biographer and, sometimes, as a logical empiricist[1] who belonged to the Vienna Circle. Despite the extent and variety of Frank's work, he has been forgotten.

To help revive Frank's significance and to reconsider his roles in philosophy and history of science, we offer this book, *The Humanistic Background of Science*, a book we believe Frank intended to publish but that lay unpublished in the archives for more than a half century. To put the manuscript in context, we offer here an overview, both biographical and philosophical, of Frank's life that pays special attention to his

---

1. Throughout the text, we use "logical positivism" and "logical empiricism" (and their inflected variants) interchangeably. While internal to their original uses they marked some intentional differences, from the viewpoint of the story told in the introduction they do not matter. For some more details about this issue, see Uebel (2013).

life in America. We do not claim that Frank's mature years were more important in forming his philosophical oeuvre. But we do believe that *The Humanistic Background of Science*, while its intellectual roots extend to Europe, should be understood largely as a product of Frank's professional and intellectual circumstances in the United States.

In section 2, we attempt to date the manuscript. This is required because the manuscript itself is not dated and provides only indirect clues. In section 3, we examine the philosophical and intellectual context of Frank's manuscript. We discuss the main theses and approach of *The Humanistic Background of Science* in its American context, in relation to its potential influence and contemporary significance, and finally in relation to Thomas Kuhn, the celebrated author of *The Structure of Scientific Revolutions*. Frank's relation to Kuhn is an important but largely unexplored area in the history of philosophy of science. Finally, in section 4, we describe the editorial process and the challenges we faced in presenting Frank's book in a form that is not only readable but interesting, challenging, and potentially fruitful.

1. Vienna—Prague—Boston: The Life of Philipp Frank

Frank played an important role in developing the Vienna Circle's scientific world conception in Vienna and later in Prague with Rudolf Carnap. He disseminated the ideas of logical empiricism and modern scientific thought to laypeople and continued this task in the United States through his institutionalization of Otto Neurath's unity of science movement and his many publications. His friend in America, the philosopher of science Paul Feyerabend, remembered, "Philipp Frank was a delight. He was widely informed, intelligent, witty, and excellent raconteur. Given the choice of explaining a difficult point by means of a story or of an analytical argument, he would invariably choose the story. Some philosophers didn't like that" (Feyerabend 1995, 103).

Frank's career may be divided into three phases, characterized by different persons and places as well as fundamental ideas and commitments: (1) The early 1900s, until 1912, in Vienna; (2) 1912–1938, in Prague; and (3) 1938–1966, in the United States, primarily Boston and Harvard University.

1.1. Vienna: A City That Breathed Physics and Philosophy of Science

Philipp Frank was born on March 20, 1884, in Vienna, then part of the Austro-Hungarian Empire His father, Ignaz Frank, originally from Heves in Hungary, was a textile merchant. He and his wife, Jenny Frank, had four children: Philipp was the oldest, followed by a younger sister, Hedwig, and two younger brothers, Rudolf and the famous architect Jozef.[2]

Frank studied mathematics and physics at the Universities of Vienna and Göttingen, where his teachers included Ludwig Boltzman, Felix Klein, and David Hilbert. He earned a doctorate in physics in 1906 at the University of Vienna and habilitated with a paper in physics in 1909 to become a private lecturer (*Privatdozent*) until 1912 (see figure I.1).

Many years later upon Frank's death, his student—the physicist Jeremy Bernstein (1966, 24)—memorialized Frank by saying that modern physics and its "ideas were part of his instinct." Bernstein's exaggeration was appropriate for the memorial meeting at which he spoke, but he was on to something substantial, for fin de siècle Vienna was perfect for anyone interested in the special sciences and the philosophical and foundational questions raised by their rapid progress.[3] When he interviewed Frank for the "Oral Histories" series of the American Institute of Physics, Thomas Kuhn asked Frank about his intellectual development, his student years, his work in Prague, and his connections to Boltzmann, Mach, Einstein, Schrödinger, and others. Frank called Prague at that time "a big school of physics."[4] Though Ernst Mach had by then retired, Frank studied with the equally important Ludwig Boltzmann. And his classmates, friends, and teachers included Erwin Schrödinger, Hans Thirring, Paul Ehrenfest, Felix Ehrenhaft, Friedrich Hasenöhrl, Karl Herzfeld, and Franz Exner—all of whom were, or would become, international leaders of their fields in mathematics and physics.

---

2. On Jozef Frank's relation to the Vienna Circle and philosophy, see Thurm-Nemeth (1998).

3. On the background and various traditions of physics and philosophy in Vienna, see Stöltzner (1999) and (2003).

4. Interview of Philipp Frank by Thomas S. Kuhn on 1962 July 16, Niels Bohr Library & Archives, American Institute of Physics, College Park, MD USA, www.aip.org/history-programs/niels-bohr-library/oral-histories/4610. Hereafter "Frank/IAP."

In 1895, Mach had been appointed to a chair in philosophy in Vienna, the same chair that would be occupied some two decades later by Moritz Schlick. Mach was then engaged in scientific and philosophical debates, one of which famously concerned the legitimacy and existence of atoms. To see the relevance of this to Frank's thinking, we need not go into the details of Mach's views and the specific nature of his so-called antiatomism. More important were the stratification and complexity of Mach's views about atomism, which included his ideas about the nature and economy of science and his understanding of theory building and experimentation. After Mach's retirement in 1901 (due to health conditions) his chair was occupied by Boltzmann, a well-known theoretical physicist who worked on statistical mechanics and, like Mach, had philosophical leanings. He accepted and publicly defended the atomistic theory of matter, though not for simple-minded or naïve-realist reasons. In fact, Boltzmann agreed that economic reasoning plays an important role in the work and nature of science; but he weighted his values and experiential data differently than Mach. Nonetheless, their debate over the nature of the atom's legitimacy shaped the history of philosophy of science.

Frank was raised scientifically in this atmosphere. When Kuhn asked him whether Mach's influence had "vanished in so far as skepticism about the atom was concerned," Frank recalled the situation as follows:

> No, it did not vanish. There was always this interesting point: what was the relation between Mach and Boltzmann[?] [In fact] Boltzmann was himself, philosophically speaking, rather a follower of Mach. Boltzmann said once to me, "You see, it doesn't make any difference to me if I say that all the atoms are only a picture. I don't mind, this. I don't require that they are absolute (rules). I don't say this." " 'An economic, description,' Mach said. Maybe the atoms are an economic description. This doesn't hurt me very much. From the viewpoint of the physicist this doesn't make a difference." Strange as it was, in Vienna the physicists were all followers of Mach *and* followers of Boltzmann. It wasn't the case that the people would hold against Boltzmann's theory of atoms any antipathy because of Mach. And I don't even think that

Figure I.1. The young Philipp Frank, taken presumably in the 1910s. The image is from Frank's *carte de visite*, courtesy of Gerald Holton.

> Mach had any antipathy. It never came to my mind that because of the theories of Mach one shouldn't pursue the theories of Boltzmann, the atomic theories.[5]

Frank learned that presentations of theories and historical issues may be very different and that even if a debate is conceptualized as realism versus antirealism, other conceptualizations are possible as well. Mach was not a full-blooded antirealist, nor Boltzmann a naive realist, but their views were situated within layers of epistemological, methodological, and logical issues. In later writings, Frank would explain how the same theory (e.g., the special or the general theory of relativity) may be interpreted differently—even diametrically, but still legitimately—by various authors.

---

5. Ibid. Emphasis added.

As a theoretically inclined and systematically minded physicist, Frank also contributed actively to the development of the physical sciences. After successfully defending his doctoral dissertation, he published important—though largely forgotten—papers on the simplification of the special theory of relativity. He also collaborated with Austrian physicist and engineer Hermann Rothe and earned a broader reputation among physicists (see Frank 1932/1998, 290–96).

Between 1907 and 1912, Frank met regularly with the mathematician Hans Hahn and the economist-sociologist Otto Neurath at the Philosophical Society of the University of Vienna and in Viennese coffee houses. (Neurath and Hahn had attended the same Gymnasium, so they knew each other well and for a long time, while Frank joined them presumably during their shared university years.) Rudolf Haller (1991) called this trio the "First Vienna Circle" that preceded the better-known circle that formed around Moritz Schlick in the 1920s.[6] Frank recalled that

> although all three of us [Hahn, Neurath, Frank] were at that time actively engaged in research in our special fields, we made great efforts to absorb as much information, methodology and background from other fields as we were able to get. Our field of interest included also a great variety of political, historical, and religious problems which we discussed as scientifically as possible. (Frank 1949b, 1)

During these years, as they pursued careers (respectively) in mathematics, economics, and physics, they were held together by philosophy and general questions about science. Facing the recent revolutionary developments and controversies of their fields, Frank, Hahn, and Neurath embraced those philosophical movements that kept up with the special sciences. The ideas they absorbed and discussed over coffee—for example, that not just individual sentences but whole theories are tested in experiments; that what we consider "pure" data may depend on our theories; and that different theories may account equally for the same data—shaped their thinking for decades. These and other influences can be seen in detail in *The Humanistic Background of Science* (hereafter: *The Humanistic Background*).

---

6. On the First Vienna Circle the most detailed and comprehensive account is Uebel (2000).

1.2. Prague: The City of Ernst Mach and Albert Einstein

Though Frank was educated in Vienna, joined his first philosophical circle amid the smoke of Viennese coffee houses, and participated later in discussions of the Vienna Circle, the longest position he held in Europe was in Prague. He taught, conducted research, and organized intellectual life in Prague for twenty-five years.

Shortly after becoming a *Privatdozent* in Vienna, Frank applied for a new job: Einstein's chair at the Department of Physics of the German University of Prague (currently Charles University), which became vacant in 1912.[7] Frank was among three finalists for the position, titled "Ordinary Professor of Theoretical Physics": a university teacher from Vienna named Emil Kohl, the theoretical physicist Paul Ehrenfest, and Philipp Frank.

A commission evaluated the three candidates in May 1912. Its members were Einstein himself, the physicist Anton Lampa, and the mathematician Georg Pick, who had once been Mach's assistant. According to the commission's review (written by Einstein), "Ehrenfest is a man of a lucid and critical mind who has few equals in his ability to extract what is essential in a theory, and who is completely independent vis-a-vis contemporary endeavors" (Einstein 1912/1993, 302). The commission had positive words for Kohl, too, but praise for Frank dominated the report. "The great amount of able scientific work that this merely 28-year-old man has already produced is something to be admired," it read. Frank "combines a rare mastery of the mathematical tools with a good grasp of the problems of physics" (Einstein 1912/1993, 302). The review also mentioned his mathematical papers and, more interestingly, Frank's "original essays of an epistemological character" (Einstein 1912/1993, 302). The report mentioned two in particular: "Kausalgesetz und Erfahrung" ("Experience and the Law of Causality" 1907/1949) and "Mechanismus oder Vitalismus?" ("Mechanism or Vitalism?" 1908). Together with his physical and mathematical articles, they demonstrated that Frank's "talents are singularly versatile" (Einstein 1912/1993, 303).

Because "Frank has been working regularly and successfully as an academic teacher for the past two years, while Ehrenfest has not habil-

---

7. Einstein started to teach in Prague in 1910, but soon after that he got an offer from Zurich where he graduated. For Einstein's time in Prague see Gordin (2020), on Frank's activities and intellectual milieu there see Hofer (2020).

itated to this day" (Einstein 1912/1993, 303), the commission suggested that Frank be ranked first for the position, followed by Ehrenfest and, finally, Kohl. In 1912 Frank was promoted to associate professor at the German University of Prague as a successor to Einstein. He later became the director of the university's Institute of Theoretical Physics until his emigration to the United States in 1938.

Frank's professional career developed alongside his relationship with Einstein. He later recalled (1962/2001, 66) that Einstein greatly admired his "Kausalgesetz" and that after his promotion they became lifelong friends and allies. In a 1917 letter to Kathia Adler, for example, Einstein recommended Frank's paper on Ernst Mach (1917/1949) instead of his own (Einstein 1916/1996). In turn, Frank wrote philosophical and popular pieces on Einstein, including, for example, "Einstein, Mach, and Logical Positivism" (1949a) for Einstein's volume in *The Library of Living Philosophers* and the concluding essay here in *The Humanistic Background*.[8]

His most important work on Einstein remains his biography of 1947, *Einstein: His Life and Times,* named recently as one of the great physicist's "authoritative biographies" (Canales 2016, 57). Frank worked on the book as early as 1939 and delivered the manuscript for translation in 1941.[9] Gerald Holton (2006, 302), first a student, then an associate of Frank, noted that the "book is still one of the best . . . even though the manuscript . . . was horribly mangled by its publisher in the English-language edition." This, Holton recalled, was because "Alfred A. Knopf [the publisher] gave the manuscript to edit to an American [George Rosen] who, Philipp told me, knew English but no science, and also to a Japanese [Shuichi Kusaka], who knew science but no English."[10]

Still, the volume was a success. Harvard historian of science I. Bernard Cohen (1948, 252) wrote that "the scientific world has long been awaiting Professor Frank's book on Einstein [and] it fully justifies our expectations." The book was published in late February 1947 and

---

8. The most detailed account of the Frank-Einstein relationship is given in Howard (2021).

9. Frank to Neurath, April 15, 1939, and September 5, 1943, ONN.

10. The German edition of the book has a "Preface" which explains that Frank started to write it in New York in 1939, then worked on it in Chicago (1940) and finished the majority of it Boston (1941). As the book was translated from German to English, a quarter of its material was cut. As Frank notes, the German edition is "the first complete edition of the manuscript" (1949c, 5).

reprinted two months later. With the success of the biography, Frank reached a wide audience and made a name for himself alongside Einstein that outlived his philosophical reputation. For as Cohen (1948, 253) warned Frank's readers, one "is hard put to tell when Einstein is speaking through Frank and when Frank through Einstein!" Einstein was at least prepared to speak through Frank's book, for he wrote a preface that Knopf omitted from the American edition and was only later published in German.[11] Perhaps owing to Knopf's decision, Einstein prepared another preface (Einstein 1950) at Frank's request for his book *Relativity–A Richer Truth*.

Besides a strong community of physicists, Prague presented ideal conditions for Frank's interdisciplinary approach to science (see figure I.2). Frank hired experimental physicist Reinhold Fürth as an assistant before Fürth was appointed as professor of physics. Frank also maintained good relations with mathematicians, including the professors Ludwig Berwald, Karl Löwner, and Georg Pick, as well as supporters and colleagues of the Jewish feminist Berta Fanta, whom Frank met at meetings of the so-called *Fantakreis*. This circle, visited previously as well by Einstein, Franz Kafka, and other cultural figures of Prague, was an extension of the German University where Fanta studied; as a woman, however, she was not permitted to get a degree (Wein 2016, 54).[12]

Frank's eclecticism extended into his personal life as well. After his classes, on the way back to his office he would step into the library and talk "about politics, about physics, about anything that he might picked up at the 'Kaffeehaus,'" said his student Peter G Bergmann (1966, 5).

---

11. See Einstein (1979). An English translation appeared recently in Rowe and Schulmann (2007, 129–131). On Einstein's preface, see Holton (2006, 303).

12. On the Frank's personal relationship to Kafka in Prague, Nina Holton tells the following story: "One evening—it must have been 1950 or 1951—we had a large party with friends of our age, and Hania and Philipp came as sort of guests of honor. In the 1950s, everyone in our circle of friends read Kafka, and on that particular evening Kafka was widely discussed. Hania pricked up her ears, and her eyes turned large with astonishment. 'Kafka?' she shouted to someone sitting on the floor near her. 'How do you know about Kafka?' The young man so addressed seemed rather embarrassed and replied: 'You see, Madame, Franz Kafka is one of the greatest writers of this century. Everybody knows his work.' Hania listened with astonishment, then she turned to her husband and said, 'Philippushka, what have we done with Franzl's letters to me?' 'You see,' Philippushka answered in his usual unperturbed way, 'they were packed in our lift to be sent from Prague in 1938, and the lift never arrived'" (Holton 2020, 171).

10 The Humanistic Background of Science

Figure I.2. In the woods of Czechoslovakia, Frank is the second from the right. Image courtesy of Harvard University Archives.

Another student in Prague, the Finnish philosopher Max Söderman, confirmed Frank's outsized cultural and social presence ("his little stories are very entertaining, his suppers delicious, his wife [Hania] charming") as well as Frank's interest in possible relations between scientific philosophy and contemporary politics.[13]

Frank focused mainly on three areas in his teaching: (1) relativity theory, (2) thermodynamics, and (3) philosophy of science, while also teaching courses on Maxwell's theory of electromagnetism, probability theory, statistical mechanics and kinetic theory, and Dirac's relativistic theory of the electron.[14] Frank intended to write a textbook on the

---

13. In a letter, Söderman pointed out Frank's political interests because he did not share them and compared Frank unfavorably to "the apolitical *purus logicus* [Karl] Reach." Max Söderman to Kaj Saxén, Nov. 21, 1936, GHWC, 714.249–50. English translation by Anssi Korhonen.

14. Based on her research, Veronika Hofer (2020, 61) lists the following courses Frank taught in Prague: "Molecular Physics," "Electrical, Light- and Heat-Radiation,"

theory of relativity to be published by Teubner in Leipzig in 1920, but it was never realized. After continuing to teach thermodynamics at Harvard during the 1940s and 1950s, he did succeed in writing a privately circulated textbook for natural scientists and engineers.[15]

As a physicist, his most important achievement in Prague was arguably the so-called Frank-Mises (Frank and von Mises 1925–1927), a major undertaking by Frank and the engineer and mathematician Richard von Mises, an old friend from Vienna.[16] In contrast to philosophers of science who envisioned a more traditional, unidirectional, and determinate route (or correspondence) between observations and theories (such as Moritz Schlick's idea of coordination), Frank and von Mises shared a view of this connection as statistical and thus offered a more refined picture of theory building in the sciences.

Frank and von Mises first worked together in the 1920s and early 1930s to revise the famous book *Differential Equations of Mathematical Physics* by Riemann-Weber.[17] The book had been revised before, but they decided to create a wholly updated and modernized version for mathematicians and physicists. Newly titled *The Differential and Integral*

"The Principle of Relativity, It's Foundations and Applications," "Atomism," "Theoretical Mechanics with Special Reference to the Theory of Relative Motion," "Hydromechanics and Aeromechanics," "Kinetic and Thermodynamic Theory of Heat," "Advanced Mechanics," "Calculus of Probability," "Theory of the Aeroplanes," "Theory of Gravitation," "Statistical Mechanics and Quantum Theory," "Partial Differential Equations of Mathematical Physics," "Introduction in the Theory of Relativity," "Huygens and Newton," "Theory of Compression of Light," "Discussions of New Papers on Quantum Theory," "Discussions of New Papers on Radiology," "Thermodynamics," and "General Relativity Theory." "Theory of Flight of Machines," "The Partial Differential Equations in Mathematical Physics " and "Introduction in Theory and History of the Exact Sciences."

15. Frank (1945) is a typescript edition published by Brown University on the basis of Frank's lecture during a summer course. It is not listed among Frank's official papers in the 1998 English translation of his causality book (Frank 1932/1998, 290–296).

16. Von Mises was an engineer and an applied mathematician who had been appointed in 1920 as full professor in Berlin, where he founded and directed the Institute for Applied Mathematics. After his emigration to the University of Istanbul in 1933, he was a professor of applied mathematics at Harvard from 1939 to 1953.

17. The story of the Riemann-Weber book goes back in the 1860s, but it does not concern us here. About the origins, details, and significance of the Frank-Mises see Siegmund-Schultze (2007).

Equations of Mechanics and Physics, their edition was finally published in 1925 and 1927: the first volume, on mathematics, was edited by von Mises. It includes several chapters by von Mises, but many of his colleagues contributed papers on introductory and advanced-level mathematics. The second volume was edited by Frank. He contributed a long section on analytic mechanics, while his colleagues from around the world covered other topics.

Until the 1950s, the Frank-Mises was often said to be "the standard encyclopedia of mathematical physics of the twenties and thirties" (Siegmund-Schultze 2007, 28). More importantly, it may be seen today as the first attempt by logical empiricists to theoretically combine and unify different scientific fields.

From the philosophical point of view, Frank's career in Prague began with his joining the *Deutsche Physikalische Gesellschaft* (German Physical Society) in 1918. The *Gesellschaft* had a long history, and there were many members outside of Germany who belonged to regional societies: for example, in Zurich, and Vienna. Prague also had its own small group, though its membership never exceeded sixty. Until this group was formally dissolved in 1934,[18] Frank was its chairman. The *Gesellschaft* provided Frank with an official setting for the first meeting on *Erkenntnislehre der exakten Wissenschaften* (The epistemology of the exact sciences), a seminal philosophical gathering that he organized in 1929 with sponsorship from the Ernst Mach Society in Vienna and the Society for Empirical Philosophy in Berlin (Stöltzner 2020).

The conference featured lectures on probability, causation, and foundations of mathematics (see figure I.3). In his opening speech, Frank (1930, 94) remarked that the goal was to "establish a purely scientific conception of the world [*rein wissenschaftlichen Weltauffassung*], to support the scientific trend of thought in contrast to the often reoccurring philosophical-metaphysical rather aesthetical [schöngeistigen] one."

Prior to relativity and quantum mechanics, Frank claimed (1930, 93), physics books contained merely casual remarks or impressions about theories and the nature of knowledge—"ornament[s], that had little to do with the content." Though Michael Stöltzner (1995) has shown to the contrary that physicists of the time had some natural philosophical tendencies, they were not drawn toward critical philosophical consider-

---

18. On the German Physical Society and Frank see Stöltzner (1995) who lists the various lectures held at the Society.

Figure I.3. The Fifth Meeting of the German Physicists and Mathematicians. Frank is fourth from the left. Image courtesy of the Vienna Circle Institute.

ations, at least not explicitly and self-consciously. Despite Prague's long traditions in natural science, philosophy of nature (*Naturphilosophie*), and the theory of knowledge (through Bernard Bolzano and Ernst Mach), Frank later recalled that

> [t]he audience, which consisted mostly of German scientists, knew little of philosophy, except that they had some sentimental ties to Kantianism. This doctrine was regarded in some intellectual quarters as a kind of substitute for the traditional forms of religion. My wife [Hania Frank] said to me after the lecture [Frank 1930/1949]: "It was weird to listen. It seemed to me as if the words fell into the audience like drops into a well so deep that one cannot hear the drops striking bottom. Everything seemed to vanish without a trace." (Frank 1949b, 40)

These observations from Frank and his wife Hania (née Gerson), a former student of his from Poland, introduce a theme Frank developed in his later writings: while many scientists deny explicit ties to philosophy and are often unwilling to consider philosophical arguments, they remain nonetheless immersed in traditional philosophical views and frameworks—in the case of these German scientists, certain forms of Kantianism and idealism. As the New York philosopher Sidney Hook (1930, 145) remarked after attending the conference during his travels in Europe, this tacit Kantianism was not a choice made by German scholars from an array of theoretical options. It was "rather a national possession, the blazing jewel in Germany's cultural crown."[19]

Though this first philosophical meeting was not as successful as he had hoped, Frank continued to organize and create international networks supporting science and scientific philosophy. In August 1928, for example, he read a paper on quantum theory at the Sixth Congress of Russian Physicists in Saratov, which he attended with von Mises and Max Born (see Joravsky 1961/2009, 267 and Pechenkin 2014, 107). He traveled to other Russian cities to give talks,[20] and he wrote two articles (on waves [Volny] and hydromechanics [Gidromekhanika]) for the first edition of the *Great Soviet Encyclopedia* (see Joravsky 1961/2009, 380, n. 6). The resulting contacts and Frank's ability to speak Russian allowed him to later write about Russian philosophy of science and compare it to logical empiricism (Frank 1936/1949) and to also draw on Marxism and Marxist theories of knowledge in *The Humanist Background*.[21]

Frank hoped to make Prague a globally renowned center for the scientific conception of the world, much like Vienna. To this end, he and others established a chair for natural philosophy at the university. To fill the position, in 1926 Moritz Schlick recommended as his first and second choices Hans Reichenbach (then in Berlin) and Rudolf Carnap (then in Vienna).[22] Though Reichenbach was disposed to go to Prague, he remained in Berlin, allowing Frank to campaign for Car-

---

19. In the 1930s several American philosophers travelled through Europe, including Albert Blumberg, W. V. O. Quine, Charles Morris, and Ernest Nagel who drew a somewhat more promising picture even of Germany. See Nagel (1936).

20. See Frank to Schlick, September 26, 1928, MSN.

21. In 1960, Frank's most important published book, *Philosophy of Science: The Link Between Science and Philosophy*, was translated into Russian; two years later the Russian introductory essay was translated into English and published in *Daedalus* (see Kursanov 1962).

22. Moritz Schlick to Hans Reichenbach, January 19, 1926, MSN.

nap's appointment. Despite five years of lobbying against "the adherents of traditional philosophy" (1949b, 45) who opposed the appointment, Frank eventually arranged for Carnap to come to Prague. In the end, he succeeded "because of a happy coincidence," Frank later recalled: in Prague, philosophers worked in the Faculty of Humanities, and the Faculty of Science was not able to provide courses in philosophy. But Thomas G. Masaryk, the president of the Czechoslovakian Republic, was himself a philosopher who "believed strongly in the educational value of philosophy. He insisted that the Faculty of Science should have a philosopher of their own." On Frank's suggestion, Carnap was appointed to the position in 1931.[23]

Frank and Carnap worked closely. They had a regular Thursday-evening colloquium (*Donnerstagabendzirkel*) that one might compare to the Vienna Circle, which was attended by local scientists, philosophers, and important figures in Prague's cultural life.[24] While they discussed classic philosophical papers, including Carnap's infamously provocative "The Elimination of Metaphysics Through the Analysis of Language" (Carnap 1932/1959), this group's *Kolloquium für philosophische Grundlagen der Naturwissenschaft* (Colloquium on the philosophical foundations of the natural sciences) focused more broadly on biology, physics, and their interrelations. In time, these issues became quite important for Frank. He gave a talk on the relation of physics to biology at the First International Congress for the Unity of Science in Paris in 1935 (Frank 1936a) (see figure I.4). A year later at the Second Congress in Copenhagen, the topic was "The Problem of Causality—with Special Consideration of Physics and Biology."[25]

---

23. Frank additionally served as Dean and Vice-Dean of the Faculty of Natural Sciences of the German part of the University from 1925–1927, as well as in the academic year 1930/31. This office may have helped Frank bring Carnap to the University (Hofer 2020).

24. These included students of Brentano, Georg Katkov and Walter Engel; the Russian educationalist Sergius Hessen; Kafka's close friend Felix Weltsch; the mathematician Karl Löwner (later known as Charles Loewner); the biologist Joseph Gicklhorn; the zoologist Paul Fortner; the mathematician Ludwig Berwal; the economist (and Rosa Luxemburg's lover) Kostja Zetkin, the biologist Felix Mainx, and the painter Trude Schmidl-Waehner (see Tuboly 2021c).

25. On the congress see Stadler (2001/2015, 178–182). Frank's philosophy of biology is taken up in Hofer (2002), (2003) and Wolters (1999), (2018).

Frank and Carnap also became personally close. Frank officially witnessed Carnap's marriage to his wife, Ina (nee Stöger), and—because Carnap did not speak Czech—acted as a translator so that Carnap could answer the ceremony's official questions. Perhaps not surprisingly, Carnap had questions of his own:

> When the procedure began, Carnap, the meticulous logician and philosopher of language, asked Frank to clarify the meaning of the verbal formulas required. As the procedure continued, Carnap kept interjecting questions as to the logical status of the particular statements he was expected to supply at each juncture. Frank finally interrupted him, saying, in effect, "Do you want to get married or not? If so, just answer and don't ask questions!" (Scheffler 2004, 66)

Though their friendship was a success, and Carnap would later help bring Frank to the United States, the Carnap-Frank Circle did not achieve the philosophical importance of its Viennese and Berliner counterparts. Neither through large numbers of participants, nor by formulating a unified or at least recognizable view, did these scientists and philosophers in Prague create an internationally respected style. Though these failings should be explained by detailed philosophical and sociological study that we cannot offer here, one obvious difference between the circles in Prague and Vienna was Otto Neurath, the "big locomotive" who prodded, organized, and often provoked his philosophical colleagues in Vienna toward continuous and productive collaboration. It is probably not a coincidence that Carnap later remarked, "My life in Prague, without the [Vienna] Circle, was more solitary than it had been in Vienna. I used most of my time for concentrated work, especially on the book on logical syntax" (1963, 33). Though Carnap's syntactical project had a great influence on Frank, Reinhold Fürth recalled that in Prague Frank "preferred to work on his own and never had a 'research school'" (1965, xiv).

During his years in Prague, Frank often visited his hometown to participate in the meetings of the Vienna Circle, which by the mid-1920s had become an evolving group of philosophers, sociologists, economists, jurists, historians, mathematicians, and physicists. The core members consisted in Schlick, Carnap, Neurath, Hahn, and Friedrich Waismann. But Kurt Gödel, Karl Menger, Herbert Feigl, Felix Kaufmann,

Figure I.4. From right to left Frank, Carnap, Susan Stebbing, Heinrich Neider, Carl G. Hempel, Eva Hempel, Jørgen Jørgensen, and Finnish logician Uuno Saarnio. The photograph was likely taken when the group traveled to a village in the Belgian Ardennes after the First Congress for the Unity of Science in Paris, 1935, a trip described in correspondence between Stebbing and Ernest Nagel (October 18, 1935, ASP EN P). Image courtesy of Harvard University Archives.

Edgar Zilsel, and others participated as well.[26] Though it was led by the physicist-turned-philosopher Moritz Schlick, it was a self-consciously cooperative enterprise devoted to discussing philosophical and scientific questions in exact terms. Though Frank is often described as a regular visitor and not as an inner member, the correspondence among Carnap, Neurath, Schlick Reichenbach and others shows otherwise. Frank was an honored and important member whose opinion always

---

26. The best and most detailed introduction and documentation of the Circle's activities is Stadler (2001/2015).

mattered and who regularly played the role of final judge in controversial matters.[27]

1.3. Harvard, Massachusetts, and Boston: The Promise of a Better Future

Amidst the rise of fascism and radical-right voices in Europe, Carnap moved to Chicago in 1935. A year later, Carnap reported to Neurath that according to Frank, "Antisemitism in Prague is again flourishing."[28] Frank was Jewish, but he remained in Prague for two more years. Only late in 1938 did he and Hania come to America to lecture at universities and colleges under the auspices of the Institute of International Education (See figure I.5; Holton 2006, 198). Frank and Hania had planned to return to Prague, but that became impossible after the Munich Agreement (which they learned about as they sailed toward the United States) and Germany's invasion of Czechoslovakia (Hofer 2020, 63, 64, 65). Frank now needed a job on American soil. Frank would eventually obtain a position at Harvard University, but this came only after he failed to get a job at the University of Chicago. The reasons for this failure are worth exploring, for they introduce some of the ongoing cultural battles manifest in *The Humanistic Background*.

1.3.1. A Refugee between Nazism and Thomism

For Frank, returning to Prague would have been foolhardy because the university had fallen under Nazi control. Besides being a Jew, Frank had once lectured in the Soviet Union, a nation that Hitler feared and loathed.[29] Frank had a temporary visa that was soon to expire. In order to obtain a permanent visa, he had to be employed for at least one year as a university professor. Frank's first hope was a temporary, one-year posi-

---

27. When he was in Prague, Frank edited a book series together with Schlick, called the *Schriften zur wissenschaftliche Weltauffassung* (*Writings on the Scientific World-Conception*). It featured, among others, Carnap's *Logical Syntax*, Frank's book on causality, von Mises' book on statistics and probability, and Karl Popper's famous *Logik der Forschung* (*Logic of Scientific Discovery*, as it was translated into English).

28. Carnap to Neurath, June 11, 1936 (ASP RC 102-52-26).

29. Frank to E. C. Kemble, Feb. 9, 1939 and Feb. 25, 1939. Frank to Harlow Shapley, April 7, 1939, all in HUA-HSP.

Institute of International Education
2 West 45th Street, New York City

announces

PHILIPP FRANK, Ph. D.
Professor of Theoretical Physics, German University of Prague

Lecturing under its auspices October and November, 1938

Dr. Frank became Professor of Theoretical Physics at the German University of Prague in 1912, appointed as successor to Dr. Albert Einstein upon his recommendation. After receiving the degree of doctor of philosophy from the University of Vienna in 1906, Dr. Frank became a lecturer in Mathematical Physics at that University in 1910. He is one of the founders of the movement towards logical empiricism in science and of the "Unity of Science Movement."

Dr. Frank is regarded as one of the leading authorities on geometric optics and dynamics as well as on aerodynamic theory. He is a member of the Committee of Organization of the new "Encyclopedia of Unified Science" published by the Chicago University Press.

In 1928 he lectured in Soviet Russia at the invitation of the scientific congress. We are told that his lectures will be most stimulating as he is considered to have one of the keenest critical minds at present in his field. He speaks English well.

In 1925 Dr. Frank collaborated in the publication of a volume entitled "Die Differentialgleichungen der Mechanik und Physik" which is considered the most circumstantial and extensive treatise on the differential equations of physics that exists. A book in English entitled "Interpretations and Misinterpretations of Modern Physics" was published in April of this year by Herman of Paris. He has also written many papers on scientific subjects, both in French and in German.

LECTURE SUBJECTS

Popular Lectures
1. The Rôle of Metaphysics in the Physics of the Twentieth Century
2. Modern Physics and Common Sense

Lectures for Graduates
1. Philosophical Interpretations and Misinterpretations of the Theory of Relativity
2. Philosophical Interpretations and Misinterpretations of the Quantum Theory
3. How to Eliminate all non-scientific Elements from the Quantum Theory

Technical Lectures
1. Recent Generalizations of Geometrical Optics
2. Bohr's Principle of Complementarity and Modern Logic

Figure I.5. Brochure advertising Frank's American lecture series, October–November, 1938. Courtesy of ASP-ENP.

tion at the University of Chicago with Carnap and Charles Morris, the American philosopher who had befriended the unity of science movement years before and had helped Frank to organize his American lecture tour.

A position at Chicago may have seemed ideal to Frank, for the university was becoming a center of the Unity of Science Movement. The first pamphlets comprising Otto Neurath's new *International Encyclopedia of Unified Science* were published by the university press the year before, and the university's philosophy department was friendly to philosophical pragmatism and a science-friendly naturalism. John Dewey and George Herbert Mead refined their philosophies there, while Morris, a student of Mead, dedicated himself to reconciling and joining logical empiricism and pragmatism. When Frank arrived in the United States, New York City was another important outpost of pragmatism, with Dewey and Ernest Nagel teaching at Columbia; Sidney Hook, Dewey's student, teaching at New York University; and Horace Kallen, a former student of William James, at the New School for Social Research.

Yet pragmatism in the United States was not without intellectual enemies. The nation had always been deeply religious and apprehensive of intellectuals and scientists who seemed to challenge religious and metaphysical orthodoxies (Hofstadter 1963). As the University of Chicago deliberated over whether or not to offer Frank a position, two powerful critics of scientific philosophy were close by: university president Robert Maynard Hutchins and his colleague and sometimes right-hand-man Mortimer Adler. With Hutchins and Adler leading one side, and Dewey and his fellow New Yorkers leading the other, debate raged about the nature of science and philosophy, their proper place in the modern world, and—amidst widespread unemployment and the rise of fascism in Europe—their roles in education and the future of democracy.

As a candidate for a job at the University of Chicago, therefore, Frank had walked into a battle royale. Hutchins had first joined the debate in 1936 with his book *The Higher Learning in America* (Hutchins 1936). Placing blame for the nation's problems on the failure of higher education and the nation's intellectual life, he blasted the rigid professionalism and departmentalization of the modern university (Hutchins 1936, 54). It had come to offer students, he wrote, a smorgasbord of disconnected fields of study—"an enormous miscellany, composed principally of current or historical investigations in a terrifying multiplicity of fields" (Hutchins 1936, 92). "The modern university," Hutchins wrote, "may be compared with an encyclopedia," albeit one without any internal, unifying structure:

> The encyclopedia contains many truths. It may consist of nothing else. But its unity can be found only in its alphabetical arrangement. The University is in much the same case. It has departments running from art to zoology; but neither the students nor the professors know what is the relation of one departmental truth to another, or what the relation of departmental truths to those in the domain of another department may be. (Hutchins 1936, 95)

Hutchins's call for intellectual unity may have sounded familiar to Frank, for in his crusade to unify the sciences through his new encyclopedia, Neurath had criticized the scientific landscape in similar (albeit overtly Marxist) terms:

> The bourgeois formation of thought is indefinite . . . proceeding here in an anthroposophic way, there mathematicising, here psychologizing, there pursuing the idea of fate. . . . The wealth of scientific detail is no longer held together by a unitary approach and in a certain sense it is left to chance whether a man thinks about some linguistic formations in Chinese or about a medieval legal text, about African beetles or about wind conditions at the North Pole. (Neurath 1928/1973, 294–95)

The new encyclopedia was an effort to unify this chaos, to "build bridges" among the sciences (as its promotional literature put it), and to cultivate understanding and collaboration across scientific fields (Reisch 2005, 70, 86).

But Hutchins and Neurath yearned for different kinds of unity that issued from different sources. While Neurath and other scientific philosophers understood that the source of any future unity would be science and scientific collaboration itself, Hutchins and his followers believed that such an enterprise was impossible. For science, Hutchins believed, collects information randomly, without any criteria of importance or value. "As a contemporary has said," he wrote in *The Higher Learning*, tacitly quoting the Catholic philosopher Jacques Maritain, the modern urge to master and control nature through scientific knowledge serves to harm us, not help us:

> Behold man the center of the world, a world all the parts of which are inhuman and press against him. . . . This morality

does not liberate man but on the contrary weakens him, dispossesses him, and makes him a slave to the atoms of the universe, and above all to his own misery and egotism. What remains of man? A consumer crowned with science. That is the last gift, the twentieth century gift of the Cartesian reformation. (Hutchins 1936, 100; see Maritain's *The Dream of Descartes*, quoted in D'Souza and Seiling 2014)

The problem, Hutchins elaborated, is that humanity does not know what facts and information to consume. It requires, therefore, a metaphysical system or framework to guide and connect areas of knowledge. Hutchins reassured his readers that "I am not here arguing for any specific theological or metaphysical system." But embracing some kind of system, hopefully "the most rational one" we can cultivate, he insisted, would be necessary to effectively reform education and "to establish rational order in the modern world as well as in the universities" (Hutchins 1936, 105).

In fact, Hutchins had a preferred system in mind—the neo-Thomism of Maritain.[30] Not unlike Dewey, who had created the Laboratory School at the University of Chicago to test and refine his pragmatic theories of elementary education, Hutchins treated his university as a kind of laboratory for his educational vision. Shortly after he became president in 1929, for example, he offered his close friend (and fellow crusader for Thomism) Mortimer Adler a handsomely paid position in the philosophy department. This caused an uproar for a number of reasons, one of which was that Adler was not a trained academic philosopher. At Columbia, he had taught psychology, and his PhD had been in music appreciation. He eventually found a home in the university's law school, but not before senior faculty in the department resigned in protest of his appointment and Hutchins's heavy-handed maneuver. This created the departmental openings that had brought Morris and, in turn, Carnap to the department (Reisch 2005, 39).

Hutchins and Adler also reformed Chicago's undergraduate education to prioritize close readings of classic texts. Their "Great Books" approach lasted until Hutchins left the university in 1950; but their advocacy and leadership helped establish and legitimize Great Books programs at other colleges and universities. With help and funding from the *Encyclopedia Britannica*, which Frank scrutinizes in *The Humanistic*

---

30. The words Hutchins quoted appear in *Notes et documents*, Issues 22–25, Institut international Jacques Maritain, 1981.

Background for its treatment of science and values,[31] Hutchins and Adler created the monumental fifty-four-volume collection *Great Books of the Western World* (1955) that stretches from Homer to Freud. Its periodic updates and its *Synopticon: An Index to the Great Ideas* (Adler 1952), a volume that traces 102 concepts that circulate throughout the collection, suggest another point of contact and competition between these philosophical camps: besides describing human knowledge as a metaphorical encyclopedia, both sides created and promoted actual paper-and-spine books designed to help unify modern knowledge and to distribute that knowledge to the educated public.

Neurath and his Unity of Science Movement had an advantage in this competition, especially in New York City where Neurath's cousin, Waldemar Kaempffert, an editor and writer for the *New York Times*, wrote glowingly of Neurath's new encyclopedia and his crusade to unify the sciences from within (Reisch 2019b). But in Chicago, Hutchins and Adler benefited from the largesse and commercial expertise of William Benton, an advertising executive and publisher of the *Encyclopedia Britannica*. Under the tutelage of Adler and Hutchins, Benton converted to neo-Thomism and supported their Thomist crusade with zeal. Apropos of Chicago's iconic Palmolive Building, Hutchins is said to have remarked approvingly that Benton and Adler were "made for each other. The two men were consummate promoters of hot products—Palmolive soap and Thomism."[32]

This competition between Thomism and scientific philosophy set the stage for several well-known conflicts and confrontations. In his autobiography, for example, Carnap (1963, 42) famously dismissed Adler as a medieval atavist. Morris tried repeatedly to convince Hutchins that he had nothing to fear from the Unity of Science Movement and Neurath's new encyclopedia, but he was rebuffed and dismissed (Reisch 2005, 73–76).

Given their advocacy of socialism in the 1930s, the New York pragmatists were especially alarmed by Hutchins and Adler's Thomistic crusade. If the rise of Nazism had not been threatening enough, by the late 1930s it had become clear that the Soviet Union was not the bastion of intelligent, humane, and scientific socialism that many New York intellectuals had taken it to be. It was instead an authoritarian dictatorship marked by command economics, farcical show trials, brutal labor camps (Stalin's infamous Gulags), and an official philosophical system—dialectical materialism—that seemed to explain and justify everything in the minds

---

31. See Pt. II. Ch. 3.
32. See Mayer (1993, 303) and Dzuback (1991, 219).

of true believers. In 1937 Hook, Dewey, Kallen, and other New York intellectuals took their grievances against Stalin public by organizing the American Committee to Defend Leon Trotsky in the attempt to promote the exiled Bolshevik as the rightful leader of worldwide socialism.

That same year, these philosophers turned to face the similar menace they saw in Hutchins and Adler's Thomist movement. Their Conference on Methods in Philosophy and the Sciences, held at New York University, was designed to shine light on "a strong and growing trend toward dogmatic authoritarianism in philosophy and the sciences." Dewey, the featured speaker, praised natural science and experimental inquiry as a method to combat social problems, while Hook attacked Adler and Hutchins as reactionaries calling for a modern "retreat from reason."[33]

Despite these tensions and hostilities, Frank remained hopeful about his prospective position at Chicago. The one-year appointment would not be expensive for the university because he had obtained independent funding from the Institute of International Education. But Frank did not get the job. He reported in early 1939 that Hutchins and his administration objected to the appointment. The substance of the objection remains unclear in Frank's correspondence about it, but he wrote, "My friends in the department suspect that these objections are rather a pretext":

> For a certain time everybody observes at the University of Chicago a tendency of the President to influence the faculties and particularly the Department of Philosophy in favour of Neo-Thomistic philosophy. President Hutchins, who is himself in all philosophical questions under the influence of Dr. Mortimer Adler, a champion of Catholic philosophy at the University of Chicago, dislikes the activity of the Carnap-Group (of Logical Empiricism) and does not want any inforcement [sic] of this Group.

Hutchins and Adler, Frank added, see this group as competition—"a stumbling block"—to their goal of making scholastic philosophy the blueprint for modern education at Chicago and beyond.[34]

---

33. See "Notes," *International Journal of Ethics* 48, No. 1 (Oct, 1937): 141–42; Gilbert (1997, 85).

34. Frank to E. C. Kemble, March 17, 1939; see also Frank to Kemble, February 25, 1939, HUA-HSP. The surviving copies of these letters do not include the year 1939 in

1.3.2. A Home at Harvard

In the wake of this disappointment, Carnap wrote to physicist Percy Bridgman at Harvard to ask whether Harvard could offer Frank a position. The situation was urgent, for Frank's visa would soon expire on May 1, 1939. Bridgman was optimistic. But within weeks Frank's hopes were dashed again, for the Harvard Corporation met to consider the matter but declined to invite Frank.[35]

Upon hearing this news, the Harvard astronomer Harlow Shapley sprang into action. For years, Shapley had worked with refugee organizations to find new homes for displaced scholars from Europe (see Jones 1984). He took the matter up with university President James Bryant Conant and wrote to Frank two days later with promising news: Conant agreed to personally ask the Corporation to reconsider its decision. Financial support would have to come from outside the university, but Shapley brokered promises from both the Institute for International Education and the Rockefeller Foundation to support Frank's year at Harvard. "I feel a great relief to know, that by your friendly efforts the situation at Harvard seems to me much better now," Frank wrote to Shapley. Shapley also helped to secure funds for Frank's second year at Harvard, from 1940 to 1941, at which time Frank became a permanent part-time member of the faculty.[36]

Unlike Hutchins at the University of Chicago, Conant was a university president who admired logical empiricism. Frank's arrival at Harvard coincided with the Fifth International Congress for the Unity of Science, which Conant convened by welcoming the participants. As Conant later put it, his philosophy of science was a "mixture of William James's Pragmatism and the logical empiricism of the Vienna circle, with at least two jiggers of pure skepticism thrown in."[37] As we explain below, the remark captures something of Frank's own philosophical ambitions as

---

their dates; but their context and proximity to other letters in Shapley's papers indicate this year, and we have dated them accordingly.

35. Frank to P. W. Bridgman, March 24, 1939; Carnap to Bridgman, March 16, 1939, HUA-HSP.

36. Shapley to Frank, April 3, 1939; April 10, 1939. Frank to Shapley, April 13, 1939, HUA-HSP. On Shapley's active career, see his autobiography (Shapley 1969). For more on European scholars emigrating to the United States, see Leff (2019).

37. Conant to John Boyer, Sept. 12, 1952, HUA-JCP.

they would develop at Conant's university, namely, to temper scientific philosophy as it was practiced by Carnap, Reichenbach, and other formalists with pragmatism and its traditional regard for the historical and sociological realities of modern science. As Frank and Hania adjusted to their new life in the United States, Frank aimed to prepare scientific philosophy for its new life in America, as well (see figure I.6).

When he intervened on Frank's behalf, Conant was also engaged in the national debates over the future of education that had indirectly helped to derail Frank's appointment at Chicago. A chemist by training, Conant shared the ambition of the unity of science movement to educate the public about science and its methods. These ambitions bore fruit several years later at the conclusion of the Second World War, when Harvard unveiled proposals and curricular guidelines for "general education" in the nation's high schools and colleges. All Americans must learn more about science, the guidelines recommended, by learning about its history and, in particular, closely examining case studies of important discoveries or theoretical breakthroughs in the past. At Harvard itself, Conant's reforms helped to legitimate the so-called marriage between history and philosophy of science by nurturing scholars

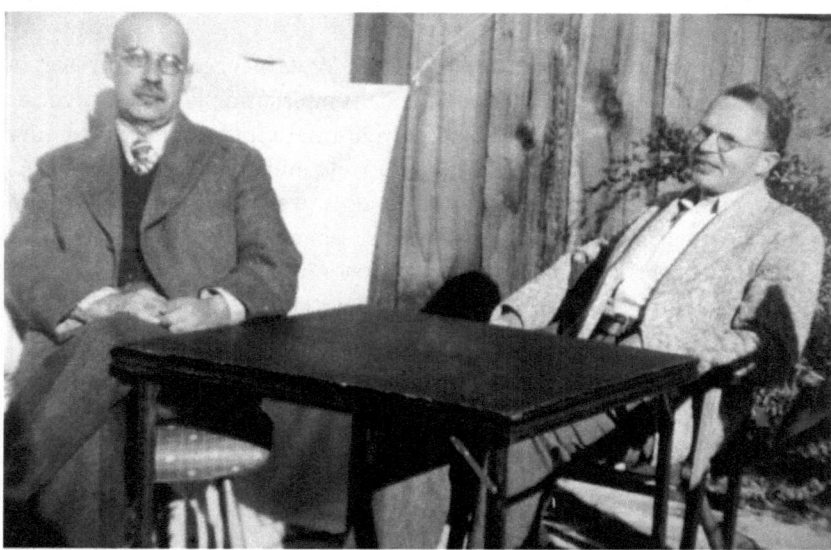

Figure I.6. Frank and Carnap in America. Image courtesy of University of Pittsburgh, Rudolf Carnap Papers.

including Gerald Holton, Thomas Kuhn, and Frank, all of whom taught in the new program.

Frank thrived intellectually at Conant's Harvard. He joined the Harvard Science of Science Discussion Group founded by Harvard psychologist S. S. Stevens, partly as an American counterpart to the weekly meetings of philosophers and scientists in Vienna and Prague (and suggested to Stevens by Carnap himself [see Hardcastle 2003]). In 1944 Frank formed a successor group, the Inter-Science Discussion Group, which similarly aimed to overcome institutional and conceptual boundaries between fields.[38]

By some measures, these were the most successful and productive years of Frank's career, for he was well supported by his colleagues and honored by several prestigious institutions. He was elected to the American Academy of Arts and Sciences in 1943, for example, as the first logical empiricist to become a member. (Additional refugee members would later include von Mises and von Neumann in 1944 and Carnap, Reichenbach, and Tarski in 1948.) Frank's support from the academy additionally allowed him to create within it an American Institute for the Unity of Science to continue the work of his Inter-Science Discussion Group and, in the wake of Neurath's death in 1945, the work of Neurath's European Institute for the Unity of Science. Supported for several years by the Rockefeller Foundation, Frank's institute sponsored lectures in and around Boston and became the official editorial home of the *International Encyclopedia of Unified Science*. Frank was also elected to join the American Physical Society in 1945 and, two years later, became the first elected president of the Philosophy of Science Association.[39]

Alongside these professional activities, however, Frank remained a part-time Lecturer in Physics and Mathematics in the Physics Department and in need of supplementing his income. He therefore lectured at other institutions (including Radcliffe and City College of New York).[40] In

---

38. Some further details of this group could be found in Gerald Holton's recollections; see Holton (1995), which also contains the reprints of a few original documents; see further Holton (2017).

39. As Fons Dewulf has shown, Frank's election may be understood as a compromise between opposing groups with different visions of the organization's future (Dewulf 2021).

40. According to the announcement of Frank's appointment in City College, Einstein supported him again. The *New York Times* quoted the following Einstein letter to college authorities: "Professor Frank has a deep insight in philosophy and epistemology,

addition, he wrote articles and books. Besides his successful biography of Einstein, he published two collections of papers (*Between Physics and Philosophy* [1941] and *Modern Science and its Philosophy* [1949]), and he wrote *Relativity and its Astronomical Implications* (1943), *The Foundations of Physics* (1946), *Relativity: A Richer Truth* (1950), and *Philosophy of Science: The Link between Science and Philosophy* (1957).[41]

1.3.3. The Conferences on Science, Philosophy, and Religion and Frank's Contribution

Relativity: A Richer Truth was one of Frank's most popular books. It emerged from his longtime affiliation with the New York–based Conferences on Science, Philosophy, Religion and their Relation to the Democratic Way of Life (CSPR). Convinced in the 1930s that democracy and other American institutions needed to band together to withstand the menace of Nazism and the threat of worldwide Communism, Rabbi Louis Finkelstein organized the conference in 1940 at his home institution: the Jewish Theological Seminary in New York City. What began as a single conference became a yearly event in which Frank participated regularly (on Frank's contributions, see Beuttler 1995, 100).

By including "religion" in their organization's title, Finkelstein and his co-organizers staked their own ground in the ongoing dispute between New York's pragmatists and Chicago's neo-Thomists. The Conference on Methods organized by Kallen, Hook, and Dewey acknowledged only "science" and "philosophy" as bulwarks against totalitarian attacks on reason and rationality. But only science, philosophy, and religion *together*—as Finkelstein explained to one of his organizers—could effectively protect Western democratic culture from "the insidious spread of Nazi and Communist doctrines in this country" (Beuttler 1995, 90).

Finkelstein chose Harlow Shapley, the astronomer who helped bring Frank to Harvard, to represent science on his planning committee. Shapley's discoveries in astronomy and his advocacy of public science education had made him one of the nation's most famous and celebrated scientists. Through Shapley, Frank quickly became involved in Finkelstein's conference. In the summer of 1940 he met personally with Finkelstein

---

and is at the same time a real physicist doing original work in this field" (*New York Times*, May 26, 1941).

41. Frank also edited *The Validation of Scientific Theories* (1957), the chapters of which were first published in *The Scientific Monthly* in 1954 and 1955.

in New York to deliver the essay on "Science and Democracy" that he would read at the first conference.[42] Frank met him again a year later and was plainly enthused about participating. By the end of the 1940s, if not earlier, Frank joined the conference's board of directors.[43] Years later he collected the lectures and talks he gave each year at the CSPR to create *Relativity—A Richer Truth* (1950).

It was at the first conference in 1940 that Frank had a front-row seat for the escalating dispute between the Thomists and pragmatists that had earlier cost him his job at Chicago.[44] He vividly described this experience to Charles Morris:

> Perhaps you have heard that I participated at the Conference on Science, Philosophy and Religion in New York last September. It was a very strange experience for me. I was the only one (besides H. Shapley) among so many people who upheld the viewpoint of positivism and empiricism. I had the opportunity of getting in contact with your colleague Mortimer Adler. I attacked him very often. However we got along personally quite well and he honored me by saying: "you are arguing very cleverly. It would be worthwhile to convert you to Thomism."[45]

In a lecture titled "God and the Professors," Adler hyperbolically denounced "positivists" who overlooked the primacy of metaphysics and mistakenly put their intellectual faith in science. Drawing from Hutchins's *The Higher Learning* and defending Hutchins's claim that only a hierarchical metaphysical system could save higher education and civilization, Adler remarked that the threat posed by wayward positivists was akin to ancient tyranny—a threat to civilization more menacing than Hitler. It would not be overcome "until the professors and their culture are liquidated" (Adler 1941, 134).

---

42. Gilbert (1997, 340) n. 53; Frank to Shapley, June 22, 1940, HUA-HSP.

43. See, for example, Bryson et al. (1950, 1964).

44. Another religious organization Frank would later participate in was the Institute on Religion in an Age of Science, founded in the early 1950s by the meteorologist Ralph Wendell Burhoe, whom Frank knew from within the American Academy of Arts and Sciences. See Breed (1990, 344). Frank's lectures remain listed on the organization's website, www.iras.org/pastspeakers.html.

45. Frank to Morris, 1 December 1940 (CMP).

Even before Adler delivered it, the lecture caused a commotion within the conference. Finkelstein had commissioned it personally and persuaded Adler to participate despite Adler's misgivings about the logical empiricists and Deweyites involved. The misgivings were mutual: Dewey refused Finkelstein's invitation, while those pragmatists who accepted prepared for fireworks. When Adler read his paper on the second day of the conference, Sidney Hook became so agitated that the conference threatened to unravel. The *New York Times* reported that Hook denounced Adler as "presumptuous" and argued that "his proposals were 'categorically false'" (*New York Times*, 1940). Behind the scenes and "under cover of the low wood partition that separated us from the audience," Hook later recalled, Finkelstein stamped on Hook's foot in an effort to keep his outrage contained (Gilbert 1997, 77, 78; see also Hook 1987, 337).

In the wake of Adler's attack, Hook, Dewey, and Ernest Nagel took the dispute to the pages of the *Partisan Review*. Under the title "The New Failure of Nerve" they chided Adler and other Thomists for merely pretending to help preserve democracy (see, e.g., Hook 1943) and for taking refuge in apriorism and supernaturalism (see Reisch 2005, 78 for a summary of this forum). Dewey and Kallen then organized another conference for 1943 to set the record straight about science and democracy. The name of the conference, they announced, was the Conference on the Scientific Spirit and Democratic Faith, purposefully excluding the word "religion" (Hook 1987, 347; Gilbert 1997, 84).

Finkelstein and other organizers had originally called for unity—a unified front from science, religion, and philosophy to defend democratic culture and democracy. As one put it at the first meeting, civilization would surely collapse unless "we can evolve some basic idea to which people of intelligence in the different spheres of life would pledge" (Gilbert 1997, 71). Finkelstein explained in his introduction to the conference proceedings that philosophers, theologians, and scientists were now as important as diplomats and standing armies because the military struggle in Europe was "but one phase, perhaps a minor phase, of a far greater conflict." At root, the current crisis was a battle of ideas and ideology—a battle "between ideas which make for the development, and those which make for the destruction of human civilization" (Finkelstein 1941, 11).

Frank endorsed Finkelstein's mission to protect democracy but disagreed with the popular notion that developments in physics, namely Einstein's relativity theories, had somehow weakened Western culture in its struggle to defend democracy and democratic values. Then writing his biography of Einstein, Frank became the conference's expert on relativity and patiently instructed his colleagues that relativity theory is not to blame for social immorality, nihilism, or the spread of totalitarianism. One problem, Frank believed, was that critics of relativity theory confused absoluteness and objectivity:

> Since the doctrine of the relativity of truth does not imperil the "objectivity of truth," it is hard to understand how the "relativism" of twentieth-century science could shock the belief in the "objectivity of human values." (Frank 1950, 39)

While relativism could question the absoluteness of values (e.g., their unconditional, or unqualified character), it did not question their contextual objectivity. Relativism, Frank explained, claimed only that statements—whether about the behavior of physical systems or about political or social values—require contextualization or, as Frank put it, "specification or qualification" (Frank 1950, 9). In *Relativity—A Richer Truth*, he provides simple examples, such as the assertion "John is wicked," said of a child who disobeys his parents. As life goes on, Frank explains, the same proposition may involve different relationships, such as teachers, fellow students, or other contextual frames within which "wicked" acquires various concrete and specific meanings. (If John's parents happen to be criminals, Frank cleverly notes, then "John is called 'wicked' if he *obeys* his parents.") His point is that just as in the case of relativity theory, using "relativized" language makes these contexts clear and, in the case of normative terms like "wicked," hardly imperils morale or denies "objective truth" (Frank 1950, 12–13).

This contextualization can be logical, sociological, cultural, or historical; what matters is that value statements be filled with *empirical* or *operational* meaning so that they become explicitly related to the world. "Without using the operational meaning," Frank explained as he generalized Bridgman's operational theory of science, "we derive from our abstract principles only abstract principles. We never get in touch with an actual human problem." (Frank 1950, 41–43, 45).[46] By making the empirical and contextual content of value statements explicit, Frank offered a way to approach what Thomas Uebel has called the

---

46. For perspective on Frank's (1950) we are indebted to Nemeth (2003), which nicely summarizes Frank's engagement with democratic theory. See further Siegetsleitner (2017) and Stamenkovic (2021), and Stadler (2021).

"intersubjective accountability" of public discourses.[47] Far from being an enemy of values, morality, and the democratic foundations of modern states, Frank's demand for contextualization would therefore make us put everything on the table so that we are better able to rationally and democratically discuss the various possibilities we must choose from.

As for Hutchins and Adler's caricature of modern research as random fact gathering, Frank argued in his lectures that values and aspirations were essential to research and the growth of science. Thomists were simply wrong to maintain that research itself was bereft of unifying principles and methods and to maintain that such principles could issue only from religion or metaphysics. On the one hand, as Frank argued frequently, the idea of relativity in physics could and should be generalized to enhance and support scientific as well as public discourses. On the other, the growth of science itself proved that researchers did not randomly gather facts. Instead, they aimed to unify and connect different areas:

> Recent advances in science have increasingly revealed unifying traits connecting the different fields. . . . We notice this trend, for instance, in chemistry. . . . There was a mathematical physics, but no mathematical chemistry. Today we scarcely can distinguish between the two fields. . . . A field of biophysics is emerging. . . . The same situation exists in the humanities and the social studies. (Frank 1950, 58–59)

Science does not need external metaphysical principles to cultivate unity. It was in order to both articulate and promote this reality of modern science that Frank organized and directed the Institute for the Unity of Science.

1.3.4. The Institute and General Education

The Institute was established officially in 1947 (Frank 1947a) and came to life in 1949 as a department of the American Academy of Arts and Sciences in Boston. As head of the Institute, Frank worked to promote the programmatic vision he defended at Finkelstein's conferences. With members including Charles Morris, Ernest Nagel, Percy Bridgman, Rudolf Carnap, Herbert Feigl, W. V. O. Quine, Hans Reichenbach,

---

47. See Uebel (2020). Though Uebel speaks only about Carnap and Neurath, we think that his views may be extended to include Frank.

Harlow Shapley, and others, the Institute also supported the *International Encyclopedia of Unified Science*, which had become moribund during the war and, with Neurath's death in 1945, had lost its main organizer and promoter. With support from the Rockefeller Foundation, Frank and his Institute sponsored lectures and colloquia, writing contests designed to recruit talent into the unity of science movement, and (as we will discuss later) promoted the sociological study of science.[48] In part, Frank aimed to fill Neurath's shoes and to organize scientists and philosophers to move beyond slogans and actually collaborate and build unifying bridges from science to science, as Neurath had urged since the 1930s (see Reisch 2005, 294–310).

At the same time, Frank allied his Institute and his own activities with Conant's postwar venture in general education. Besides teaching courses in the new program for undergraduates who were likely to become professionals, business leaders, or politicians, Frank addressed other areas of the academy and spoke to growing anxieties about (and a widening gulf between) the "arts" and the "sciences." In the late 1950s, these concerns would crystallize around C. P. Snow's famous essay "The Two Cultures and the Scientific Revolution," in which Snow described between scientists and humanists 'a gulf of mutual incomprehension—sometimes (particularly among the young) hostility and dislike, but most of all a lack of understanding" (Snow 1961, 4). Years before, however, Frank's Institute officially "work[ed] for an integration between the sciences and the humanities" (Frank 1951, 6), and Frank himself spoke frequently at colloquia and staged debates dedicated to general education and interdisciplinary relations among the sciences and the humanities.[49] In semipopular venues, including *Daedalus* and *The Bulletin of the Atomic Scientists*, and popular magazines such as the *Saturday Review* (one issue of which featured him on its cover), Frank reviewed books and addressed

---

48. For more on the unity of science movement, see Galison (1998) and Reisch (2003a, 2003b, 2005).

49. Examples of topics to which Frank spoke include "Education at Harvard—Means and Methods," (*Harvard Crimson*, April 18, 1949); "The Scholar and Society," (*Harvard Crimson*, March 6, 1948); "Is the Concept of Science Different in Biology from What It Is in the Physical Sciences?" (with Conant) (*Harvard Crimson*, December 9, 1952); "Ethics for Modern Man," (*Harvard Crimson*, February 19, 1954); "Poetry, Philosophy, and Science: Approaches to Reality," (*Harvard Crimson*, May 13, 1954); Frank was also involved in the General Education program where he led a seminar on "Criteria of Validity in the Physical Sciences" (*Harvard Crimson*, March 16, 1948).

issues of the day for readers outside of physics and philosophy, if not also outside the academy altogether.

For many admiring students and colleagues, however, it was Frank himself they remember most vividly as he personified his humanistic approach to science and learning. Besides his vast historical knowledge, he spoke (besides his native German) French, English, Czech, Russian, Persian, and some Arabic (Bernstein 2001, 71). As it had been in Vienna and Prague decades before, Frank's erudition was public, well known, and much in demand at Harvard. He was a regular fixture on campus, whether reading a newspaper alone or looking for someone—a student or prominent scholar, it did not matter—to discuss anything from social problems to physics and culture. His famous lectures on philosophy of science were attended by 250 students (Holton 2006; Ridenour 1949, 9), not least because of Frank's famous ability to explain complex problems clearly and simply as his own clarity of understanding shone forth.[50] Even Frank and Hania's home in Cambridge was a venue for lively discussions in the style of the "Vienna Coffee House," as suggested by one surviving photograph (see figure I.7).

1.4. The Final Years

Frank retired from Harvard in 1954. Along with P. W. Bridgman, who retired at the same time, Frank was honored with a conference in 1956, titled "Science and the Modern World View: Toward a Common Understanding of the Sciences and the Humanities" (see figure I.8). Over 250 participants, including prominent luminaries as diverse as Suzanne Langer, Detlev Bronk, Lewis Mumford, and Sidney Hook convened to honor Frank and Bridgman's "eloquent and persistent exposition of a broad view of science," as Gerald Holton put it in an issue of *Daedalus* dedicated to the conference (Holton 1958, 6–7; see also the *Crimson*, May 7, 1956).

---

50. Bernstein 2006, 9; see also Bernstein 2020, 139–44. Frank's predilection for teaching elementary and introductory courses was remembered as late as 1989 in a report about the quality of American education that contrasted European and American conventions: "In Vienna, where Frank had previously taught, beginning courses were considered the greatest honor—one to be bestowed on only those who had mastered their fields sufficiently to be able to generalize" (Cheney 1989, 8). But in the United States, where graduate assistants and nontenured faculty taught elementary courses, Frank was at first not allowed to teach the introductory and elementary courses he wanted to teach.

Figure I.7. Undated photograph of Frank and Hania presiding at a "Vienna Coffee House" at their home in Cambridge. Courtesy of Harvard University Archives.

Figure I.8. Frank and Bridgman's joint retirement conference, 1954. Front row: Gerald Holton, I. I. Rabi, Detlev Bronk, P. W. Bridgman, W. V. O. Quine, Giorgio de Santillana. Back row: Charles Morris, Nathan M. Pusey, J. Robert Oppenheimer, John Burchard, Perry Miller, Howard Mumford Jones, Hartford Brown. Courtesy of Gerald Holton.

After his retirement, Frank remained active. In 1956, he taught a guest course in MIT's School of Humanities and Social Studies. He also continued to publish, to attend conferences, and to be honored by his colleagues. In 1962, for example, fifty years after Frank's appointment to the chair of theoretical physics at Charles University, the Boston Colloquium for the Philosophy of Science held an anniversary dinner.[51] That same year, he was interviewed by Thomas Kuhn about his roles in the history of physics. But Frank's fortunes soon declined as he succumbed to dementia and could no longer live independently. Reports about Frank's as well as Hania's decline shocked and saddened friends and colleagues. By the mid-1960s, Nina Holton (2020, 172) recalled, Frank was eighty-two years old and "became quite often confused and forgetful, but always tried to be cheerful." Early in 1965, Kuhn acknowledged Frank's "plight" and prepared to forfeit royalties from his *Structure of Scientific Revolutions* (originally commissioned as a part of the *International Encyclopedia of Unified Science*) as he feared that Frank was reduced to "living, penuriously at that, on the royalties of the *Encyclopedia*."[52] Both Frank and Hania were "quite senile and very weak and pathetic," Carl Hempel wrote to Ernest Nagel in a letter that described a heartbreaking visit Paul Feyerabend had paid to his fellow Austrian. During the visit, Feyerabend reported, Frank asked him "whether he knew a philosopher called Feyerabend."[53]

In September 1965, Philipp and Hania were moved into the Cambridge nursing home where Robert Cohen would visit to show Frank

---

51. Robert S. Cohen organized the event and solicited from Niels Bohr a congratulatory telegram that applauded and admired Frank's "rich life work devoted to the elucidation of the foundations of science and its place in contemporary culture" (Cohen to Bohr, May 11, 1962, and Bohr to Cohen. Bohr-Frank correspondence, NBSC-NBA, folder 95). Frank's appointment at MIT was announced by MIT's News Service: "For release in morning papers of Wednesday, March 14, 1956" (MIT Archives and Special Collections, News Office (AC0069).

52. Thomas Kuhn to Maurice English, January 21, 1965, TKP-MIT, box 25, folder 55.

53. Hempel to Nagel, September 2, 1966, ASP CH 28-02; see also Feyerabend 1995, 103. Frank had an important impact on Feyerabend's thought, and he explicitly noted in both in his autobiography and in his notorious *Against Method*: "Frank argued that the Aristotelian objections against Copernicus agreed with empiricism, while Galileo's law of inertia did not. As in other cases, this remark lay dormant in my mind for years; then it starts festering. The Galileo chapters of *Against Method* are a late result" (Feyerabend 1995, 103. See Feyerabend 1988, 277).

the newly printed volume two of the *Boston Studies in the Philosophy of Science*—a *Festschrift* dedicated to Frank. He died six months later on July 21, 1966.

2. Dating the Manuscript

It is difficult to know the precise date of the composition of *The Humanistic Background* because the bulk of Frank's personal papers were lost and never acquired by Harvard or some other library or repository. One indication that the book was among Frank's last projects is the circumstances by which Harvard University Archives acquired the manuscript. In early 1963 an unknown member of the faculty wrote to inform Frank about a manuscript and other papers and letters of his "tucked away in compartment 521"—a faculty research carrel, evidently—"which you were using for a period" (see p. ii). Whether it was Frank, Hania, one of Frank's students, or this unknown professor who secured the manuscript and documents for the archives (where the note survives today, along with two small boxes of archived documents), it appears that Frank's carrel had been reassigned by early 1963, a likely *terminus ad quem* for the manuscript as well as Frank's productive academic career.[54]

As for a *terminus a quo*, National Science Foundation reports suggest that *The Humanistic Background* may have grown out of the work that culminated with Frank's conference on *The Validation of Scientific Theories*. In his introduction to the published volume Frank noted that the American Academy of Arts and Sciences and the National Science Foundation funded the conference that took place during the Christmas vacation of 1953 (Frank 1956, viii). A 1954 report of the foundation mentions the conference and additional funding that had been provided to Frank. Under the heading "History, Philosophy, and Sociology of Science," it mentions Frank's plan

> to undertake an historical analysis of the validating grounds of scientific theories together with the social and psychological atmosphere in which theoretical ideas originate, develop, and become accepted, rejected or modified. This investigation is

---

54. Unknown to Frank, March 27, 1963, HUA-PFP. The note's typist initialed it "RHH," providing one clue to who inherited Frank's compartment 521.

aimed at clarifying the relations and interactions of various fields of science on each other and in demonstrating what influences have contributed to the progress of modern science. (Anonymous 1954, 51)

These social and psychological—broadly, cultural—interests roughly match the aims and goals of Frank's manuscript. This suggests that Frank had conceived the project by 1953 (in order to receive support announced the following year).

There are also chronological markers in the manuscript itself. In §1.2.3., Frank referred to Viscount Herbert Samuel's then-current role in the British House of Lords, one he occupied from 1944 to 1955. This suggests that Frank was at work on the manuscript by 1955, at the latest—a dating confirmed by other markers: similarities between the manuscript's concluding "Appendix" and a talk Frank gave in November 1955 at Harvard titled "Science and Religion in the Mind of Einstein"; reference to a talk by Adlai Stevenson given in 1955 (Pt. 1. Ch. 1. Sect. 1); and Frank's remark (in Pt. 1. Ch. 1. Sect. 5) that important issues surrounding relations between science and the humanities can be found in nearly any daily newspaper. He illustrated his point by quoting from a newspaper of February 4, 1956.

It is less clear when Frank finished writing. The text does not refer to events or people from the 1960s, so it remains possible that he considered it finished by the end of the 1950s. Given reports of Frank's declining health, therefore, we are inclined to date the manuscript's creation to the last half of the 1950s, from roughly 1954 to 1960.

Despite lack of evidence that Frank attempted to publish the manuscript, there is no doubt that he envisioned the manuscript as the "book"—or, often, "the present book"—that he mentions in its pages. As Frank originally conceived and wrote it, that book was very long. The existing manuscript pages were repaginated three times in a manner indicating that it went through two major reductions (of 100 and 416 pages, consecutively) and then an expansion of 36 pages that brought the manuscript to 342 typed pages (excluding occasional handwritten insertions). It is possible, therefore, that the preliminary manuscript of 857 pages consisted of chapters from which Frank selected those that he published separately as his last published book, *Philosophy of Science: The Link Between Philosophy and Science*. While only an examination of the manuscript pages for *Philosophy of Science* could confirm this, we speculate that Frank originally set out to produce a synoptic exposition of his mature philosophy of science and then divided that project into two complementary books. While it contains some themes in common with *The Humanistic Background*, including Frank's interest in how interpretations of science "have actually served as supports of moral, religious and political creeds," (1957, iv), Frank's *Philosophy of Science* of 1957 is a more focused investigation into the sciences themselves. It discusses traditional philosophical topics, such as induction, geometry, causality, atoms, and relativity. What remained of the original manuscript were chapters exploring different backgrounds—both historical and contemporary—that inform contemporary cultural and philosophical debate about science. They formed the book that Frank titled *The Humanistic Background of Science*.

3. The Multilayered Significance of *The Humanistic Background*

It cannot be emphasized enough that Frank's recognition and his influence are among the most curious issues in the history of twentieth-century philosophy of science in general and the history of logical empiricism in particular. Frank undoubtedly had a very basic and general institutional recognition in the United States: many reviews of his books appeared; he was often invited to conferences, seminars, workshops, and even art galleries.[55] Nonetheless, most of these were related to his local contexts in Boston and New York, and to his well-known biography of Einstein. Among mainstream analytic philosophers in the 1950s and early 1960s pursuing relatively formal studies of theories and methods, Frank was neglected, and his reputation declined, as illustrated by dismissive and sometimes acerbic reviews of his books (see Reisch 2005, 324–30 and

---

55. After Frank's retirement, it appears, Frank wrote advertising copy for a New York gallery show of paintings by Fraydas, an illustrator of books whom we believe was Stan Fraydas, Hania Frank's nephew. Frank applauded the largely abstract paintings on display for the "ideas about the world and about human behavior" they manifest, for drawing on "the growth of the behavioral sciences like psychology, anthropology and sociology with their new interpretation of human behavior," and for endeavoring—like modern science, he emphasized—"to convert the stubborn facts of nature in a beautiful and useful structure" (Brochure, "Fraydas," at the Ruth White Gallery, HUA-FFP).

Tuboly 2017, 272–74). His eclectic and synthetic approach to understanding science remained dominant only among those in New York and Boston who knew him—including Robert S. Cohen and Marx W. Wartofsky. Besides dedicating the second volume of *Boston Studies in the Philosophy of Science* to Frank as a Festschrift, they organized and chaired the Boston Colloquium for the Philosophy of Science that "construes the philosophy of science broadly, as [Frank] had advised us to do." Their credo continued:

> We try to discuss open problems in the foundations of science, and, wherever relevant, to bring material from the history and cultural relations of science to bear upon such problems. We try also to talk with each other across all boundaries of discipline, to include scholars from philosophy, logic and mathematics, the physical and biological sciences, history and the social sciences, and the humanities as well. (Cohen and Wartofsky 1965, vii)

This broad view of philosophy of science is nowadays called "integrated history and philosophy of science" (or *&HPS*). As there is no generally accepted methodology of *&HPS* (see Arabatzis and Schickore 2012; Arabatzis and Howard 2015), *The Humanistic Background* may be of particular interest today for the insights and methods it illustrates. More broadly, the history of positivism that Frank presents seems poised to improve our understanding of how contemporary science studies, shaped by historical as well as intellectual factors, inherited its current practices and disciplines.

That said, *The Humanistic Background* is not a systematic treatise. Instead it shows Frank making integrative sociological and historical points as he appeals to a uniquely broad and eclectic range of primary and secondary sources, including personal and scientific correspondence, biographies, textbooks, handbooks, unpublished materials, journals, and newspapers. He had a sense of where a given story might lead, and the fact that he did not back up his investigations with detailed archival work suggests this general interpretation of his lifework: Frank laid down a new approach to understanding science that emphasized equally its epistemic and social aspects and that science is primarily a human undertaking.

On this view, Frank remained relatively obscure because he did not have a fully executed and detailed research program that could unite and inspire his colleagues. He additionally lacked organizational skills and often did not follow through on projects (such as a book series in philosophy of science that Frank was to edit for Harvard University Press in the early 1950s).[56] Though his name was often mentioned in the philosophical literature,[57] outside of Boston the success of Carnap, Feigl, Hempel, and others overshadowed Frank's work. His influence was real but not evident for those who had not known him or had not carefully examined the literature of the 1940s and 1950s.

Although this professional neglect may be explained by various factors (Reisch 2005, chap. 15; Tuboly 2017), the question remains about how *The Humanistic Background* might have been received by pragmatists, philosophically sensitive sociologists, and historically inclined scientists had it been published in Frank's lifetime. Frank had made some of the book's main points in scattered papers he wrote in the 1940s and 1950s,[58] but we cannot but wonder whether the collected, sustained, and often provocative treatment of these themes in *The Humanistic Background* might have helped to preserve Frank's reputation for later generations— and possibly kept on philosophy's table some of his interdisciplinary and cultural ambitions.

---

56. Frank asked Niels Bohr to write a book on the interpretation of quantum mechanics for this series, but we see no evidence that his request came to fruition (Frank to Bohr, June 4, 1952, NBSC-NBA, folder 95). See also Reisch (2005, chapter 15).

57. There is, for example, a master thesis from 1969 dealing with Frank's philosophy of science in a systematic fashion which remains one of the longest and detailed presentations of his views. See Synnestvedt (1969). A recent exception is Amy Wuest's dissertation (2015; for an extract see Wuest [2017]).

58. Frank's *Between Physics and Philosophy* (1941) and *Modern Science and its Philosophy* (1949d) collect many of his philosophical writings from the 1930s and 1940s in which appear themes covered in *The Humanistic Background*, as well. These include the Copernican revolution, school philosophies and their relations to science, metaphysics and common sense, education and the humanities, and the popularization of science. The same goes for Frank's last published book, *The Philosophy of Science* (1957), which discusses these issues and utilizes some of the terminology used here (e.g., "metaphysical interpretations"). In notes we often refer the reader to Frank's earlier writings when relevant or similar. Nonetheless, *The Humanistic Background* often goes beyond them with, for example, its sustained and systematic treatments of the history of positivism, scientific philosophy's place within the history of pragmatism, and the politics of "metaphysical interpretations." We note again Frank's use of divergent and sometimes unlikely sources in *The Humanistic Background*.

3.1. The Main Theses and Approach of *The Humanistic Background*

In the broadest outline, the book comprises in Part I an exposition of Frank's mature philosophy of science and, in Part II, applications of that philosophy of science to a number of issues and problems. Given Frank's somewhat mosaic-like style of writing and assembling of chapters, exceptions to this organizational scheme abound, including, for example, seven manuscript pages introducing a Part III, titled "Positivism and Ideologies," that Frank evidently excluded when he consolidated and repaginated the manuscript. As a result, different readers may well identify different issues and problems as central. Indeed Frank himself, in the first chapter of Part I, identifies at least four separate (though related) goals for "the present book":

> (1) "[The] chief topic of the present book will be the pragmatic approach to present day science including the line of descendance from which our present science has originated."

> (2) "[W]e have also to study the meaning of these [scientific] symbols as expressions of human aspirations. Thus, the variety of meanings which have been attributed to scientific symbols is a main topic in the present book."

> (3) "We shall discuss, in the present book, the ways in which these philosophical groups attempt to trace their 'genealogy,' back to scientific theories."

> (4) "We shall learn in the book that when one knows which philosophical approach should be supported, one will find a way to get this support out of several physical or biological theories."

The first two topics indicate Frank's interest in philosophical pragmatism (of Dewey, Peirce, Bridgman, and others) and in the philosophical study of symbols in human life (then strongly represented by the work of Charles Morris). The third and fourth topics capture the relative uniqueness of *The Humanistic Background* in so far as Frank extends and applies these interests to the sociology of philosophical schools or, as he put it here, "philosophical groups" and to the interpretive uses (and abuses) of scientific theories by leaders and representatives of social and national movements who seek scientific legitimacy for their agendas. "We shall discuss in the present book," as Frank put it in yet another formulation, "the role which the symbols of scientific discourse have played in the struggle for moral and political goals" (Pt. 1, chap. 1, sect. 7).

To illustrate these struggles, Frank singles out three philosophical schools for extended analysis. The first two are Thomism, endorsed by Hutchins and Adler at the University of Chicago, and pragmatism in its various forms, which Frank gathers under the umbrella of "positivism." The third, dialectical materialism, takes Frank not only outside of academic philosophy but also into the ideological core of the primary US Cold War enemy: the Soviet Union. Frank treats each of these schools and the factions within them sympathetically—surprisingly so in the cases of Thomism and Soviet philosophy, especially given the nation's popular suspicion of Soviet ideology. With something of a sociological or anthropological objectivity, Frank quotes extensively from each school's internal literature and portrays them as trying to organize and improve society and to "guide human conduct" in ways that inherit the authority and prestige of modern science.

By dedicating long stretches of the book to this subject, Frank attempted to critically enlighten the very publics that advocates of these schools aimed to convert to their respective worldviews and to enlist in their movements. In this aspect, Frank's project joined the CSPR and the humanist movement (discussed in the next section) in aiming to educate and enlighten the public about science and its philosophical interpretation and to continue Neurath's lifelong efforts to educate and inform the public about science and its roles in the modern world.

Frank's hopes for *The Humanistic Background* to join ongoing discourse about science and democracy were necessarily joined to an internal critique of academic philosophy of science. This critique was surrounded and made urgent by public anxiety about science after the Second World War. Writing soon after the public learned of the atomic bomb and the threat of nuclear annihilation,[59] Frank understood how

---

59. During the late 1930s, Frank was commissioned to write a monograph for Neurath's *Encyclopedia* on the "foundations of physics." While Frank was known for his slow progress in his works, the final delay in late 1945 was explained to Neurath as follows: "[The delay was] caused by the explosion of the atomic bomb. I had to add to the galley proofs a new chapter on the 'atomic power' without which no American would have ever bought this pamphlet." Frank to Neurath, December 15, 1945, ONN. See Frank (1946, chapter 8). This might have been motivated by Conant (1947/1951, xii) who

this anxiety was often blamed on the nuclear physicists and chemists of the Manhattan Project whose Promethean hubris could only lead to tragic consequences: "A great many people would wish that the vultures get at the livers of the nuclear scientists," as well, Frank writes in his introduction. But philosophers of science, including those in the Vienna Circle and allied groups, had also contributed to science's postwar reputation. Philosophical accounts of modern theories as formalizable systems of statements devoid of emotion or value encouraged critics of science (including neo-Thomists) to argue that modern science is essentially harmful to cherished human values or to cultural progress.

The most notable American critique of Vienna Circle philosophy in this vein had come from John Dewey, who contested logical positivism's strict cognitivism and its disregard of norms and values. In the late 1930s, Otto Neurath, aware that Dewey's stature could only be a boon to his encyclopedia project, visited Dewey personally at his home and persuaded him to contribute to the *International Encyclopedia of Unified Science* (Reisch 2005, 84–85). Frank's *The Humanistic Background* can be seen through a similar lens: an effort to persuade American philosophers of science that in fundamental ways Dewey's critique was important and not to be underestimated if postwar philosophy of science were to have credibility and influence in the modern world.

The task for philosophy of science was not simply to reform itself along naturalist, pragmatist lines or to replace texts by Carnap or Reichenbach with texts by Peirce or Dewey. Frank urged instead the adoption of a historical perspective within which American pragmatism as well as European scientific philosophy could be seen as allied descendants of the original positivist movement. Alongside its analyses of philosophical schools, *The Humanistic Background* also surveys Comtean positivism and its development—including predecessors such as La Mettrie in France, and descendants such as Mach in Austria and Peirce in the United States. This synoptic picture provides Frank a way to unify European logical positivism and American pragmatism within a single "positivist" framework.[60] Thus *The Humanistic Background* contains a long chapter

---

said in his very popular *On Understanding Science* that "to write a book about science in the year 1946 without some consideration of the atomic bomb may seem the academic equivalent of fiddling while Rome burns."

60. Frank took a similar approach in his contribution to Einstein's *Library of Living Philosophers* volume; see Frank (1949a).

(Part II, chap. 5) about how to integrate logical empiricism (or logical positivism, as Frank often called it) into American pragmatism and how their common sources could be made explicit. This unification preserves important innovations of Vienna Circle philosophy (primarily its rejection of "picture theory" epistemology and recognition of the basic unity of the sciences) and yet tempers the formalism that was quickly beginning to dominate academic philosophy of science in the 1950s. Logical positivists who wield logical symbols, as well as artists and humanists intimidated by those symbols, can see within this larger framework that logical and scientific proofs as well as poems and works of art meaningfully draw on a larger humanistic background.

Frank's impulse to portray philosophy of science as a powerful, unified project can perhaps be understood as his response to the growth and especially the continuing diversification of scientific philosophy in the postwar United States. Frank observed, for example, the now well-documented debates and discussions organized by Alfred Tarski at Harvard during the 1940–41 academic year and attended by Carnap, W. V. O. Quine, and Hempel.[61] Frank described these meetings in a letter to Morris:

> We have now at Harvard . . . an "inflation" of logicians and positivists. There are very interesting debates. But it turns out that even among the "founders" of Logical Empiricism there is no agreement on the most fundamental issues e.g. on the question, whether the statements of mathematics and logic are empirical. Carnap and Feigl deny it, Mises and myself are inclined not to overrate the difference between empirical and so-called "analytical" statements. Even pure logicians like Quine and Tarski seem to favor a kind of "empirical" view on logic.[62]

What Frank believed should be a unified philosophical front was potentially splintering in different substantive and stylistic directions. Here

---

61. For details of the Quine-Tarski-Carnap workshop, see Frost-Arnold (2013). On debates including Frank about the future of logical empiricism, see Reisch (2005, 303–306) and Tuboly (2021a).

62. Frank to Morris, 1 December 1940, CMP, box 1. We have corrected Frank's spelling in this quotation.

Frank noted one of these developments—Quine's now-famous argument that the positivist distinction between analytic and synthetic statements was untenable—and appropriately dedicated parts of *The Humanistic Background* to his own more holistic and behavioristic view of theories and their components (Part II, chap. 5). This view (to echo Frank's comment to Morris) does not "overrate" this distinction, much less regard it as a foundational dogma on which scientific philosophy either stands or falls.

We believe this impulse to portray philosophy of science as a still-powerful and unified project led Frank to often minimize (if not sometimes ignore) what are today recognized as important differences and disagreements among notable philosophers and to portray other philosophers in a very pragmatist-friendly light.[63] On the other hand, Frank did not hesitate to spar with his colleagues and perhaps step on some toes. Although he shows that Carnap's criticism of metaphysics could be read and interpreted along pragmatist lines (much as he did to a lesser extent in his contribution to Carnap's *Library of Living Philosophers* volume [Frank 1963]) Frank does not hesitate to suggest that Carnap's formalism ignores at its peril the historical and sociological contexts that shape scientific theory and reasoning.

Even Neurath, whose antiformalist sensibilities Frank shared, comes in for a quiet yet firm reprimand on the issue of metaphysics. To be sure, Frank remained a proponent of science and a critic of antiscientific, metaphysical claims. But he finds inspiration in Peirce, Dewey, and Pierre Duhem to suggest that the logical positivist movement was mistaken—in at least two different senses—to categorically dismiss metaphysics as "unscientific" or meaningless noise as it had in the 1920s and early 1930s. On the one hand, such a dismissal is historically insensitive to the historical evolution of knowledge. The same theoretical claim may be seen as cutting-edge science awaiting confirmation, everyday common sense, or antique, outdated knowledge depending on its context and overall place in the historical evolution of the sciences. According to Peirce (Part II, chap. 3, §3), Frank explains, metaphysical knowledge—properly understood—emerges from commonsense knowledge and experience. Frank thus turns the tables on his colleagues to imply that Carnap's ordinary "thing language" and Neurath's "universal jargon"—upheld as empirical, objective, and nonmetaphysical platforms for science—are

---

63. See, for example, Frank's treatments in *The Humanistic Background* of Moritz Schlick, Ernst Mach, Pierre Duhem, and even some neo-Thomists.

themselves metaphysical, albeit not in a way that invalidates science or harms its progress. As soon as Frank arrived in the United States in 1938, he wrote to Neurath and reported surprising observations about metaphysical thinking on the part of his students that lead Frank to reconsider metaphysics.[64] Some of these developments are visible in one of Frank's early publications (Frank 1941/1949), but they are worked out in more detail in *The Humanistic Background*.

By examining a wide array of American philosophers, all of whom are more sympathetic to metaphysics than most logical positivists, it becomes clear to the reader that neither a critical analysis of metaphysical concepts (a la Carnap) nor prohibitions on metaphysical terms (a la Neurath's infamous *Index Verborum Prohibitorum*) will serve scientific philosophy well in the American intellectual context that Frank had come to know well by the 1950s.[65] They were handicapped by a blindness to the historical, sociological, and practical needs and purposes served by metaphysical beliefs and metaphysical interpretations of modern science, so they could never succeed in productively engaging (much less eliminating) metaphysics as the movement had pledged to do, for example, in the Vienna Circle's manifesto, *Wissenschaftliche Weltauffassung*, or in early antimetaphysical writings by Carnap in particular. Frank's *Humanistic Background* suggests, therefore, that "the elimination of metaphysics through the logical analysis of language" was a more complex task than logical empiricism had first envisioned.

3.2. The Humanistic Background in the American Scene

Frank's choice for a title also helps to situate his unpublished book in the postwar US intellectual milieu. The word "humanism" had gained increased currency in the 1940s with the humanist movement, its magazine, and its original manifesto of 1941. The manifesto, written by the philosopher Roy Wood Sellars and the Unitarian minister Raymond Bragg, called for a humanist religion built upon a naturalist, evolutionary worldview and dedicated to "the complete realization of human personality" within "a

---

64. See Frank's letter to Neurath, December 10, 1943, ONN. On Frank's changing conception of metaphysics and its criticism, see Uebel (2011) and Tuboly (2021b).

65. Metaphysically relevant differences between European and American culture were pointed out by American philosophers, as well. See, e.g., A. Cornelius Benjamin's review of Frank's 1949 collection of essays (Benjamin 1950).

socialized and cooperative economic order" (Sellars 1933). Reflecting the socialist ideals that many Americans had warmed to during the Depression era, the manifesto was signed by a roster of scientists, theologians, and philosophers. The most notable philosopher to sign was Dewey, whose anti-Thomist article "The New Failure of Nerve" (coauthored with Hook and Nagel [Hook 1943]) echoed the manifesto's call "to elicit the possibilities of life, not flee from them." The movement also influenced Frank's colleague Charles Morris who offered his own humanist prescriptions in his book *Paths of Life: Preface to a World Religion* (Morris 1942).

In 1956 Frank was interviewed in the pages of *The Humanist*, where he discussed themes dominant in *The Humanistic Background*—science's thoroughly naturalistic worldview, the compatibility of modern science and contextual, nonabsolute ethics, and essential roles for values in the scientific enterprise. The interviewer, Edwin H. Wilson, joined Frank in appreciating logical empiricism as an ally of the movement ("I knew, through the interest of such men as Rudolf Carnap, Herbert Feigl and Charles Morris, as well as Frank himself," he wrote as he introduced Frank to his readers, "that logical positivism is one of the various philosophical methods that arrives at an ethical position essentially compatible with humanism") and concluded the interview with a ringing endorsement of logical positivism as "essentially humanistic."[66]

As historians of philosophy now acknowledge, however, both philosophical pragmatism as well as public political engagements of the sort that Dewey, Hook, Hutchins, and others routinely undertook in the 1930s and early 1940s declined rapidly in the years after the war.[67] With few exceptions, the advent of the Cold War and the nation's prosperity

---

66. Wilson (1956, 191). In regard to Feigl, Wilson may have had in mind Feigl's "Naturalism and Humanism" (Feigl 1949).

67. Part of the campaign against university teachers and scholars consisted in FBI investigations from which logical empiricists, such as Rudolf Carnap and Frank, were not exempt. Gerald Holton, who was then Frank's assistant, later recounted the following anecdote: "Frank one day received a visit at his home from two FBI men. They had come to investigate his background and orientation, which seemed to them to have been suspiciously on the liberal side. Frank, no doubt with his usual quizzical smile, inquired whether they thought he might be a spy for the Russians, and to answer his own question, he went to his bookcase, fished out the copy of Lenin's book, and opened it to the passage where Lenin attacked him personally. As Frank ended this story, the two FBI men practically saluted him, and left speedily and satisfied" (Holton 1992, 44, n.48). For further details of these investigations based on internal FBI documents, see Reisch (2005, especially 268–71).

(relative to the depression of the 1930s) inaugurated for most scholars a new professionalism that prized internal, scholarly research and debate and that minimized (if not stigmatized) public advocacy and even scholarly engagements with controversial, politically charged subjects such as Marxism or atheistic humanism.[68]

Against this backdrop of increasing professionalism and depoliticization, *The Humanistic Background* stands in bold relief and documents Frank's sustained interest in politically perilous topics (such as Marxism and dialectical materialism) and his relative lack of interest in the professional and disciplinary boundaries then growing stronger in the American academy. This includes, for example, boundaries between philosophy and history, literature, religion, and the then-nascent field of Russianism, not to mention within philosophy: between analytic, continental, pragmatic, and sociological approaches to knowledge. At a time when most established philosophers of science were narrowing their disciplinary methods and goals, Frank's manuscript glides easily—too easily, contemporary readers may find—from discussions of important philosophers (Carnap, Quine, Whitehead, Dewey, Peirce, Neurath, and others) to sequential expositions of subjects like Thomism, Marxism, sociology of knowledge, historicism, theology, and even interpretation of the Bible.

Owing to Frank's multilingualism, *The Humanistic Background* is also unique for the quotations it contains and the sources it might have brought to wider attention had it been published in its time. These include passages from George Lukács's *History and Class Consciousness*, an influential book first published in English in 1971. It also includes quotations and summaries of French writers, including Édouard Le Roy, Edmond Goblot, Émile Littré, Henri Bergson, and Abel Rey, as well as quotations from writings by Frank's colleagues Schlick, Hahn, and Neurath that were not yet translated into English. Frank's knowledge of Russian allowed him to translate and quote writings by the Soviet philosophers and physicists Sergei Vavilov, Abraham Ioffe, and Mark Borisovich Mitin.[69] Any reader of *The Humanistic Background* who did

---

68. For a general account of the McCarthy era's effects on the American academy, see Schrecker (1986, esp. 338–41). On philosophy, see Howard (2003), McCumber (2001, 2016) and Reisch (2005). On the decline of public intellectuals in American culture, see Jacoby (1987).

69. It is known from Carnap's diaries (November 26, 1932; ASP RC 025-75-10) that Frank translated articles about "the Machians" from the official Russian Marxist journal *Under the Banner of Marxism* (*Pod znamenem marksizma*) and translated for

not read Russian would also have learned about Russian university textbooks (scientific and philosophical) and the *Great Soviet Encyclopedia*.

3.3. THE HUMANISTIC BACKGROUND, THOMAS KUHN AND THE SOCIO-HISTORICAL APPROACH TO SCIENTIFIC KNOWLEDGE

Sustained attention to the sociology of scientific knowledge in *The Humanistic Background* will legitimately lead to comparisons with Thomas Kuhn's influential book, *The Structure of Scientific Revolutions*, traditionally credited with inaugurating interest in the historical, sociological, and psychological study of science. This comparison is not abstract, for Frank's mature philosophy of science, his activities on behalf of the unity of science movement, and Frank himself—the friendly, talkative fixture in and around Harvard Yard—belonged to the intellectual landscape in which Kuhn became a historian of science and began to write *Structure*. As an undergraduate, Kuhn arrived at Harvard in 1940, one year after Frank, and studied with him, most likely in the Physics Department (Kuhn 2000, 268). After completing his PhD in physics, however, Kuhn became a historian of science and, beginning in the late 1940s, taught alongside Frank within President Conant's general education program. Besides their proximity to each other and their shared interests in physics and philosophy, both Frank and Kuhn had important relationships to Conant. Frank, we noted, owed his position at Harvard to Conant, and so did Kuhn (Reisch 2017, 2019a). Conant and his then new general education program offered Kuhn a welcome opportunity to leave physics and to teach history of science. Behind both Frank's and Kuhn's theorizing about science, moreover, lay Conant's book *On Understanding Science* (1947/1951), a book that introduced Kuhn to the case-study approach to teaching history of science that he implemented in *Structure* and whose central concept of "conceptual schemes" tacitly circulates in both *Structure* and *The Humanistic Background* (e.g., Part I, chap. 5).

The path that would lead Kuhn toward writing *Structure* also involved Frank and the Unity of Science Movement. For the book was originally commissioned in the early 1950s as a pamphlet in Neurath's *International Encyclopedia of Unified Science*. Frank was at this time an official editor of the encyclopedia (succeeding Neurath after his death), but no evidence exists showing that Frank and Kuhn discussed his contribution. It was rather Frank's coeditors, Charles Morris and (to a

---

Carnap Russian movies in the cinema (March 2, 1935; ASP RC 025-75-13).

lesser extent) Rudolf Carnap who shepherded *The Structure of Scientific Revolutions* to its eventual publication.

In *Structure* itself there is scant reference to philosophical writings by Frank.[70] There is some evidence, however, that Kuhn formed his ideas partly through collaboration and discussion with Frank. One archival document, for example, is an invitation from Frank to collaborate within Frank's Institute for the Unity of Science on a new committee to promote and organize research in sociology of science.[71] The committee already had on board two of Frank's close friends, namely the sociologist Robert Merton and the philosopher Ernest Nagel,[72] and later produced the document "Possible Research Topics / Sociology of Science" that lists themes including the interpretation of data and the metaphysics behind verbal differences; the relation between conceptual innovations and experiments; scientists' resistance to discoveries; the factor of scientists' age; and roles and effects of specialization.

There is little evidence that Frank's new committee was active or fruitful in its stated aims, but it is obvious that Frank took these topics seriously and worked on them for years. *The Humanistic Background* can perhaps be profitably read as the final result of the research done in this "sociology of science" group from a logical empiricist point of view. However much or little this group may have influenced Kuhn's developing

---

70. We note, however, that Charles Morris encouraged Kuhn to at least refer in *Structure* to Frank's work: "Do you know of Philipp Frank's book, *The Validation of Scientific Theories* (Boston: Beacon, 1957)? Frank is the editor. There are papers by Frank, Hempel, Guerlac, Koyré, R. Cohen, and others.... Frank has played an important role in the EUS [so it] would be good to have at least a reference to this book." (Morris to Kuhn, copy to Carnap, March 26, 1962, ASP RC 088-47-03). Frank's own contribution to that volume, "The variety of reasons for the acceptance of scientific theories" (1956), covers issues similar to those Kuhn treats in *Structure*. But in *Structure* Kuhn notes only Frank's biography of Einstein (Frank 1947b).

71. Frank to Kuhn, December 2, 1952, TKP-MIT, box 25, folder 53. This interesting document and Kuhn's response to Frank's invitation has only recently come to light in secondary Kuhn-literature. See, for example, Wray (2015, 171), Fuller (2003, 210), Isaac (2012, 220), and Reisch (2019a).

72. The friendship between Frank and Nagel was philosophical and personal, partly since Frank spent every second semester in Nagel's New York City. While organizing the Institute for the Unity of Science in mid-1946, for example, Frank encouraged Nagel to become the institute's managing director (Frank to Nagel, June 26, 1946, ASP-ENP). In *The Humanistic Background*, Frank discusses Nagel's philosophy of science (Pt. II, Ch. 5, Sect. 7).

ideas, given Kuhn's aim in *Structure* to reform (if not dramatically refute) logical positivist orthodoxy—and given Frank's stature at Harvard as a Vienna Circle logical positivist—it seems likely that this encounter guided and encouraged Kuhn as he began to theorize the nature of science as a professional historian of science.

A second piece of evidence is anecdotal and holds that Kuhn absorbed the now-famous notion of "Kuhn loss" from Frank. As Robert Butts tells it,

> In 1954, Philipp Frank gave a lecture for the American Academy of Arts and Sciences in Boston. In the audience, seated side-by-side, were Tom Kuhn and Adolf Grünbaum. Frank was discussing change in science, progress in science. . . . Often a newly accepted theory has lost some of the explanatory power of its earlier rival. Examples are plentiful. The loss of explanatory power of a theory is now widely referred to as "Kuhn loss." Years later, shortly before his death, Adolf reminded Tom of Frank's remarks. Tom Kuhn was shocked. (Butts 1999/2000, 197)

Whether or not Grünbaum's Freudian interpretation is persuasive ("Tom had absorbed the lesson offered by Frank, had repressed conscious memory of it, and had, by some mental trickery, called it into consciousness as his own idea" [Butts 1999/2000, 197]) the anecdote confirms what we should expect given their overlapping intellectual and professional affiliations: Frank and Kuhn knew each other, taught in the same program, and attended conferences and lectures together through the 1950s. Kuhn likely paid at least some attention, therefore, to the ideas articulated in *The Humanistic Background* and Frank's other writings.

For those interested in the origins and germination of Kuhn's influential ideas and, more broadly, the midcentury history of philosophy of science in America, *The Humanistic Background* may repay careful reading. For there can be little doubt that it illustrates a kind of integrated philosophy-history-sociology of science a decade or more before it became popular in the 1970s.[73]

---

73. From this perspective, Thomas Uebel's comparison of Frank and Neurath's sociology of science with the later strong program (Uebel 2000) is especially constructive. According to Uebel, Frank agreed that the natural sciences fall within

3.4. Evaluating *The Humanistic Background* Today

Perhaps the greatest strength of *The Humanist Background* is the way Frank's external and internal critiques join and call for a common remedy. In broadest strokes, the reason why complaints and misperceptions about science issuing from dictators (Joseph Stalin), popular theologians (Fulton Sheen), Thomist philosophers (Jacque Maritain), and poets (Archibald MacLeish) were so influential at midcentury—the external critique—had much to do with the enduring formalism and epistemological purism—the internal critique—that helped to professionalize (but publicly marginalize) scientific philosophy in this postwar landscape. In this largest aspect, then, *The Humanistic Background* can be seen as Frank's attempt to Americanize the philosophical movement he had helped to create in Europe so that it may yet achieve the jointly intellectual and cultural goals of the Enlightenment that it pursued decades before.

That said, *The Humanistic Background* is no lost masterpiece. However one assesses Frank's programmatic vision for midcentury philosophy of science, the scholarship behind it is occasionally sloppy. At its worst, it sometimes lacks clarity and coherence. Whether or not this is best understood as due to Frank's incipient health problems, his last book occasionally presents the reader with puzzles, false dichotomies, and overstatements—some of which Frank himself confesses are "flippant." Frank's eagerness to reconcile American pragmatism and logical empiricism seems to get the better of him, for example, when he writes, "That our pictures of the physical universe are not based upon intellectual research, but are influenced by our moral and political ideas, has been strongly upheld and lucidly presented by John Dewey" (Part I, chap. 6, §5). Or, in his enthusiasm for the sociology of science, Frank occasionally loses sight of empiricism: "We have learned, however, that the ultimate decisions between hypotheses in astronomy or physics are determined by sociological arguments" (Part II, chap. 3, §4).

---

the scope of sociology of science, that sociological explanations are aptly formulated for the validation of natural scientific theories, and Frank "deemed it appropriate to furnish sociological explanations not only when the hypotheses under consideration that were accepted are false, but also when they are empirically adequate" (Uebel 2000, 144–45) or, one might say, when they are true or accepted as true. We agree with Uebel's conclusion and see *The Humanistic Background* supporting his claims about Frank and the strong program of the sociology of science.

Many readers may be puzzled not only by the book's unconventional themes and subject matter but also its two-part architecture to which Frank added a relatively short tribute to Albert Einstein that he labeled "appendix." We present it as the book's "conclusion," even though it does not contain doctrinal summaries and proclamations one might expect to find at the end of a philosophical book about science. One way to make sense of the appendix is to suppose that Frank chose not to conclude his book by taking a stand on one or more doctrinal positions or theses. He chose instead to introduce Einstein—as he did in *The Humanist* magazine[74]—as an exemplary philosopher of science whose personal "cosmic religion" draws on the array of epistemological, methodological, and cultural issues covered in *The Humanistic Background*. The concluding thrust of Frank's book, that is, is not to embrace a doctrinal position (a la Thomism or dialectical materialism) but to find one's own way in life, or in science, as Einstein had, with an awareness of the intellectual as well as practical resources available within our shared humanistic background.

This is what Paul Feyerabend meant, we suggest, when he recalled that "given the choice of explaining a difficult point by means of a story or of an analytical argument," Frank "would invariably choose the story" (Feyerabend 1995, 103). Frank endorsed stories not because he was unable to produce sophisticated and sharp formal arguments, or because he believed that they are unimportant. He believed rather that the stories we tell ourselves allow better access to science's humanistic background. That background, in turn, guides our understanding of the world, our place in it, and the potentialities it offers. The stories Frank provided and applauded may be sometimes inconsistent or filled with tensions that pull in different directions. But as his experiences of the Mach-Boltzmann debate and the interpretations of Einstein's theories had taught him, theoretical mosaics can be put together in many different ways.

4. Editorial Preparation and Remarks

We faced several editorial challenges in preparing *The Humanist Background* for publication (see figure I.9). Most daunting was an absence

---

74. "Was Einstein a Humanist?" Frank was asked. He replied that he was "in the sense that he was not a supernaturalist. He liked to speak of God, but it was a God identified with the laws of nature, not a God who cared for human beings or interfered in human affairs. His God was that of Spinoza, not that of the church" (Wilson 1956, 191).

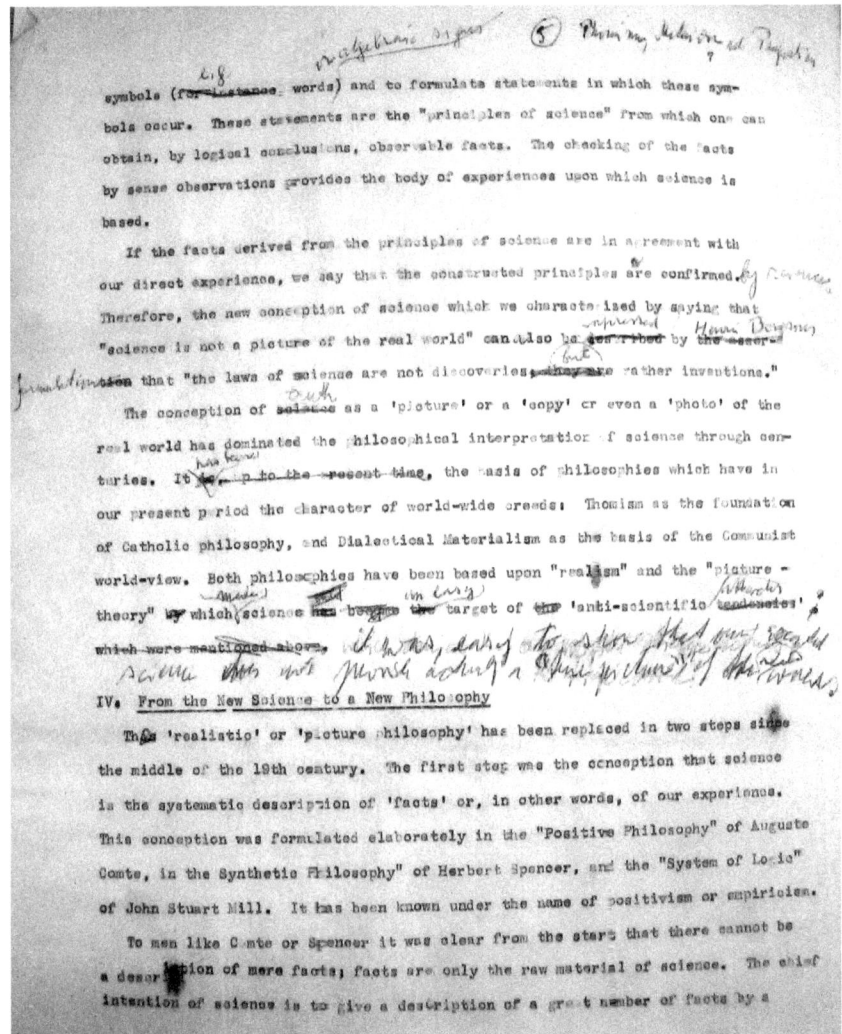

Figure I.9. A representative page from Frank's manuscript.

of references and citations to match the footnote numbers in Frank's text. With online text searches often proving fruitless given Frank's distinctive translations, we relied on the help of colleagues and scholars to determine Frank's sources for many quotations. As we finalized the manuscript, however, we located a file containing the manuscript's missing references. Still, in many cases, Frank gave little or no information about his sources. Those we were never able to identify are noted in the relevant footnotes.

When we did locate and examine Frank's sources, we discovered that he often quoted sources from memory, or perhaps from notes he had taken. Given discrepancies between Frank's quotations and their published sources, we chose to preserve Frank's inexact quotations in the main text and to provide their published counterparts in footnotes. In cases where multiple editions or translations exist, we used the one most similar to Frank's own.

Given these uncertainties and contingencies, all in-line references using the author-date system, and all footnotes containing bibliographical references should be considered as editorial notes either given or reconstructed by ourselves. In a smaller number of cases, Frank's notes contained consequential or potentially interesting information about texts, historical or philosophical circumstances, or Frank's preferences and opinions. We reproduce these notes verbatim and introduce them with a phrase such as "Frank's note here reads."

Given that Frank spoke and wrote English fluently but not always idiomatically and concisely, *The Humanistic Background* required extensive editing. In all of his books, Frank thanks editors and colleagues, often several, who prepared his manuscripts for publication. We presume that they encountered similar issues, including awkward idiomatic expressions, Frank's Germanic tendency to write very long sentences, and organizational problems owing to his occasionally meandering curiosity.

To treat the first two kinds of problems, we copyedited Frank's writing to make it more readable by contemporary standards. Though we avoided introducing words into Frank's sentences unless it was grammatically necessary, readability was improved often by adjusting Frank's idiomatic phrases and removing inessential words such as "very," "however," and standard phrases such as, "as a matter of fact." We additionally sometimes changed Frank's sentences from passive to active voice.

Organizational problems—some likely resulting from the way Frank removed and added pages from the manuscript as it evolved—required higher-level editing. The manuscript occasionally reintroduced figures or concepts that had already been introduced, and sometimes took up themes or philosophers that had never been introduced (at least in the pages that remained to constitute *The Humanistic Background*). As much as possible, we remedied these defects by removing redundant text, or by preserving but relocating Frank's own text in order to serve the flow of his argument or discussion.

In several circumstances, we added text to the manuscript. In the course of reorganizing the manuscript, Frank sometimes divided sections without providing titles for the newly created sections. In these cases we supplied descriptive section headings. Though Frank sometimes penciled in revised chapter titles, we occasionally adjusted his choices to better match what the chapter in question covered. In one case (Part II, chap. 4), Frank's chapter meandered in the first few pages before settling down to examine an identifiable topic, in this case "Science, Democracy, and the New Wave of Positivism"—a title that we supplied on the basis of Frank's reference in that chapter to "a new wave of positivism" in the late nineteenth century. Finally, at the ends of his chapters, Frank tended neither to summarize what had been discussed nor to indicate what the subsequent chapter held in store. In some cases, we inserted short paragraphs to better orient the reader who might be dazzled by abrupt changes of subject between chapters.

Finally, in his manuscript Frank provided on separate sheets basic information (affiliation, important books, and important views) about those figures he discussed or quoted. Frank employed a similar convention in the "Notes" section of his book *Philosophy of Science* (Frank 1957). In our case, we included these notations in footnotes only when Frank discussed figures who are likely to be unknown to contemporary readers or when his notations contained idiosyncratic views or potentially useful information.

5. Acknowledgments

We are indebted to many scholars and colleagues, especially for helping us track down some of the references. We would like to thank (in alphabetical order) David Bakhurst, John Connelly, Hans-Joachim Dahms, David R. Grinnell, Michael Grüttner, Klaus Hentschel, Gerald Holton, Don Howard, Anna Jedynak, Christian Joas, Anssi Korhonen, Christoph Limbeck-Lilienau, Gary Lance Lugar, Andrey Maidansky, Juan V. Mayoral de Lucas, Volker Peckhaus, Sebastian Rand, Alan D. Schrift, Friedrich Stadler, Krzysztof Szlachcic, Toby Thacker, Thomas Uebel, Eike Wolgast, K. Brad Wray, Amy Wuest, and Joanna Zaleska. We are especially grateful to Andrey Maidansky for providing us the page numbers for all the original Russian texts and for making new translations for us that helped reveal discrepancies and idiosyncrasies in Frank's versions. Also, a special thanks goes to Eike Wolgast for making available to us Heinrich

Kunstmann's book. We are also indebted to two anonymous referees who offered helpful comments, to State University of New York Press for judicious queries and copyediting of the text, and to Randall Auxier and John Shook, editors of the series American Philosophy and Cultural Thought for their generous support and encouragement.

We are indebted to the Harvard University Archives for granting permission to publish Frank's manuscript and several photographs included in this introduction. They exist among the Papers of Philipp Frank, 1943–1995, HUG 4406.10, and HUG 4406.2. We are also grateful to the Vienna Circle Institute and to Gerald Holton for granting permission to publish photographs of Frank.

During the editorial process Adam Tamas Tuboly was supported by the MTA Lendület Morals and Science Research Group and by the MTA Premium Postdoctoral Scholarship. During the finalization of the book, George Reisch and Adam Tamas Tuboly were supported by the MTA Lendület Value-Polarizations in Science Research Group.

List of Archives and their Abbreviations

ASP CH   Archives of Scientific Philosophy, Carl Gustav Hempel Papers, 1903–1977, ASP.1999.01, Archives & Special Collections, University of Pittsburgh Library.

ASP ENP   Archives of Scientific Philosophy, Ernest Nagel Papers, 1925–1982, ASP.2020.01, Archives & Special Collections, University of Pittsburgh Library.

ASP RC   Archives of Scientific Philosophy, Rudolf Carnap Papers, 1905–1970, ASP.1974.01, Archives & Special Collections, University of Pittsburgh Library.

CMP   Charles Morris Papers, Institute for American Thought, Indiana University–Purdue University Indianapolis.

GHWC   Georg Henrik von Wright Collection, Finnish National Library, Special Collections.

HUA-HSP   Harvard University Archives, Papers of Harlow Shapley, 1906–1966, (HUG 4773).

HUA-JCP   Harvard University Archives, Papers of James Bryant Conant, 1862–1987 (UIA 15.898).

HUA-PFP   Harvard University Archives, Papers of Philipp Frank, 1943–1995 (HUG 4406.10; HUG 4406.2).

MSN   Moritz Schlick Nachlass, Wiener Kreis Archiv, Rijksarchief in Noord-Holland, Haarlem, The Netherlands.

NBSC-NBA   Niels Bohr Scientific Correspondence, 1903–1962, Niels Bohr Archive.

ONN   Otto Neurath Nachlass, Wiener Kreis Archiv, Rijksarchief in Noord-Holland, Haarlem, The Netherlands.

TKP-MIT   Thomas S. Kuhn Papers, MIT Institute and Archives.

References

Adler, Mortimer J. 1941. "God and the Professors." In *Science, Philosophy and Religion: A Symposium*, 120–38. New York: Conference on Science, Philosophy and Religion in Their Relation to the Democratic Way of Life, Inc.

———. 1952. *The Great Ideas. A Syntopicon of Great Books of the Western World*. Chicago: Encyclopedia Britannica.

Anonymous. 1954. *Fourth Annual Report for the Fiscal Year Ending June 30, 1954*. US Government Printing Office: Washington, DC.

Arabatzis, Theodore, and Don Howard. 2015. "Introduction: Integrated History and Philosophy of Science in Practice." *Studies in History and Philosophy of Science Part A* 50: 1–3.

Arabatzis, Theodore, and Jutta Schickore. 2012. "Introduction: Ways of Integrating History and Philosophy of Science." *Perspectives on Science* 20: 395–408.

Barber, Bernard. 1961. "Resistance by Scientists to Scientific Discovery." *Science* 134: 596–602.

Benjamin, A. Cornelius. 1950. "Review of Frank's Modern Science and its Philosophy." *Philosophical Review* 59: 387–89.

Bergmann, Peter, G. 1966. "Philipp Frank in Prague." In *Expressions of Appreciation as Arranged in the Order Given at the Memorial Meeting for Philipp Frank, October 25, 1966*, edited by Gerald Holton, 4–6 Cambridge, MA.: Harvard University Press.

Bernstein, Jeremy. 1966. "Philipp Frank as a Teacher in America." *Expressions of Appreciation as Arranged in the Order Given at the Memorial Meeting for Philipp Frank, October 25, 1966*, edited by Gerald Holton, 23–26. Cambridge, MA.: Harvard University Press.

———. 2001. "Recollections of Frank." In *Ernst Mach's Vienna 1895–1930: Or Phenomenalism as Philosophy of Science*, edited by J. Blackmore, R. Itagaki, and S. Tanaka, 70–72. Dordrecht, The Netherlands: Springer.

———. 2006. *Secrets of the Old One: Einstein, 1905*. Dordrecht, The Netherlands: Springer.

———. 2020. *Quantum Profiles*. 2nd ed. Oxford: Oxford University Press.

Beuttler, Fred W. 1995. "Organizing an American Conscience: The Conference on Science, Philosophy and Religion 1940–1968." PhD diss., University of Chicago.

Breed, David R. 1990. "Ralph Wendell Burhoe: His Life and His Thought. I. Perceiving the Problem and Envisioning its Solution, 1911–1954)." *Zygon: Journal of Religion & Science* 25(3): 323–51.

Bryson, L., Finkelstein, L., MacIver, R. M. & McKeon, R. (eds.). 1964. *Symbols and Values: An Initial Study; Thirteenth Symposium. Conference on Science, Philosophy and Religion in their Relation to the Democratic Way of Life*. 9th ed. Columbia University, New York.

Butts, Robert E. 2000. "The Reception of German Scientific Philosophy in North America: 1930–1965." In *Witches, Scientists, Philosophers: Essays and Lectures*, 193–204. Dordrecht, The Netherlands: Kluwer.

Canales, Jimena. 2016. "Einstein's Bergson Problem: Communication, Consensus and Good Science." In *Cosmological and Psychological Time*, edited by Yuval Dolev and Michael Roubach, 53–72. Cham, Switzerland: Springer.

Carnap, Rudolf. 1932/1959. "The Elimination of Metaphysics through Logical Analysis of Language." In *Logical Positivism*, edited by A. J. Ayer, 60–81. New York: Free Press.

———. 1963. "Intellectual Autobiography." In *The Philosophy of Rudolf Carnap*, edited by Paul A. Schilpp, 3–84. La Salle, Illinois: Open Court.

Cheney, Lynne V. 1989. "A Core Curriculum for College Students." *Humanities* 10: 4–10.

Cohen, Bernard I. 1948. "Review of Frank's *Einstein, His Life and Times*." *Isis* 38: 252–53.

Cohen, Robert S., and Marx W. Wartofsky. 1965. "Preface." In *Boston Studies in the Philosophy of Science. Volume Two: In Honor of Philipp Frank*, edited by Robert S. Cohen and Marx W. Wartofsky, vii. New York: Humanities.

Conant, James B. 1947/1951. *On Understanding Science*. New York: New American Library.

Dewulf, Fons. 2021. "The Institutional Stabilization of Philosophy of Science and its Withdrawal from Social Concerns after the Second World War." *British Journal for the History of Philosophy* 29, no. 3.

D'Souza, Mario, and Jonathan R. Seiling, eds 2014. *Being in the World: A Quotable Maritain Reader*. Notre Dame: University of Notre Dame Press.

Dzuback, Mary Ann. 1991. *Robert M. Hutchins: Portrait of an Educator*. Chicago: University of Chicago Press.

Einstein, Albert. 1912/1953. "Report to the Philosophical Faculty of the German University on a Successor to the Chair of Theoretical Physics." In *The Collected Papers of Albert Einstein*. Vol. 5: *The Swiss Years: Correspondence, 1902–1914* edited by M. J. Klein, A. J. Kox, and R. Schulmann, 470–73. Princeton, NJ: Princeton University Press.

———. 1916/1997. "Ernst Mach." In *The Collected Works of Albert Einstein*. Vol. 6: *The Berlin Years: Writings, 1914–1917* (English translation supplement), 141–45. Princeton, NJ: Princeton University Press.

———. 1950. "Preface." In *Philipp Frank: Relativity—A Richer Truth*, v–viii. Boston: Beacon.

———. 1979. "Vorwort." In *Philipp Frank: Einstein, Sein Leben und seine Zeit*, v–vii. Braunschweig/Wiesbaden., Germany: Friedr. Vieweg & Sohn.

Feigl, Herbert. 1949. "Naturalism and Humanism: An Essay on Some Issues of General Education and a Critique of Current Misconceptions Regarding Scientific Method and the Scientific Outlook in Philosophy." *American Quarterly* 1: 135–48.

Feyerabend, Paul. 1988. *Against Method*. 5th ed. London, New York: Verso.

———. 1995. *Killing Time: The Autobiography of Paul Feyerabend*. Chicago and London: University of Chicago Press.

Finkelstein, Louis. "The Aims of the Conference." In *Science, philosophy and religion: A symposium*, 11–19. New York: Conference on Science, Philosophy and Religion in Their Relation to the Democratic Way of Life, Inc., 1941.

Frank, Philipp, and Richard von Mises, eds. 1925–1927. *Die Differential- und Integralgleichungen der Mechanik und Physik*. Teil I-II. Braunschweig: Friedr. Viewig & Sohn.

Frank, Philipp. 1907/1949. "Experience and the Law of Causality." In *Modern Science and Its Philosophy*, 53–60. Cambridge, MA: Harvard University Press.

———. 1908. "Mechanismus oder Vitalismus? Versuch einer präzisen Formulierung der Fragestellung (Besonders im Hinblick auf den Neovitalismus von Hans Driesch)." *Annalen der Naturphilosophie* 7: 393–409.

———. 1917/1949. "The Importance for our Times of Ernst Mach's Philosophy of Science." In *Modern Science and Its Philosophy*, 61–78. Cambridge, MA: Harvard University Press.

———. 1930. "Eröffnungsansprache." *Erkenntnis* 1: 93–95.

———. 1932/1998. *The Law of Causality and its Limits*. Dordrecht, The Netherlands: Springer.

———. 1936. "L'abîme entre les sciences physiques et biologiques vu à la lumière des theories physiques modernes." In *Actes du Congrès international de philosophie scientifique*, fasc. 2: *Unité de la science*, 1–3. Paris: Hermann & Édituers.

———. 1936/1949. "Logical Empiricism and the Philosophy of the Soviet Union." In *Modern Science and its Philosophy*, 198–206. Cambridge, MA: Harvard University.

———. 1941. *Between Physics and Philosophy*. Cambridge, MA: Harvard University Press.

———. 1941/1949. "Why Do Scientists and Philosophers So Often Disagree about the Merits of a New Theory?" In *Modern Science and its Philosophy*, 207–15. Cambridge, MA: Harvard University Press.

———. 1945. *Thermodynamics*. Providence: Brown University.

———. 1946. *Foundations of Physics*. Chicago: University of Chicago Press.

———. 1947a. "The Institute for the Unity of Science." *Synthese* 6: 160–67.

———. 1947b. *Einstein: His Life and Time*. New York: Alfred A. Knopf.

———. 1949a. "Einstein, Mach, and Logical Positivism." In *Albert Einstein: Philosopher-Scientist*, edited by Paul A. Schilpp, 269–86. New York: Tudor.

———. 1949b. "Introduction: Historical Background." *Modern Science and its Philosophy*, 1–52. Cambridge, MA.: Harvard University Press.

———. 1949c. *Einstein: Sein Leben und seine Zeit*. Munich: List.

———. 1949d. *Modern Science and its Philosophy*. Cambridge, MA: Harvard University Press.

———. 1950. *Relativity—A Richer Truth*. Boston: Beacon.

———. 1951. "Introductory Remarks: Contributions to the Analysis and Synthesis of Knowledge." *Proceedings of the American Academy of Arts and Sciences* 80: 5–8.

———. 1956. "Introduction." In *The Validation of Scientific Theories*, edited by Philipp Frank, vii–xi. Boston: Beacon.

———. 1957. *Philosophy of Science: The Link between Science and Philosophy*. Englewood Cliffs, NJ: Prentice-Hall.

———. 1962/2001. "T. S. Kuhn's Interview." In *Ernst Mach's Vienna 1895–1930. Or Phenomenalism as Philosophy of Science*, edited by J. Blackmore, R. Itagaki, and S. Tanaka, 61–6. Dordrecht, The Netherlands: Springer.

Frank, Philipp, ed., 1957. *The Validation of Scientific Theories*. Boston: Beacon.

Frost-Arnold, Gregory. 2013. *Carnap, Tarski, and Quine at Harvard*. Chicago: Open Court.

Fuller, Steve. 2003. *Kuhn vs. Popper: The Struggle for the Soul of Science*. London: Icon.

Fürth, Reinhold. 1965. "Reminiscences of Philipp Frank at Prague." In *Boston Studies in the Philosophy of Science*. Vol. 2: *In Honor of Philipp Frank*, edited by Robert S. Cohen and Marx W. Wartofsky, xiii–xvi. New York: Humanities.

Galison, Peter. 1998. "The Americanization of Unity." *Daedalus* 127: 45–71.

Gilbert, James. 1997. *Redeeming Culture: American Religion in an Age of Science*. Chicago: University of Chicago Press.

Gordin, Michael D. 2020. *Einstein in Bohemia*. Princeton, NJ: Princeton University Press.

Haller, Rudolf. 1991. "The First Vienna Circle." In *Rediscovering the Forgotten Vienna Circle. Austrian Studies on Otto Neurath and the Vienna Circle*, edited by Thomas Uebel, 95–108. Dordrecht, The Netherlands: Kluwer.

Hardcastle, Gary. 2003. "Debabelizing Science: The Harvard Science of Science Discussion Group, 1940–41." In *Logical Empiricism in North America*, edited by Gary Hardcastle and Alan W. Richardson, 170–96. Minneapolis: University of Minnesota Press.

Hofer, Veronika. 2002. "Philosophy of Biology around the Vienna Circle: Ludwig von Bertalanffy, Joseph Henry Woodger and Philipp Frank." In *History of Philosophy of Science. New Trends and Perspectives*, edited by Michael Heidelberger and Friedrich Stadler, 325–33. Dordrecht, The Netherlands: Springer.

———. 2013. "Philosophy of Biology in Early Logical Empiricism." In *New Challenges to Philosophy of Science*, edited by Hanne Andersen et al., 351–63. Dordrecht, The Netherlands: Springer.

———. 2020. "Philipp Frank's Civic and Intellectual Life in Prague: Investments in Loyalty." In *The Vienna Circle in Czechoslovakia*, edited by Radek Schuster, 51–72. Cham, Switzerland: Springer.

Holton, Gerald. 1958. "Perspectives on the Issue 'Science and the Modern World View.'" *Daedalus* 87, no. 1: 3–7.

———. 1995. "On The Vienna Circle in Exile: An Eyewitness Report." In *The Foundational Debate: Complexity and Constructivity in Mathematics and Physics*, edited by Werner Depauli-Schimanovich, Eckehart Köhler, and Friedrich Stadler, 269–92. Dordrecht, The Netherlands: Springer.

———. 2006. "Philipp Frank at Harvard University: His Work and His Influence." *Synthese* 153: 297–311.

———. 2017. "Philipp Frank and the Wiener Kreis: from Vienna to Exile in the USA." *Studies in East European Thought* 69 (3): 207–13.

Holton, Nina. 2020. "On Hania Frank." In *The Vienna Circle in Czechoslovakia*, edited by Radek Schuster, 165–72. Cham, Switzerland: Springer.

Hook, Sidney. 1930. "A Personal Impression of Contemporary German Philosophy." *Journal of Philosophy* 27: 141–60.

———. 1943. "The New Failure of Nerve." *Partisan Review* 10: 2–23.

———. 1978. *Out of Step: An Unquiet Life in the 20th Century*. New York: Harper and Row.

Howard, Don. 2003. "Two Left Turns Make a Right: On the Curious Political Career of North American Philosophy of Science at Midcentury." In *Logical Empiricism in North America*, edited by Gary Hardcastle and Alan Richardson, 25–93. Minneapolis: University of Minnesota Press.

———. 2021. "The Philosopher Physicists: Albert Einstein and Philipp Frank." In *Logical Empiricism and the Physical Sciences: From Philosophy of Nature to Philosophy of Physics*, edited by Sebastian Lutz and Adam Tamas Tuboly, 121–56. New York and London: Routledge.

Hutchins, Robert M., ed. 1955. *Great Books of the Western World*. Chicago: Encyclopedia Britannica.

Hutchins, Robert M. 1936. *The Higher Learning in America*. New Haven, CT: Yale University Press, 1936.

Isaac, Joel. 2012. *Working Knowledge: Making the Human Sciences from Parsons to Kuhn.* Cambridge, MA: Harvard University Press.
Jacoby, Russell. 1987. *The Last Intellectuals.* New York: Basic Books.
Jeff, Laurel. 2019. *Well Worth Saving: American Universities' Life-and-Death Decisions on Refugees from Nazi Europe.* New Haven, CT: Yale University Press.
Jones, Bessie Zaban. 1984. "To the Rescue of the Learned: The Asylum Fellowship Plan at Harvard, 1938–1940." *Harvard Library Bulletin* 32: 205–38.
Joravsky, David. 1969/2009. *Soviet Marxism and Natural Science: 1917–1932.* London and New York: Routledge.
Kuhn, Thomas. 2000. *The Road since Structure: Philosophical Essays, 1970–1993, with an Autobiographical Interview.* Edited by James Conant and John Haugeland. Chicago and London: University of Chicago Press.
Kursanov, G. A. 1962. "Philipp Frank and His Philosophy of Science." *Daedalus* 91: 617–41.
Mayer, Milton. 1993. *Robert Maynard Hutchins: A Memoir.* Los Angeles: University of California Press.
McCumber, John. 2001. *Time in the Ditch.* Evanston, IL: Northwestern University Press.
———. 2016. *The Philosophy Scare.* Chicago: University of Chicago Press.
Morris, Charles. 1942. *Paths of Life: Preface to a World Religion.* New York and London: Harper and Brothers.
Nagel, Ernest. 1936. "Impressions and Appraisals of Analytic Philosophy in Europe. I and II." *Journal of Philosophy* 33, no 1. and no. 2, 5–24, 29–53.
Nemeth, Elisabeth. 2003. "Philosophy of Science and Democracy. Some Reflections on Philipp Frank's 'Relativity—A Richer Truth.'" In *Wissenschaftsphilosophie und Politik/Philosophy of Science and Politics,* edited by Michael Heidelberger and Friedrich Stadler, 119–38. New York: Springer.
Neurath, Otto. 1928/1973. "Personal Life and Class Struggle." In *Otto Neurath: Empiricism and Sociology,* edited by Marie Neurath and Robert S. Cohen, 249–98. Dordrecht, The Netherlands: D. Reidel.
New York Times. 1940. "Religion of Good Urged by Einstein." September 11, 1940.
———. 1941. "To Teach at City College: Dr. P. G. Frank Named Visiting Professor of Philosophy." May 26, 1941.
Pechenkin, Alexander. 2014. *Leonid Isaakovich Mandelstam—Research, Teaching, Life.* Dordrecht, The Netherlands: Springer.
Quine, W. V. O. 1986. *The Time of My Life: An Autobiography.* Cambridge, MA: MIT Press.
Reisch, George. 2003a. "Disunity within the International Encyclopedia of Unified Science." In *Logical Empiricism in North America,* edited by Gary Hardcastle and Alan W. Richardson, 197–215. Minneapolis: University of Minnesota Press.

———. 2003b. "On the International Encyclopedia, the Neurath-Carnap Disputes, and the Second World War." In *Logical Empiricism: Historical and Contemporary Perspectives*, edited by Paolo Parrini, Wesley Salmon, and Merrilee H. Salmon, 94–108. Pittsburgh, PA: University of Pittsburgh Press.

———. 2005. *How Cold War Transformed Philosophy of Science: To the Icy Slopes of Logic*. New York: Cambridge University Press.

———. 2017. "Pragmatic Engagements: Philipp Frank and James Bryant Conant on Science, Education, and Democracy." *Studies in East European Thought* 69: 227–44.

———. 2019a. *The Politics of Paradigms: Thomas S. Kuhn, James B. Conant, and the Cold War "Struggle for Men's Minds."* Albany: State University of New York Press.

———. 2019b. "What a Difference a Decade Makes: The Planning Debates and the Fate of the Unity of Science Movement." In *Neurath Reconsidered: New Sources and Perspectives*, edited by Jordi Cat and Adam Tamas Tuboly, 385–411. Cham, Switzerland: Springer.

Ridenour, Louis N. 1949. "The Man Who Will Know the Answers." *Saturday Review*, July 30, 1949, 9–10.

Rowe, David E., and Robert Schulmann, eds. 2007. *Einstein on Politics: His Private Thoughts and Public Stands on Nationalism, Zionism, War, Peace, and the Bomb*. Princeton, NJ: Princeton University Press.

Scheffler, Israel. 2004. *Gallery of Scholars: A Philosopher's Recollections*. Dordrecht, The Netherlands: Kluwer.

Schrecker, Ellen. 1986. *No Ivory Tower: McCarthyism and the Universities*. New York: Oxford University Press.

Sellars, Roy Wood. 1933. "Humanist Manifesto." *New Humanist* 6: 58–61.

Shapley, Harlow. 1969. *Through Rugged Ways to the Stars*. New York: Scribner.

Siegetsleitner, Anne. 2017. 'Philipp Frank on Relativity in Science and Morality." *Studies in East European Thought* 69: 215–25.

Siegmund-Schultze, Reinhard. 2007. "Philipp Frank, Richard von Mises, and the Frank-Mises." *Physics in Perspective* 9: 26–57

Snow, C. P. 1961. *The Two Cultures and the Scientific Revolution*. Cambridge: Cambridge University Press.

Stadler, Friedrich. 2001/2015. *The Vienna Circle: Studies in the Origins, Development, and Influence of Logical Empiricism*. 2nd ed. Dordrecht, The Netherlands: Springer.

———. 2021, forthcoming. "Philipp Frank and his Contributions to the 'Conference for Science, Philosophy and Religion,' 1940–1968." In *The Vienna Circle and Religion*, edited by Esther Ramharter. Cham, Switzerland: Springer.

Stamenkovic, Philippe. 2021, forthcoming. "Philipp Frank's 'Doctrine of the Relativity of Truth': Appreciation and Critique." In *The Socio-Ethical*

Dimension of Knowledge: The Mission of Logical Empiricism, edited by Christian Damböck and Adam Tamas Tuboly. Cham, Switzerland: Springer.

Stöltzner, Michael. 1995. "Philipp Frank and the German Physical Society." In *The Foundational Debate: Complexity and Constructivity in Mathematics and Physics*, edited by Werner Depauli-Schimanovich, Eckehart Köhler, and Friedrich Stadler, 293–302. Dordrecht, The Netherlands: Kluwer.

———. 1999. "Vienna Indeterminism: Mach, Boltzmann, Exner." *Synthese* 119: 85–111.

———. 2003. "Vienna Indeterminism II: From Exner to Frank and von Mises." In *Logical Empiricism: Historical and Contemporary Perspectives*, edited by Paolo Parrini, Wesley Salmon, and Merrilee H. Salmon, 194–229. Pittsburgh, PA: University of Pittsburgh Press.

———. 2020. "Scientific World Conception on Stage: The Prague Meeting of the German Physicists and Mathematicians." In *The Vienna Circle in Czechoslovakia*, edited by Radek Schuster, 73–95. Cham, Switzerland: Springer.

Synnestvedt, Justin. 1969. "Philipp G. Frank: Critic of Modern Science." PhD diss., Loyola University Chicago.

Thurm-Nemeth, Volker, eds. 1998. *Konstruktion zwischen Werkbund und Bauhaus. Wissenschaft—Architektur—Wiener Kreis*. Vienna: Springer.

Tuboly, Adam Tamas. 2017. "Philipp Frank's Decline and the Crisis of Logical Empiricism." *Studies in East European Thought* 69: 257–76.

———. 2021a. "To the Icy Slopes in the Melting Pot: Forging Logical Empiricisms in the Context of American Pragmatisms." *HOPOS: Journal of the International Society for the History of Philosophy of Science* 11, no. 1.

———. 2021b. "Understanding Metaphysics and Understanding Through Metaphysics: Philipp Frank on Scientific Theories and Their Domestication." In *Logical Empiricism and the Physical Sciences: From Philosophy of Nature to Philosophy of Physics*, edited by Sebastian Lutz and Adam Tamas Tuboly, 401–21. New York and London: Routledge.

———. 2021c. "Building a New Thursday Circle: Carnap and Frank in Prague." In *Young Carnap in an Historical Context: 1918–1935*, edited by Christian Damböck and Gereon Wolters, 243–64. Dordrecht, The Netherlands: Springer.

Uebel, Thomas. 2000. "Logical Empiricism and Sociology of Knowledge: The Case of Neurath and Frank." *Philosophy of Science* 67: 138–50.

———. 2011. "Beyond the Formalist Meaning Criterion: Philipp Frank's Later Antimetaphysics." *HOPOS* 1: 47–72.

———. 2013. "'Logical Positivism'—'Logical Empiricism': What's in a Name?" *Perspectives on Science* 21: 58–99.

———. 2020. "Intersubjective Accountability: Politics and Philosophy in the Left Vienna Circle." *Perspectives on Science* 28: 35–62.

Wein, Martin. 2016. *History of the Jews in the Bohemian Lands*. Leiden, The Netherlands: Brill.
Wilson, Edwin H. 1956. "Editorial Interview: Philipp Frank—A Modern Positivist." *Humanist* 4: 189–91.
Wolters, Gereon 1999. "Wrongful Life: Logico-Empiricist Philosophy of Biology." In *Experience, Reality, and Scientific Explanation. Essays in Honor of Merrilee and Wesley Salmon*, edited by Maria Carla Galavotti and Alessandro Pagnini, 187–208. Dordrecht, The Netherlands: Springer.
———. 2018. "'Wrongful Life' Reloaded: Logical Empiricism's Philosophy of Biology 1934–1936 (Prague/Paris/Copenhagen)." *Philosophia Scientiæ* 22, no. 3: 233–55.
Wray, K. Brad. 2015. "Kuhn's Social Epistemology and the Sociology of Science." In *Kuhn's Structure of Scientific Revolutions—50 Years On*, edited by William J. Devlin and Alisa Bokulich, 167–83. Dordrecht, The Netherlands: Springer.
Wuest, Amy. 2015. *Philipp Frank: Philosophy of Science, Pragmatism, and Social Engagement*. PhD diss., University of Western Ontario.
———. 2017. "Simplicity and Scientific Progress in the Philosophy of Philipp Frank." *Studies in East European Thought* 69: 245–55.

PART I

Chapter 1

Introduction

Science, Facts, and Values

1. Science and Poetry

Before the eighteenth century, it was understood that human values—or, in other words, the rules of human behavior—were provided by theology and philosophy. In *The Existential Revolt*, Kurt Reinhardt writes:

> In the course of the nineteenth century and with the adoption of the 'scientific method' by historians, jurists, sociologists and 'humanists,' the Western mind began to reject the guiding principles provided by these two disciplines (philosophy and theology). 'Truth' henceforth was to be found exclusively by those sciences which analyzed and described extended and measurable physical reality.[1]

However, it has been repeated again and again that science can only teach us "facts" but cannot provide any "values." Hence, by basing the rules of human behavior on the sciences, the influence of moral principles disappears, and our lives become empty of spiritual guidance. This distinction between facts and value has also served in the nineteenth and twentieth centuries to bring about a "truce" between science and religion. While science was to provide "facts," the task of providing "values" was assigned to religion. By this method, a division of labor could be established. When institutions of higher learning put the emphasis on science, they failed to develop a sense of values among the students. Distinguished speakers at commencement celebrations, such as

Adlai Stevenson at Smith College in 1955, deplore the contemporary environment in America as "an environment in which 'facts,' the data of the senses, are glorified, and value judgments are assigned inferior status as mere 'matters of opinion' . . . philosophy is not only neglected but deemed faintly disreputable because 'it never gets you anywhere.' "[2]

The overemphasis on facts and the neglect of values in our education have been made responsible for a wide variety of evils on our contemporary scene; for the irresponsible way in which scientists produced the atomic bomb, for the decline of religion among the educated, and even for juvenile delinquency. This kind of argument has almost become commonplace among college graduates and readers of magazines. However, this sharp distinction is based on a rather vague concept of "fact." It is interesting to note that Archibald MacLeish, a prominent contemporary poet, who certainly has not been partial to science, accuses science of underemphasizing fact. In his article "Why Do We Teach Poetry?," he asserts that "science" is very remote from "facts" if by "facts" we mean "facets of reality."

He points out that "science" never speaks about real facts that occur in nature. For instance, when science describes an atomic explosion, it describes neither the human sensations nor the human emotions connected with it but rather a set of abstract symbols like electric forces, chemical affinities, and so on. Science "interprets" facts but does not "present" them. MacLeish tries to formulate the basic difference in the way in which science and poetry approach reality.

> Science can abstract ideas about apple from apple. . . . But poetry, we know, does not abstract. Poetry presents. Poetry presents the thing as the thing. . . . It should be possible to *know* the thing *as the thing it is*—to *know* apple *as* apple—this, the true child of the time will assure you, cannot be done. To the true child of abstraction, you can't know apple as apple. You can't know tree as tree; you can't know man as man. All you can *know* is a world dissolved by analyzing intellect into an abstraction—not a world composed by imaginative intellect into itself.[3]

His main point is that science does not teach us the real facts but an "impoverished" picture of them. He writes: "Young men and young women graduate from American schools and colleges by hundreds and thousands every year to whom science is the only road to knowledge, and to whom poetry is little more than a subdivision of something called 'literature.'"⁴ According to MacLeish, the emphasis on science in education and the neglect of poetry as a way of "knowledge" leads to a worldview based essentially upon "abstractions." He continues:

> This sort of thing has consequences. . . . Abstractions have a limiting, a dehumanizing, a dehydrating effect on the relation to things of the man who must live with them. The result is that we are more and more left, in our scientific society, without the means of knowledge of ourselves as we truly are, of our experience as it actually is. We have all the tools—we are suffocating in tools—but we cannot find the actual wood to work on or even the actual hand to work it. We begin with one abstraction (something we think of as ourselves) and a mess of other abstractions (standing for the world) and we arrange and rearrange the counters . . . but we simply do not know what we are doing.⁵

While in the nineteenth century it was a generally accepted truth that science was the beacon that guided mankind toward a bright future, in the twentieth century there have been quite a few critical voices. Science has not only been accused of failing to provide a sense of values among its students, it was even claimed that science cannot provide a knowledge of facts if one means by that a knowledge of the real world. We learned this clearly from MacLeish, but we find similar opinions expressed by other representatives of the arts. We may quote the words of Aldous Huxley who claims in *Science, Liberty and Peace* that the scientists and their followers "tend to accept the world picture implicit in the theories of science as a complete and exhaustive account of reality" and "to regard those aspects of experience, which scientists leave out of account, because they are incompetent to deal with them, as being somehow less real than the aspects which science has arbitrarily chosen to abstract from out of the infinitely rich totality of given facts."⁶

2. Charges against the Monopoly of Science

The confidence in science in the nineteenth century was twofold. First, one believed that science would provide us with the "truth about the universe." This knowledge and insight would improve along with advances in science. Second, advances in scientific knowledge would increase the satisfaction and happiness of human beings. Scientific knowledge and personal happiness would increase simultaneously. Toward the end of the nineteenth century, quite a few authors had cautioned against this confidence. One started to doubt whether scientific knowledge actually gives us insight into the nature of the universe and whether the advances of the nineteenth century actually had enhanced the happiness of men. These doubts were eloquently voiced by Friedrich Nietzsche in the preface to his *Birth of Tragedy* (1886): "What I got hold of that time was something terrible and dangerous, a problem with horns, not necessarily a bull, but certainly a problem; the problem of science itself, science comprehended for the first time as something problematical and highly questionable."[7]

These doubts about science were encouraged by the mounting conviction that science in its nineteenth-century form—with its belief in "progress"—had actually failed to achieve its promise. Technical progress had rather produced suffering and despair among the people who were supposed to serve as the guinea pigs of technical advancement. Toward the end of the nineteenth century and particularly around the turn of the twentieth, two strong political and social groups—or, more generally speaking, two schools of human conduct—came to fruition: on the one side a certain longing for the society and the philosophy of the Middle Ages and, on the other, the strong belief in Marxist socialism. Both groups strongly opposed the political and religious way of life that was based on what one had called "objective science." They were, above all, opposed to the belief that science, by its own methods and without any help from political or religious creeds, could "objectively" derive a way of life that would guide the world to happiness.

The right-wing movement against "scientism," (i.e., this overrating of science) has been known by names like Christian Socialism, National Socialism, fascism, Religious Revival, and, occasionally, New Conservatism. The left-wing movement has been known as Marxism, or, in a more philosophical parlance, dialectical materialism. Both movements, left and right, have been strongly opposed to what has been called the "science" of the "liberal bourgeoisie," which is taught in Western universities. This science is allegedly infiltrated by the political and religious ideas of the liberal bourgeoisie to which groups the faculties mostly belong. Specifically it is infiltrated by the ideas of "free thinking" and "free economy." The rightist and leftist groups of which we spoke stress that the belief in free thinking and free economy is not the result of some "objective" science but of the political and religious predilections of scientists. Science claims to provide objective truth but is actually a manifestation of the political and religious creeds that dominate our universities.

It has been a conspicuous phenomenon in our twentieth century that these "anti-scientific tendencies" have been advocated with great fervor. The heat that drives these movements certainly has its source not in any longing for "ultimate knowledge of the universe" but in the belief that by demoting science as the only source of truth one can promote human happiness. This drive against science accumulated strength after science enabled mankind to liberate nuclear energy and produce the atomic bomb. In a significant way, our twentieth century repeats the attitude of Greek paganism by which Prometheus was accused of having "liberated atomic energy": it enabled man to produce at will the "reaction between carbon and oxygen" and to pave the way for the destructive forces of fire. According to Greek tradition, Prometheus was punished because he had "stolen" the method of fire making from the gods to do a favor for his fellow man. He was tied to a rock in the Caucasian Mountains where his liver was eaten by vultures.[8]

We must think of this tradition when we note how scientists of today are accused of "stealing" forces of nature, the nuclear forces from God, and of having taught their manipulations to their fellow mortals. A great many people would wish that the vultures would get at the livers of the nuclear scientists.

As a matter of fact, myths about the jealousy of the gods against the technical and scientific advances of men are ubiquitous. We have only to remember the story of the Tower of Babel in the Old Testament, where the Lord intervened personally against the construction of the first skyscraper. These ancient myths reflect an antagonism that has existed between two activities of mankind: research in the natural sciences and the search for moral and spiritual advice by superior beings.

What we experience today as the conflicts between science and the humanities[9] in education are the effects of these myths upon the human mind. In particular, the rise of antiscientific tendencies we spoke of is due to the sometimes subconscious reminiscences of ancient myths. These traditions stress that science cannot reveal to us the real nature of the universe, so we must expect it from other sources.

3. Twentieth-Century Science and Philosophy

While antiscientific tendencies have attempted to undermine the prestige of science, there has been *within science* a critical movement advocating a new conception of science. It approves some of the objections raised against the claims of science but does not admit that other sources can do better in revealing to man the real nature of the universe. This new conception no longer defines the task of science as giving a true picture of the universe. However, it denies that a true picture of the real world can be given at all, and, more specifically, denies that the expression "true picture of reality" has any meaning.

This would mean that statements like "our physical universe consists, in reality, of matter," or "consists, in reality, of mind" cannot be confirmed or refuted by science: they lack scientific meaning. This means, in turn, that philosophical creeds like materialism or idealism are not actually supported by science. Moreover, the new conception of science has been opposed, in general, to the doctrine of "realism" according to which science explores the "real world." The new conception of science stresses the point that there is no language with which the "real world" can be described directly. What science actually does is construct a system of symbols (e.g., words or algebraic signs) and to formulate statements in which these symbols occur. These statements are the principles of science from which one can obtain, by logical conclusions, observable facts. The checking of the facts by sense observations provides the body of experiences upon which science is based. If the facts derived from the principles of science agree with our direct experience, we say that the constructed principles are confirmed. Therefore, the new conception we characterized by saying that science is not a picture of the real world can also be expressed by Henri Bergson's formulation that "the laws of science are not discoveries but rather inventions."[10]

The conception of truth as a picture, a copy, or even a photo of the real world has dominated the philosophical interpretation of science for centuries. It has been the basis of philosophies that in our present period are worldwide creeds: Thomism as the foundation of Catholic philosophy and dialectical materialism as the basis of the Communist worldview.

This realistic or "picture" philosophy has been replaced in two steps taken beginning in the mid-nineteenth century. The first step was the conception that science is the systematic description of facts or, in other words, of our experience. This conception was formulated elaborately in the "positive philosophy" of August Comte, in the "synthetic philosophy" of Herbert Spencer, and the "system of logic" of John Stuart Mill. It has been known under the name of positivism or empiricism.

To men like Comte and Spencer, it was clear from the start that there cannot be a description of mere facts, which are only the raw material of science. The chief intention of science is to describe a great number of facts through a unifying idea. Mathematicians, physicists, and historians of science, in the second half of the nineteenth century, investigated the way in which this unifying idea is formed. They worked out a conception of science in which the human mind builds up a framework or pattern in order to describe the facts. This turn in positivism was dubbed "New Positivism" by the French historian of science Abel Rey.

Positivism such as Comte's and Spencer's had been connected with the conviction that Euclidean geometry and Newtonian mechanics were the final word in our knowledge of nature. The rise of non-Euclidean geometry and non-Newtonian physics had a profound influence upon the rise of the New Positivism. Its representatives, including Ernst Mach, Henri Poincaré, and Pierre Duhem, emphasized more and more the role of human creativity in the production of new scientific theories. They advocated the conception of science as a flexible, conceptual frame that is constructed by the human mind in order to accommodate new facts.

In the same period, the end of the nineteenth century, a new conception of science originated in the United States that was, in many respects, similar to the New Positivism. We refer here to the school of Pragmatism, known best by the works of the psychologist William James. However, Charles Sanders Peirce, a mathematician and logician, was the first to introduce the word and concept of "pragmatism."[11] He applied the pragmatic views to science. Since, for him, science was a background of action, he strongly rejected the old picture theory with its attempt to make science a copy of the real world.

In the twentieth century, new physical theories emerged in which the role of human imagination and the formation of new languages became even more conspicuous. The New Positivism (sometimes called "neo-positivism") developed more and more into a theory of language, and philosophical interpretation became linguistic analysis.

The history of science was now being studied as a history of factual discoveries accompanied (and sometimes even guided) by the invention of new languages. All advances in science are closely connected, according to the new view, with advances in language. This has to be kept in mind if we want to understand the philosophical groups into which the New Positivism branched out in the twentieth century.

In this introduction I will briefly describe these philosophical groups that have overcome the "realistic" or "picture theory" of science. We note in the United States, beginning from 1920, the "operationalism" of Percy W. Bridgman. We note in Central Europe, at the same time, the opposition to the prevalent German idealistic and realistic philosophy. This opposition started in the first decade of the twentieth century under the name of "scientific world-conception" advanced by the Vienna Circle. From this group there developed what in the United States became known as "logical positivism" and "logical empiricism." In England, we note in the twentieth century the school of analytical philosophy that like the New Positivism was based on the analysis of language. The prevalent idea of this English philosophy was, above all, to study the language by which one expresses our everyday life experience and called "natural language." In all these groups, one has regarded metaphysical problems as confusions that have arisen from the realistic and picture theory and by neglecting the analysis of natural language.

4. The "Real World" Is Not Describable

If we want to describe the so-called real world, we soon become aware that there is no adequate language for a direct description. Therefore one will use symbols that have no clear-cut connection with the natural language. Most of the British philosophers in our period hope that by analysis of "natural" language the eternal metaphysical puzzles can be eliminated. If one "explicates" the meaning of metaphysical statements by using statements in natural language they become a consistent system of statements that can be checked by "experience" in the ordinary sense of this word. British analytical philosophy has its origin in Bertrand Russell and G. E. Moore and is now mainly based upon the teachings of the Austrian Ludwig Wittgenstein who has been the link between the Vienna Circle and British analytical philosophy.

Besides the branches of New Positivism, several schools have developed that attempt, in a practical way, to actually improve the language used in philosophical, religious, and political discourse. In the United States, the most important and successful group has been the school of "general semantics" founded by Alfred Korzybski. A similar group is the school of Significs in Holland.[12] All these more-or-less neo-positivistic groups share the opinion that by abandoning the picture theory and introducing a very subtle analysis of language one can dissolve the insurmountable difficulties that have tormented traditional philosophy.

Nowadays, we no longer conceive science as a picture of reality but as a system of symbols that helps us to bridge the gap between the science and humanities, or, in other words, between facts and values. According to the new conception, science consists of an instrument and the directions telling us how to use it. The instrument is a system of propositions that are, in turn, relations between symbols. In geometry, for instance, the system consists in the axioms of geometry. The symbols are "straight line," "point," etc., while the propositions are of the form "two points are coinciding with one straight line only." The directions that tell us how to use the system are the operational definitions, for instance, the statement that a straight line should be realized by a light ray or the edge of a rigid body. From the propositions, one can logically derive statements about observable facts. Such statements are the basis for the production of technical devices. However, neither these observations nor success in producing technical devices determine the general principles of science unambiguously. Since the earliest times of scientific and technical endeavor, the general principles have been used to provide a general view of the universe and, in this way, guidance for human conduct. The Ptolemaic system was preferred to the Copernican one by some scientists, although everyone agreed that it was not the more practical instrument. But the Ptolemaic system was in better agreement with the religious and philosophical traditions and thus a better basis for making good citizens.

According to the new conception of science, we cannot simply say that a certain system of principles provides a "true picture of the real world" but only that it is useful from a technological, moral, or religious viewpoint. From the start, the definite form of scientific principles is determined by their purpose. Technological as well as moral and political goals have their influence upon the final shape of these principles. We can no longer say with certainty that science as a system of statements about the world is independent of those who make these statements. Since science is a manmade invention, it is invented for a certain purpose. It is, therefore, dependent upon the purpose that has been selected. If we speak rather perfunctorily we may say that the principles of science depend upon whether we want an instrument for technological purposes or an instrument for supporting a certain moral or political conduct. Anyway, the fact-finding and evaluating tasks of science cannot be completely separated from each other.

5. The Humanities Are Trailing behind the Natural Sciences

One has often deplored a situation in which the advance of the humanities—namely philosophy, ethics, and religion—has lagged sadly behind the advance of science and technology. Such complaints have become commonplace today. To find an instance, I merely pick at random the contemporary issue of a Boston daily newspaper and a letter to the editor: "Science and technology have given us material benefits almost beyond imagination, but ethics and moral values have fallen far behind our rapid advancement in technological knowledge. It is the time—here and now—to close this gap."[13]

We meet the same complaints at the summit of intellectual culture. Professor I. I. Rabi, Nobel Prize–winning physicist of Columbia University addressed (on Feb. 1, 1956) the American Institute of Physics on the small effect of scientific advances upon general culture. He pointed out that "the impact of scientific thought on the culture of the time becomes less and less even as science advances to greater pinnacles of understanding and discovery. As the importance of science in this country increases, its dignity seems to be diminishing."[14] The speaker obviously referred to the ratio of "technological importance" to "moral dignity." If one describes this situation not in moral terms but in a more sociological vocabulary, one could say that our society is technologically well advanced, but the evolution of social institutions is lagging far behind the technological advancement.

While religious-minded groups have deplored this lagging of moral advance behind technological progress, advocates of a materialistic conception of history have similarly deplored the fact that our present social structure no longer fits the highly developed state of today's technology. Karl Marx and Friedrich Engels, the founders of revolutionary socialism, saw in this incongruence the very root of future social revolutions. This lagging of moral and cultural progress behind scientific and technological advancement can also be formulated by claiming that advances in science have not contributed much to the advance of our knowledge about social improvement, or, generally, about human values. In other words, the study of science has been completely isolated from the study of the humanities. Aldous Huxley, again in *Science, Liberty and Peace*, related this shortcoming of science to the fact that "some scientists, many technicians and most consumers of gadgets have lacked the time and the inclination to examine the philosophical foundations and background of the sciences."[15] However, according to Huxley, "The scientific picture of the world is inadequate for the simple reason that science does not even profess to deal with experience as a whole, but only with certain aspects of it in certain contexts."[16] Huxley thinks that the scientific, positivistic worldview leads practically to totalitarian forms of government with their regimentation of human life. "As theory," he writes, "pure science is concerned with the reduction of diversity to identity."[17] The most familiar example is the "mechanistic world view" by which the various qualities appearing in our sense experience are reduced to the positions and velocities of material particles. Another example is the "electromagnetic world view" by which all qualities are reduced to electric and magnetic forces. The term "reduces" is certainly misleading because it has disparaging connotations; it would hint that mechanical and electromagnetic forces are regarded by science as "more real" than human qualities or values. Actually, according to the practice of science, these mechanical or electromagnetic quantities belong to a system of symbols from which statements about our human experience can be derived. Nobody would doubt that this system should be as simple as possible, provided it allows us to derive the world of our experience. Exactly speaking, therefore, science is not a "reduction of diversity to identity" but a method to derive the immense diversity of human experience from a simple conceptual scheme.[18] Huxley claims nonetheless that "as a praxis, scientific research proceeds by simplification. These habits of scientific thought and action have, to a certain extent, been carried over into the theory and practice of contemporary politics."[19] According to Huxley, the restriction of individual liberty in authoritarian countries has been justified by the methods of science. "Philosophically," he writes, "this ironing out of individual idiosyncrasies is held to be respectable, because it is analogous to what is done by scientists, when they arbitrarily simplify an all too complex reality."[20] He points out that "a highly . . . regimented society . . . is felt by the planners and even . . . by the planners to be more 'scientific' . . . than a society of freely cooperating individuals."[21] This view belongs not only to Huxley but has been repeated again and again by contemporary authors. However, as we have seen, it has its roots in an obsolete "philosophy of science."

6. The "Special Sciences" Don't Exhaust "Science"

When modern science was born in the seventeenth century and grew up in the eighteenth and nineteenth centuries, the central idea was to build up "special sciences" such as astronomy, physics, and biology. Each of these special sciences was based upon a conceptual scheme that contained a small number of symbols. This specialization and simplification were certainly very successful and resulted in the construction of marvelous buildings in the form of modern physics, chemistry, and so on. But one must not forget that no concrete problem put to us by nature—in other words, by life—can be solved by applying one of the special sciences.

Let us consider this "simple" problem: what will be the potato crop of this fall? Or, more pointedly, how many bushels of potatoes will we reap on a certain plot? To solve this problem "scientifically," we have, of course, to apply botany. Since insects may influence the crop, we must also use zoology and, since the structure of the soil will be relevant, mineralogy and geology. Because of the influence of rain and storms, we will also need the results of meteorology or cosmic physics. But all these "special sciences" don't allow us to predict the crop, for there may be labor unrest that we can only predict from economics and sociology; there may be even a war that we can only predict by knowledge of foreign affairs. There may be epidemic diseases that can prevent us from digging potatoes, and if we believe in traditional religion, the crop will depend on the favor of the gods, which we can only predict from theology or possibly astrology. The predictions of all these doctrines are interwoven in their application to reaping an individual potato crop.

We note that we must deal not only with "sciences" in the ordinary sense of this word, including physics, chemistry, and botany but also with "humanities" and "social sciences" such as theology, history, foreign affairs, and sociology. Generally speaking, marvelous "buildings" like physics and biology are beautiful and impressive in their proud isolation. But in order to solve an actual problem, even predicting a potato crop, we have to make use of these "special sciences" together as building blocks for a larger edifice. If each of the special sciences contained unique symbols and concepts, they could never form a combination of botany, history, and religion from which to derive statements about the potato crop. In order to draw such conclusions, botany, economics, and theology must have concepts in common. Ultimately, all observable facts can be described in terms of our everyday language. If these terms did not occur in all the special sciences, from mechanics to theology, we could not derive from science statements about observable facts (e.g., potato crops).[22]

7. Semantic and Pragmatic Components of Science

As a matter of fact, the opinion that science deals with an impoverished world that is deprived of all human elements is based on an obsolete conception of science and ignores contemporary attempts to build up a science that reflects the whole of our experience of the world. This new conception of science and its relation to man is often called "unified science"; it has been advocated and elaborated by the *International Encyclopedia of Unified Science*. The traditional scientific approach to the physical world consisted in reducing this world that was rich in qualities to a mathematical skeleton. This "desiccated world" was described and presented itself by a "science" that was an impoverished picture of the genuine rich world. Science was based upon a scheme that consisted of two components, "reality" and "picture," a "dyadic scheme."

Toward the end of the nineteenth century, it had turned out that it would be much more expedient to base the presentation of science on a richer scheme that considers three factors: (1) objects directly observed by the scientists as physical phenomena; (2) the scientist as a person; and (3) the signs invented and used by the scientist. C. S. Peirce spoke of a "triadic" scheme, and we find these three factors in the works of all the great philosophers of science who were working at the turn of the century (1900), in particular, in the works of the great

French scientists, Henri Poincaré and Pierre Duhem. This scheme can be illustrated by a triangle.

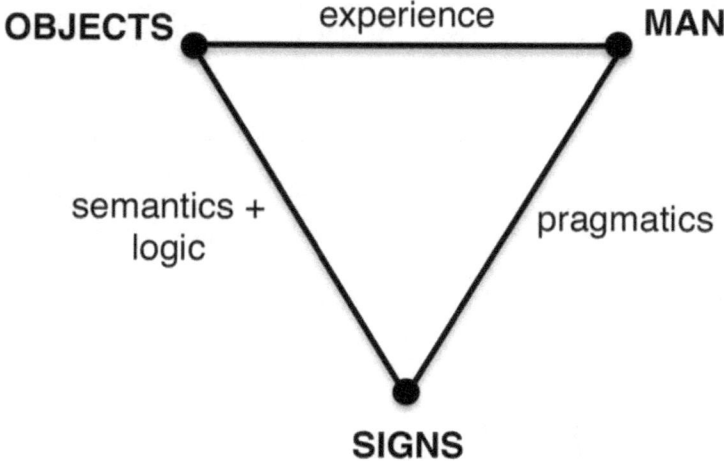

In the *International Encyclopedia of Unified Science*, this triadic scheme was elaborated and clearly defined by the leaders of the Unity of Science movement (e.g., Rudolf Carnap and Charles Morris).[23] To give examples of this scheme, we may choose a very new and a very old one.[24] The contemporary one may be taken from the explosion of the atomic bomb. The first factors, physical objects, consist in the famous mushroom-shaped cloud, the pressure exerted upon buildings or people, and the injuries produced by burning, among others. The second factors are the nuclear scientists who have attempted to understand these phenomena. The third factor of the scheme that we call "nuclear physics" consists in the signs and symbols that have been introduced by these scientists (e.g., the concept of electron, nucleus, neutron, cross section, isotope, etc.). What now does the scientist *actually* do? First he studies the way in which scientists can actually observe the physical facts: that is, he does experimental research. Second, he studies the correlation of the physical facts to the scientific symbols and signs. He studies, for instance, how the mushroom cloud can be understood as a swarm of electrons, radiations, neutrons, etc. He invents the symbols, electrons, neutrons, etc., by which these physical phenomena can be understood in a simple and comprehensible way, or in other way, he studies the "physical meaning" or "operational meaning" of symbols like nucleus, electron, neutron, etc. These studies are often referred to as "semantics."

There is, however, a third kind of investigation. One can investigate by what factors the scientist is driven to invent certain signs or symbols. Pragmatics is concerned with the predilections of scientists and their public for certain symbols. These predilections are obviously determined by the role these symbols have played in human life. This means that these predilections or "values" can only be ascertained by studies in the "behavioral" sciences, like psychology, sociology, and so on. Hence, by the "triadic scheme" of science, a large field of humanistic and social studies is introduced even in the most "inhuman" sciences like mathematics and physics.

One often says that "science" treats only a very poor skeleton of the world because human personality and human values are eliminated by a far-reaching process of abstraction.[25] But this is only true if we restrict ourselves to logic and semantics, thereby neglecting the pragmatic component. Pragmatic studies of science include the influence of social, philosophical, and religious factors on the invention of scientific symbolism. In order to understand this point well, we have only to remember the passages in Plato's *Republic* that stress the value of astronomy for supporting Greek religion and good citizenship.[26] In the same category belongs the insistence of the Roman Church that Galileo's presentation of the Copernican System, in which our earth is freely moving in the world's space, is harmful to traditional Christian religion. A similar result can be drawn from the attitude of the present Soviet philosophy to Einstein's theory of relativity. The symbolism by which this theory had been presented seemed to be a help to philosophical systems (like Agnosticism, idealism, etc.) which were regarded as harmful to the political philosophy of the present government. By using the pragmatic approach to scientific theories of a certain period, we are impelled to study the human values of this period.

For this reason a chief topic of the present book will be the pragmatic approach to present-day science including the line of descendance from which our present science has originated. If we want to study this approach, we have to study the symbols of science not just from a technological angle. We cannot study the actual meaning of symbols merely for the purpose of predicting future facts or for producing gadgets; we must also study the meaning of these symbols as expressions of human aspirations. Thus, the variety of meanings that have been attributed to scientific symbols is a main topic in the present book. Scientists and technologists have always attempted to interpret these symbols in order to serve "practical" purposes in the technological sense of this world. However, at the same time, philosophers, educators, ministers, and politicians have attempted to give these symbols "moral" interpretations by which the statements of science would become helpful in the support of human aspirations and goals. Every new term in science was accompanied by a lively discussion about the possibility of employing this new term for improving human conduct in a line that would have a desirable religious, moral, or political effect. This was the case when Galilean and Newtonian science replaced medieval scholasticism, and it was equally the case when twentieth-century physics (relativity and quantum theory) replaced the Newtonian concepts of strict determination[27] by the initial position and speed of particles.

We shall discuss in the present book the role that the symbols of scientific discourse have played in the struggle for moral and political goals. In this way, the humanities and social sciences are tied almost inseparably to the physical and biological sciences.

8. Philosophical Schools Woo the Support of Science

As far back as any recorded history of science and philosophy goes there have been attempts to build up symbolic systems that would embrace all "adventures of the mind," all efforts to master nature by effective prescriptions, and to master human conduct by establishing systems of values. These structures of concepts have been mostly referred to as "philosophical" systems. They have been advocated by "philosophical schools," which have become actual social groups that have occasionally cooperated with religious and political parties.[28] We have seen, for example, the cooperation of Thomistic philosophy with Catholic religious and political organizations. The philosophical schools of British Empiricism, American Pragmatism, and Western and Central European Positivism have intensively cooperated with democratic and liberal groups of the Anglo-American political type. The most instructive example of our period is perhaps the close cooperation between the Communist Party in the Soviet Union and the philosophical groups advocating dialectical materialism. All these philosophies have looked for the support of science; again and again new achievements of science like Newton's mechanics and Einstein's relativity have been interpreted as supporting or refuting philosophical schools.

We shall discuss the ways in which these philosophical groups attempt to trace their "genealogy" back to scientific theories. One and the same scientific theory (e.g., the theory of relativity) is often used to support antagonistic philosophical schools. This comes obviously from the fact that no philosophical school can claim to be derived from science. The shoe is on the other foot; the primary fact is the wish to justify the philosophical approach that supports favored political or moral ideals. We shall learn in the book that when one knows which philosophical approach should be supported, one will find a way to get this support out of several physical or biological theories.

Scientific theories play a double role. On the one hand, they are statements about observable physical facts that can be logically derived from those theories. But they have also been interpreted as statements about the nature of the universe. In order to do so, one had to interpret terms like "matter," "energy," "force," "determinism," not by the "operational meaning" by which they have been defined in science but by some analogy in commonsense language. If scientists speak of "mass," they are referring to the results of physical operations by which they assign to the word a certain number of pounds: the "quantity" of mass. If they speak of "kinetic energy" or "radiating energy," they are referring to the results of other physical operations by which they assign to these terms "quantities of energy," which are measured in "ergs" or "watts," etc. In this context "mass" and "energy" have the same status. However, when one speaks of "mass" and "energy" in a metaphysical context, one does not replace "mass" and "energy" with the operational definitions of physics. One attempts rather to embed or to integrate expressions like "mass" and "energy" directly into the language in which one describes the facts of our everyday life.

In this language, "mass" is something clumsy and inert, but "energy" is something much more subtle and active. To a very high degree, the philosophical schools have made use of these metaphysical interpretations of science. A main thrust of their intellectual argument has been an attempt to prove that a certain school has the blessing of science. Scientists have often branded this type of philosophy as "school philosophy" in order to stress the fact that the driving force behind these philosophies has not been the search for solutions to scientific problems but the urge to serve an ethical, political, or religious purpose embraced by particular philosophical schools. The term "school philosophy" has a slightly derogatory connotation. The scholastic philosophy of the Middle Ages has been referred to as the "philosophy of the schoolmen" because this particular school for centuries has been regarded as *"the school."* Its central work was, perfunctorily speaking, the adjustment of Aristotelian and Platonic philosophy from pagan to Christian faith. Hence, in the eighteenth and nineteenth centuries "school philosophy" had the connotation of a philosophy that was not adjusted to the results of modern science. In this book, not all philosophical schools are going to be discussed. We shall only examine those schools that embody the political and religious creeds of the twentieth century and at the same time look to adjust to modern science.

According to J. B. S. Haldane, the British biologist and advocate of Marxist philosophy, there are three schools in our twentieth century that are to be considered from these viewpoints. They are scholastic philosophy, modern science, and Marxist philosophy.[29] By regarding modern science as a philosophical school, the author refers, of course, to the group that has attempted to "purge" modern science of any metaphysical interpretation and has been known as "positivism." The attempts to avoid all such interpretation have led to new types of interpretations and the formation of new "schools" with all the characteristics of social organizations that work to shape human behavior.

However, my present goal is not a history of philosophical schools but rather a discussion of the social phenomena one calls "philosophical schools." We shall present as examples first the Thomistic interpretation of science, as supporting Christian—or specifically Catholic—ways of life. Second, we shall present the positivistic interpretation of science in the widest sense of this word, which includes the pragmatist interpretation. As a third example, we shall present the materialistic interpretation of science, in particular the doctrine of dialectical materialism, which has been the official philosophy of the Soviet Union and its affiliated states. This philosophy has been regarded as a support for the way of life advocated by Marxism. All these philosophical schools have ardently claimed that they agree with "science" and moreover that each of these is the genuine philosophy that fits with science.

9. Principles of Science and Human "Values"

We have heard from quite a few scientists' utterances that would claim that a certain philosophy like Thomism or dialectical materialism is "antiscientific." As we shall see, there have even been authors who have maintained that positivism is antiscientific because it denied the "reality" of the physical world upon which, allegedly, belief in science is based. As a matter of fact, none of these philosophical schools will ever admit to contradicting science because, to speak flippantly, no government of any ideology could survive without an atomic bomb. This means that no ideology can afford to alter the technological results of the sciences. The slightest deviation from the recognized laws of science could frustrate the production of nuclear weapons. That is, the results of science have never actually depended upon the ideology professed by a certain government. Materialist, Christian, or Buddhist bombs have the same effects. This may seem to be true if we regard science only as an instrument to produce technological rules which, in turn, are helpful in predicting observable phenomena and constructing gadgets that serve man's pleasure—including his victory in battles with other men.

However, science has never restricted itself to the role as an instrument for the production of technological rules. When the Epicurean school, in antiquity, claimed that the celestial bodies are made of the same material as the terrestrial ones, it proclaimed a radical change in man's interpretation of the physical universe. The new physical theory made it more difficult to believe that there are spiritual beings apart from or above human beings. Plato chastises this physical theory as a subversive view about human society because it endangered the traditional religion upon which good citizenship seemed to be based. We see a similar phenomenon in the attitude of authorities to the Copernican theory. It would be a mistake to think that this hostility was representative of a special backwardness of the Catholic Church. For even "liberals" like Francis Bacon or John Milton recommended caution, at least, in accepting such changes because these would reflect on our beliefs on human society.

It would be equally wrong to believe that this interference by authorities with natural science was a characteristic only of the so-called Dark Ages. The attitude of the Soviet government toward Einstein's theory of relativity and Mendel's genetics is well known. Since we would hardly believe that an immensely practical government like the Soviet Union would risk backwardness in nuclear technology or agriculture, we must assume that these attitudes had other origins. They were all based upon the fact that every fundamental change in theory about the physical universe was used to support some change in man's view about the desirable type of human society. We note a clear example of this connection in the interpretation of the new formulations of determinism and causality that have been given to us in the twentieth century by atomic and nuclear physics. The view was abandoned that all particles, independent of their size, follow the same laws, namely Newton's Laws of Motion. The laws by which the motions of very small particles were determined would then be statistical laws. If statistical laws were the ultimate laws of nature, one would say that there is no "determinism" in the physical universe. "Indeterminism" would be the characteristic of our world. However, there are also scientists who surmise that one could perhaps regard these statistical laws as merely superficial laws behind which the actual laws are hidden but as yet unknown. If these laws could be found, our physical universe would be governed again by determinism. It seems as if science alone cannot come to a decision whether "determinism" or "indeterminism" is ruling our world.

Many authors have maintained that the belief in indeterminism supports the belief in the freedom of will, while the belief in determinism may lead to a belief in an iron causality that is incompatible with human freedom and moral responsibility. The American philosopher of law Justice Jerome Frank says in his book *Fate and Freedom*[30] that the belief in an iron causality, if applied to human society, is incompatible with the belief in the free choice of every citizen, which is, in turn, essential for the belief in the "American way of life." According to Justice Frank, the belief in indeterminism is much more favorable to the belief in the American way of life than the belief in determinism. Since science cannot provide a clear-cut decision between these beliefs, he maintains, one should choose the theory that is more favorable to the American way of life—and this belief is indeterminism. We see, therefore, that even today in our own country, as in the times of Plato and Galileo, scientific theories are preferred according to their fitness in supporting a way of life: that is, according to their ability to encourage certain kinds of human behavior.

All these considerations show that ultimately the acceptance of the general principles of a science, say physics, has never been decided by this "special science" alone; there have always been arguments of social science that had an influence. One can visualize this situation by looking back to the "hierarchy of the sciences," or the "pyramid of the sciences." This idea has played at all times a role in the attempts to overcome the isolationist view that the whole of our knowledge consists merely of an aggregate of "special sciences." When Auguste Comte, in 1830, attempted to build up a synthesis of all the special sciences ("positive philosophy"), he established a certain hierarchy among them. He started from the following sequence: mathematics, astronomy, physics, chemistry, biology, and sociology. The main point was that every science presupposes the sciences that were prior in this series but not the posterior ones. "Astronomy," for example, needs "mathematics" but not biology, while "biology" needs "physics" and "chemistry" but not "sociology" as its presuppositions.

From this order, it follows in particular that none of the natural sciences like astronomy, physics, or chemistry need sociology as a basis. However, we have learned (and shall continue to learn) that the most general principles of natural science, for instance, of astronomy and physics, cannot be unambiguously derived without considering the social predilections of a certain period. As we noted, for instance, we can learn from Plato's *Republic* why the astrophysical statement that the sun is made of the same material as the earth is not compatible with Plato's moral and religious value system. In the same way, the Copernican system was not regarded as compatible with the value system of the Middle Ages.

Hence, we have to admit that the principles of astronomy and physics are not independent of social evaluation as they should be according to Comte's scheme of the sciences. As a matter of fact, an immediate disciple of Auguste Comte, E. Littré, proposed an important alteration to this scheme: the order of the sciences should not be linear but circular. One could perhaps, in a rather perfunctory way, propose a scheme of the sciences arranged on a circular (closed) curve in which they follow in the order—mathematics, astronomy, physics, chemistry, biology, sociology, and then again mathematics, astronomy, etc. This circular scheme may help us to visualize the role of pragmatics among the sciences.

Chapter 2

The Longing for a Humanization of Science

1. Dissatisfaction with Nineteenth-Century Science

It has become obvious that the instrument we call "science" is extremely helpful in harnessing nature for human purposes. This success gives to the scientist the satisfaction a man gets out of the construction of a machine that uses simple means to successfully complete a complex job. It is the ingenuity of the human mind that feels elated and proud of this accomplishment. However, when we investigate how men with various kinds of backgrounds and various desires react to the accomplishments of modern science, we record some widespread and typical expressions of dissatisfaction. Particularly men who are deeply interested in the problem of how to guide human conduct toward a desirable way of life express their feeling that modern science is not always helpful and is occasionally even detrimental to the realization of their goals. They feel that older science, before the nineteenth century, had been much more valuable in supporting standards of desirable human conduct because it provided direct information about the "real world." Modern science, according to these men, consists, on the one hand, of a system of symbols without any immediate meaning for human life and, on the other, the observable facts that are logically derived from this system and confirmed by sense observation. Modern science can be regarded either as a system of humanly senseless symbols, as a mathematical or logical structure, or as a system of statements about sense observations. It is, therefore, accused by some authors of being too "abstract" and remote from human experience but by other authors of being too "sensate" and speaking to us about sense impressions only. Both groups accuse modern science of giving to man no real understanding of the universe—no human angle—according to which man could orient himself in his conduct. Hence, modern science, according to these critics, would be of little use in guiding men toward desirable goals.

If we look at ancient and medieval laws of motion, we can well understand what these critics are talking about. They blame modern science for having abandoned the search for the "real causes" of physical phenomena and for having dropped all attempts to provide reasonable explanations for observed phenomena. It is quite understandable that we find this dissatisfaction with modern science frequently among those who are primarily interested in the guidance of human conduct, such as workers in the fields of education, politics, and religion.

2. Emerson on the Changing Role of Science

This dissatisfaction manifested itself clearly after the mid-nineteenth century. We may quote, for example, passages from the essays of Ralph Waldo Emerson, who was not a scientist but was deeply interested in the effect of science teaching upon education. He discusses the difference between the role of science in the period of its unity with philosophy and modern nineteenth-century science in its state of divorce from philosophy. Emerson writes:

> The motion of science was the extension of man, on all sides into Nature . . . (until) through his sympathy, heaven and earth should talk with him. But this is not our science. . . . The formulas of science are like the papers in your pocket-book, of no value to any but to the owner. Science in England, in America, is jealous of theory, hates the name of love and moral purpose. There is a revenge for this inhumanity. What manner of man does science make? The boy is not attracted. He says, I do not wish to be the kind of man as my professor is.[1]

Emerson blames the failures of science education upon the isolationist attitude of the special sciences and, in particular, on the separation of science from philosophy. He writes that "there is a science of stars, called astronomy; a science of quantities, called mathematics; a science of qualities, called chemistry, so there is a science of sciences which is the intellect discriminating the false and the true."[2]

Emerson stresses that the special sciences collect and record facts indiscriminately, without much attention to whether these facts are relevant for our general view of the universe or not. Emerson writes that "the sciences, even the best—mathematics and astronomy—are like sportsmen, who seize whatever prey offers, even without being able to make use of it."[3]

The special sciences, according to Emerson, cannot bring to bear their full educational value without the assistance of a "science of sciences." Emerson was also fully aware that the general laws of science, by revealing the "nature of the universe," have been interpreted frequently as rules for human conduct. To the scientist these laws were merely formulas that served the purpose of coordinating observable physical phenomena; but the philosopher sees an analogy between the laws of science and human conduct. Emerson wrote bluntly:

> The axioms of physics translate the laws of ethics. Thus, 'the whole is greater than its parts'; 'reaction is equal to action'; 'the smallest weight may be made to lift the greatest'; 'the difference in weight being compensated by time'; and many like propositions, which have an ethical as well as physical sense. These propositions have a much more universal sense when applied to human life than when confined to technical use.[4]

The scientists' attitudes regarding science as a system of practical formulas, and of not searching for truth, real causes, and ultimate reality, are branded by the prominent and imaginative social scientist Pitirim Sorokin as follows:

> Decadent sensory science even declares that it is not concerned with any reality. It offers merely certain propositions based upon sensory observations which appear to be convenient and therefore speciously true. Such a formulation of the task of sensory science is equivalent to burying the truth, reality, and science itself.[5]

He attacks most severely modern science's regard for itself as a system of symbols that serve as an instrument for producing and utilizing sensory propositions.

Since a sensory proposition is not concerned with any reality, and does not represent the statement of the cognizing mind about this reality, then just what is it? In this way sensory truth eventually digs its own grave.... The theories of science [become] purely utilitarian and conventional fictions.

According to Sorokin, they become "empty and irresponsible."[6]

3. Lord Herbert Samuel for Modern Science

If we wish to study how contemporary educators and statesmen react to modern science, it is instructive to take as a representative sample some writings of Viscount Herbert Samuel. The author is, as of this writing,[7] the leader of the liberal party in the House of Lords, and through half a century has played an active role in the political, administrative, and diplomatic life in Great Britain. Although not a professional in science and philosophy, he has studied intensively these two fields of knowledge and, in particular, their mutual interdependence. He had strong feelings about the great bearing that philosophy has upon the individual and social behavior of man and especially the importance of science as a support of philosophical doctrines. Like many educational and political leaders, he held that modern science had abandoned its historical role as a guide to philosophy and had isolated itself in order to remain "pure" and aloof from the controversies of philosophical, political, and religious doctrines. Lord Samuel writes:

> This is to desert what has been regarded hitherto by modern science as its fundamental duty. Had this agnostic attitude prevailed five centuries ago and since, the discoveries of Copernicus, Galileo, [etc. to] Einstein ... and Rutherford would never have been achieved.[8]

The great physicists of the past believed firmly that "philosophy and science must never desist from the search for the true causes of phenomena. [Today's] mathematical physics ... is shy of the word Causation."[9] Lord Samuel continues: "If theoretical physicists are often shy of the Cause, they are repelled by the word Explain."[10]

This attitude of modern physics can be shown, according to Lord Samuel, by so simple a case as the current representation of the law of inertia or, in other words, the conservation of momentum. If we recall the presentation of the law of inertia in modern science, the motion of masses in the universe can be conveniently described by the superposition of two motions. One of them is prevalent if the observed mass is very large compared with the masses it is interacting with; then this large mass will traverse a straight line with constant speed relative to an inertial system in the first approximation relative to the constellations of our fixed stars. More precisely, there is in the world an "inertial system" relative to which these large masses will traverse rectilinear trajectories with constant speeds. Along these trajectories, the linear momentum of the masses will remain constant.

Lord Samuel denounces the unwillingness of theoretical physicists to investigate the cause: "When we ask for an explanation of Momentum, or Gravitation, or whatever it may be, they demur at once. They say—What do you mean by Explanation?"[11] According to the views of modern science the law of inertia is justified by the fact that it provides, in combination with other laws, a convenient description of the motions that we observe in nature and in the laboratory. If we ask why a mass, even without external forces acting upon it, keeps moving in a straight line, we are usually told that "inertia" keeps it moving. But Lord Samuel would object to this, asserting not only that gravity and atoms are *factual* but that concepts like momentum, mass, inertia, force, space, and time are *fictional*.

If we persist in asking what is actually happening in the universe to bring a shunted truck to a standstill or to carry a thrown ball across the room, to refer us to these fictional abstractions does not satisfy us. They may justify each other in their mathematical universe, but they can *do* nothing in the phenomenal universe.

By referring to "velocity" or "impetus" at one instant as the cause of the movement afterward, the physicists have abandoned the search for real causes. Lord Samuel objects that velocity is coincident with movement and not prior to it. It can therefore have no operative effect of any kind. Momentum and impetus are words that *describe* what is happening, but they cannot *account* for it.

Einstein's theory of gravitation accounts for the phenomenon of gravitation by the curvature of space. Here again Lord Samuel objects to introducing a mathematical concept like curvature as the real cause of physical phenomena. He calls Einstein's theory of gravitation "an attempt to treat linguistic terms [like velocity or curvature of space] as though they could be factors in a sequence of physical events."[12]

4. Dehumanization of Science

If we wish to understand the deep dissatisfaction with science among those who have been interested primarily in how to guide human conduct toward desirable ends, we must investigate in what way the final goal of modern science differs from the aims of the older period. We will see in sections 10–12 in this chapter that Aristotle saw the final goal of science being to reduce our sense observations to "intelligible principles" that were to be "intrinsically clear" to the human mind. In this way, science attempts a penetration of the physical universe by the human mind, which is, in a certain way, a "humanization of the universe." This was the only type of knowledge that seemed to make sense. In modern science, however, the observable phenomena are derived from principles that are no more "intelligible" and no more "intrinsically clear" to the human mind than the sensory observations themselves. The theories of modern physics do not seem to contribute much toward the conquest of the universe by the human mind.

This is certainly the feeling of many philosophers and educators who deplore the fact that modern science undermines belief in a metaphysical and somehow mystical approach to the search for truth about the universe. We may quote the Russian philosopher Nicolas Berdyaev, an eloquent representative of this group who has been a strong advocate of a mystical approach and a severe critic of science in the modern sense. He writes that

> knowledge, indeed, is *humanization* in the deepest ontological sense of the word. . . . Humanization is at its lowest degree in scientific knowledge and particularly in the physico-mathematical sciences. Contemporary physics demonstrate the dehumanization of science, for their researches are leading them outside the human universe as understood by man.[13]

Our author points out, however, that "physicists are blind to the fact that the very researchers of dehumanized physics symbolize the power of human knowledge."[14] It is certainly true that science in the modern sense can be described as "dehumanized" only if we concentrate our attention upon the finished systems of symbols produced by the physicists and disregard the action of the inventor. We shall see more and more that the presentation of modern science can be *rehumanized* if we focus our attention upon the interdependence between the scientist and the scientist's environment: in other words, if we treat the invention and acceptance of scientific theories (or symbol systems) as phenomena treated by psychology and sociology. We shall learn then that to abandon the search for real causes and real explanations—briefly speaking, to abandon all efforts toward humanizing the universe—has been interpreted as a deliberate and evil attempt of one or more "enemy group" to destroy belief in the real world. These groups have been accused of frustrating the attempts of the human mind to penetrate and conquer the universe and of replacing these heroic efforts with symbol systems that do not represent any cosmic reality. In this way, those "enemy groups" have been held responsible for strangling the belief that we can know the "real world" and, hence, are responsible for strangling all efforts to improve this world.

Under the Third Reich in Germany, for example, the official philosophy explicitly accused the "inferior races" of advocating the interpretation of science as a system of symbols while, according to the doctrine approved by the ruling party, "Aryan" scientists, not satisfied by symbols without meaning, attempted to grasp the truth about the world by merging their souls with the soul of the universe.

5. Soviet Philosophy and Modern Science

In a somewhat similar way, the official philosophical doctrine of the Soviet Union approved by the ruling party, condemns the conception that science is a system of abstract symbols. This interpretation of science is known by the derogatory term "formalism" and is regarded as a typical view of "bourgeois scientists." The official philosophy maintains that the presentation of science by a formal system disguises the real character of the world and makes people lose interest in the fight for social improvement.

In the book *Dialectical and Historical Materialism*, introduced as a textbook in the universities in the Soviet Union, the author Mark

Borisovich Mitin connects the emphasis on relations between symbols prevailing in modern science with the "new logic" advocated by Bertrand Russell and his school. "The New logic," he writes,[15] "expels things and their qualities and deals only with 'pure relations' (i.e., not with relations between things, but with pure relations without content and without objects that are related)." He concludes, "In this way contemporary idealists purge science of its content, its meaning, its truth."[16]

These statements about modern science are very similar to Sorokin's statement that the theories of science grow progressively "thinner" and "shallower." If we wish to understand the great bearing that modern science has on men's orientation in the world, and upon his rules of conduct, it is important to appreciate the fact that a fervent advocate of idealistic philosophy like Sorokin and an official representative of "Soviet Materialism" like Mitin agree on a very important point: they are dissatisfied with the interpretation of science as a system of symbols to be judged only for its logical coherence and the agreement of its results with experience. Idealists and naturalists agree in their dissatisfaction with a science that limits itself to the logico-empirical or positivistic philosophy.[17]

6. The Birth of Modern Science Was the Birth of Dissatisfaction

It would be wrong to believe that this dissatisfaction with modern science originated in our present period. It is probably true that this dissatisfaction has become particularly vociferous in the twentieth century because the physical theories of today are much more abstract and symbolic than those of earlier centuries. But essentially the allegation of dehumanization, of pure formalism, and of deserting the search for truth have accompanied modern science to a certain degree since its birth.

The separation of science in the modern sense from philosophy started first in astronomy. We can therefore expect that the first complaints about abandonment of the search for real causes, for real truth, for the real nature of a phenomenon would arise in the discussion about new astronomical theories. If we glance at the period circa 1600 when the conflict between the Ptolemaic and the Copernican world systems was at its peak, we find quite a few examples of the type of complaint that we find in modern times in the writings of authors as varied as

Ralph Waldo Emerson, Pitirim Sorokin, Nikolai Berdyaev, Lord Samuel, and Mitin.

Everyone who has even a slight familiarity with the history of philosophy will know that at about that time Francis Bacon, a British philosopher and statesman, was regarded as the man who attempted to refute medieval philosophy and introduce a new "empirical" philosophy that would better fit with the rising tide of modern science. He called the Ptolemaic doctrine of eccentrics and epicycles an absurd theory; but he added that the absurdity of these notions has "thrown men upon the extravagant idea of the diurnal motion of the earth, an opinion which we can demonstrate to be most false."[18] He strongly disapproved of accepting an astronomical theory just because it allows for a convenient computation of the future position of planets; he disapproved the inclination of scientists to devote all their efforts to the invention of convenient mathematical theories of motion. He continued:

> But scarce any one has inquired into the physical causes of the substance of the heavens . . . the different accelerations of motion in the same planet; the sequences of their motion from east to west: the progressions, stations, and retrogressions of the planets. . . . Inquiries of this kind have hitherto been hardly touched upon, but the pains have been chiefly bestowed in mathematical observations and demonstrations which indeed may show how to account for all these things ingeniously, but not how they actually are in nature: how to represent the apparent motions of the heavenly bodies, and the machines of them, made according to particular fancies; but not the real causes and the truth of things.[19]

These lines could just as well have been written by Lord Samuel in 1951 as 350 years ago. And in the same way as Emerson, Sorokin, and Berdyaev have expressed regrets that sciences like astronomy have lost their dignity because they no longer claim to tell us the truth about the universe, Bacon wrote as long ago as 1605:

> Astronomy, as it now stands, loses its dignity by being reckoned among the mathematical arts, for it ought in justice to make the most noble part of physics. . . . We, therefore, report the physical part of astronomy as wanting, in comparison of which

the present animated astronomy is but as the stuffed ox of Prometheus—aping the form, but wanting the substance.[20]

7. Bacon on the Copernican System

It is important to note that Bacon not only blames Copernican astronomy for producing only a mathematical scheme without attempting to find the real physical causes; he also launches the same objection against the old Ptolemaic theory of eccentrics and epicycles. The orthodox Aristotelians had never admitted that the theory of epicycles was "true" in the philosophical sense. From these historical facts, it seems to follow that whenever a new theory was advanced from which experimental facts could be precisely derived, this theory was at once accused of being "purely mathematical" and telling us nothing about the real causes of the phenomena. Neither the Ptolemaic nor the Copernican system was recognized by Bacon as being a sound physical theory.

In order to find out what has actually been done to make science more satisfactory to the human mind, it is perhaps advisable to find out why Bacon does not recognize the Copernican theory as a satisfactory physical theory and what he recommends for the improvement and humanization of science: in this particular case, astronomy. By what argument did he demonstrate that the Copernican theory was "extravagant" and "false"? Bacon wrote:

> The separation of the sun from the planets [in the Copernican system] with which it has so many affections is a harsh step; and the introduction of so many immovable bodies into nature, as when he [Copernicus] makes the sun and the stars immovable, the bodies which are peculiarly lucid and radiant, and his making the moon adhere to the earth in a sort of epicycle, and some other things he assumes, are proceedings which mark a man who thinks nothing of introducing fictions of any kind into nature provided his calculations turn out well.[21]

Another reason advanced by Bacon in favor of the earth as immovable center is "the evidence of sight and inveterate opinion." In favor of the sun as center, on the other hand, Bacon invokes the consideration that that body which has the chief office in the system should occupy that place from which it may best act on the whole system and communicate its influence. And since the sun is that which seems most to vivify the world by imparting heat and light, it appears to be altogether right and in order that it should be placed in the middle of the world.[22]

If we look at Bacon's argument from the point of view of modern science, we would say that neither the plea in favor of the earth nor the one in favor of the sun contains reasoning that could be called "scientific" in the strict sense this word is used in today. No investigation is made to find out whether observational statements that would contradict each other can be logically deduced from the two assumptions, nor are we told which of these statements agrees with actual observations. What Bacon really investigates is the question of which theory, the heliocentric or the geocentric, agrees with the most familiar facts of our everyday life, or with what could be called commonsense experience. He invokes, for example, the "evidence of sight": this means that the closer the statements of the theory are to familiar statements about what our eyes have seen, the more probable the theory. Since childhood we have been accustomed to look at the motion of the sun and the moon in relation to the walls and the floor of our rooms. Then, of course, every child would say "men move in the room," and not that "the room moves around men who are at rest." Therefore, when we look at the "sunrise" in the morning, we say, according to speech habits acquired as children, "the sun is moving and the horizon is at rest." However, if we train ourselves, on looking at the sunrise we can just as well "see" the horizon sinking and the sun resting. Everyone has experienced the fact that at the start of a train ride we can "see" our train moving just as well as the train on the adjacent track; but we shall, in any case, say that our train is moving and the station is at rest. What Bacon actually proved is that the Ptolemaic system is more in agreement with the way we describe our most familiar experiences in the language of our daily life. We can say, in brief terms, that it is more in agreement with commonsense parlance.

All advances in science have been accompanied by a growing departure from this commonsense parlance in the formulation of scientific theories. These remarks may also serve to correct an appreciation of the Copernican system that has been repeated again and again in books and lectures, yet which is very misleading. The Copernican system has been praised for having refuted the judgment of our sense observations and for having proved the superiority of reason over the senses. As a matter of fact, the Copernican system is in no point contradictory to the judgment of our sense observations. As we said above, our senses do not tell us whether the sun rises or the horizon sinks but only that the distance between the sun and the horizon increases. We can only say that the statements of the Copernican theory are more remote from the statements of commonsense language than are the statements of the Ptolemaic system. Copernicus did not "refute" the judgment of our senses, but he showed that the phenomena of planetary motions can be derived more conveniently from principles that are remote from commonsense language than from principles that are close to it. Which is, from the "scientific aspect," everything that can be said about this point. By the remoteness of the general principles that are the backbone of modern science, however, the impression is produced that these principles are somehow dissatisfactory and only serve as useful fictions.

From the "humanistic aspect," which is ours in this book, we must recognize, following the testimony of a great many prominent authors, that modern science has left a gap in the minds of men who are eager for truth. We notice in Bacon's appreciation of the Copernican system that, frankly speaking, he misses the "truth" in it. He regards the Copernican system as a "fiction"—in the best case a "useful fiction"—but a "fiction" in any case. By analyzing Bacon's attitude toward the Copernican system, we noted that what he actually blames is the disagreement of the Copernican system with the statements of our commonsense experience. If we are not satisfied with the usefulness of a theory for the description of phenomena and with logical consistency—if we, in fewer words, are not satisfied by the logical-empirical criteria for the acceptance of a theory—we must introduce a further criterion of truth. From our analysis of Bacon's criteria for acceptance of the Copernican theory we learned that what he was actually after was agreement of the general principles of astronomy with commonsense experience. If an astronomical theory is the best possible one by purely scientific, logico-empirical criteria, it is by no means certain that it is "nearest to the true nature of the universe." It is very instructive to learn that Bacon outlined a kind of peace treaty between philosophy and science (astronomy) in which the rights and duties of both parties are stipulated in a way that reminds us of actual pacts between hostile powers. Also in our case the pact means "armistice" rather than peace. We can easily point to some ambiguous paragraphs that may provide good food for legal minds. Around 1612 he wrote:

> Let it then be arranged, if you will, between philosophy and astronomy, as by a convenient and legitimate compact, that astronomy shall prefer those hypotheses which are most suitable for compendious calculation; philosophy those which approach nearest the truth of nature.[23]

What follows is a paragraph that really sounds like an agreement between diplomats. "The hypotheses of astronomy shall not prejudice the truth of the thing, while the decisions of philosophy shall be such as are explicable on the phenomena of astronomy."[24] Obviously Bacon wanted to say that one and the same phenomenon, for instance, planetary movement, can be presented by astronomical theories and by philosophical hypotheses, each of which has its own purpose. While scientific hypotheses are guided by the logico-empirical criteria only, philosophy attempts to bring out the "real course," the "explanation" of the phenomenon. The "philosophical interpretation" of the scientific law would cure the dissatisfaction that has been felt by educators as well as by political and religious workers when confronted by modern science. What is needed in order to relieve the uneasiness described by Emerson, Sorokin, and others is the addition to the scientific hypothesis from which we can derive logically the observed facts of a piece of "philosophy" that provides an *understanding* of these hypotheses; this would counteract the dehumanization of science, which is the result of what we have called the "purely scientific aspect," or the "logico-empirical" aspect.

8. How Science Has Been "Humanized"

In antiquity and the Middle Ages, no need for humanizing science existed. According to Aristotle and his school, the principles of intermediate generality (for example, the orbits of planets on this sphere and the trajectories of launched stones on the earth) could be checked by experiments fairly directly; but they were derived from more general principles that were "intelligible" or "intrinsically clear" to the human mind. The laws of motion were derived from "organismic" principles, according to which every terrestrial body tried to return to its "natural place" as quickly as possible, and according to which the nobler celestial bodies moved eternally in perfect and noble orbits (i.e., in circles). In what sense did these principles seem to be intelligible to the human mind? Obviously, the human mind considers these terrestrial bodies somehow similar to itself and understands by analogy that such a being does not feel sheltered before it has reached its "natural place." In the same way he sees in human society more exalted beings who walk along their paths for a long time without becoming tired and without being deviated by the influence of other less exalted beings. Man understood the laws of the physical universe by identifying himself or other human beings with the parts of the universe.[25] Astronomy was based upon humanizing the physical world.

Because Newton's laws of motion did not lend themselves easily to a derivation from intelligible principles, men have always tried to introduce "intelligible" principles from which these laws of motion could be derived. Newton himself at two points introduced organismic arguments that again mean the interjection of the human mind and its action into the physical universe. If we look, for instance, at Newton's law of inertia, it does not seem "intelligible" that a body is moving in a straight line relative to absolute space. This was later reformulated into: "There is an inertial system with respect to which a body is moving in a straight line with constant speed." However, we do not understand why, for instance, the system of fixed stars is approximately an inertial system, while other mechanical systems are not. Newton identified absolute space and hence the inertial system with the "sensorium of God" that filled the whole world space. By "sensorium" he meant the organ by which God perceives all events in the world. There could be some doubt as to whether this means an actual "filling the space" analogous to the way in which a piece of wood "fills space." We can solve this doubt by looking into the diary of David Gregory, a good friend of Newton. In 1705 he wrote concerning the question of what fills the space that is called empty space:

> The plain truth is that he believes God to be omnipresent in the literal sense; and that as we are sensible of objects when their images are brought home within the brain, so God must be sensible of every thing, being intimately present with every thing; for he supposes that as God is present in

space when there is no body, He is present in space where a body is present.²⁶

Since obviously God cannot be present in space in the sense in which geometry speaks of the presence of a body in space, the "omnipresence" of God is meant in an "analogical" sense.²⁷ Newton defines "inertial system" not by reference to a physical body that is an object of our sense observations; he refers rather to a spiritual being to which only by analogy with a human being can the quality of presence in space be ascribed.

As a matter of fact, Newton replaced the inertial system by an object of commonsense experience, a human being with the purpose of reducing the law of inertia to an intelligible principle. The law became more intelligible because a human being was injected into the physical world and made it more "intrinsically clear" to the human mind. It was again an attempt to "humanize" physical science by introducing analogies taken from our commonsense experience.

9. Analogies as Humanizing Elements

It is well known that attempts to "explain" or "understand" Newton's law of gravitation use analogies with attraction exerted by one human being upon another. The meaning of attraction was in some cases "pulling with a rope," but in other cases the analogy was with psychological attraction, as in the expression "an attractive girl." As the words "force" and "energy" suggest, there have been repeated attempts to "understand" the laws of mechanics by interpreting them as analogies to commonsense experience gained by observing human beings under familiar conditions. In the theory of relatively the "humanizing element" was injected by the "observer." If an observer is "inside" a moving system, he cannot make any observation outside the system, and then it seems plausible that the laws of nature can be formulated to contain only the observations of this one observer.

In Bohr's theory of spectral lines, every line was due to the jump of an electron from one orbit to another; but the theory predicted only how many jumps there would be on the average every second. It did not predict at what moment any individual electron would jump. This new type of regularity was formulated by saying that each individual electron is "free" to jump when it "likes."

The statistical laws of atomic physics have been made "intelligible" by analogy with the commonsense experiences of human beings with "free will." However, the commonsense experience of mankind does not only consist in the behavior of living organisms, as in many cases the behavior of inanimate bodies is also a part of that experience. To the oldest experience of mankind certainly belongs the behavior of colliding solid bodies. It has been known since time immemorial that when a stone hits a second stone, the small one will be pushed and moved. Therefore, there has been a general tendency to "explain" the principles of science by analogy with the phenomena of pushing and pulling, for instance, to account for electromagnetic phenomena by the push and pull of hidden material bodies. This type of "mechanistic" picture must also be regarded as an explanation by analogy taken from our everyday experience. It seems therefore that the complaint that modern science has become dehumanized—that the search for real causes and explanations has been abandoned—has been met by introducing analogies with everyday experience. By adding these analogies to the general principles of science, we cause these principles to become "intelligible," "intrinsically clear," and "humanized."

10. "Humanization," "Metaphysics," and the "Inner Eye"

We stressed that in ancient and medieval science there was no dissatisfaction with science because of "dehumanization." The reason was simply that, as stated earlier, the principles of Aristotelian and Thomistic science, for instance, were formulated in a language that was very close to the commonsense level and therefore seemed plausible to everybody who accepted the experience of everyday life. The philosophies of that period derived the plausibility of these principles not from their proximity to commonsense experience but from the fact that they were "intelligible," or "intrinsically clear." In ancient and medieval philosophy a statement was "intelligible" if man could recognize its truth by his intellect directly, as man recognizes the truth of an observational statement like "this flower is red" by his senses. When the belief in this ability of the intellect started to fade away, as it did in the thought of men like Francis Bacon around 1600, the plausibility of the most general principles of science came to rest no longer on the role of the intellect as an "inner eye" but upon the analogy of those principles with commonsense experience, as we learned from our analysis of Bacon's views on Copernicus (Part I, chap. 2, §6–7). If we follow this analysis, we learn that "seeing with the inner eye" and "believing in common sense experience" yield about the same results. When, however, science rose beyond the commonsense level, as in Newton's mechanics (and even more in relativistic and quantum mechanics), the general principles could not be supported by commonsense experience and could not be seen either by the inner eye or by common sense. Then, in order to "humanize" science, general principles had to be interpreted by pointing out analogies to common experience. To these analogies belong the introduction of "observers" and "free will of electrons" into science. We call these analogies "philosophical interpretations" of science.

To understand what the ancient and medieval philosophers actually meant by "seeing with the intellect as an inner eye," we can consider Thomas Aquinas who, following Aristotle, wrote that "truth is defined by the conformity of intellect and thing. . . . As the good denotes that towards which the appetite tends, so the true denotes that towards which the intellect tends."[28] In our usual language the conformity between a mental object like intellect and a physical thing seems to make little sense. In order to grasp this sense correctly, we must understand that for St. Thomas "everything is true according as it has the form proper to its nature."[29] The terms "form" and "nature" will be more elaborately discussed in the chapter on Thomism (Part II, chap. 7). At this point, we need only keep in mind that "form of a physical body" meant not only geometrical shape but all the properties that determine the reaction of the body to its environment (e.g., the elastic and optical properties of the body). As a thing is true if it has the form proper to its nature, "the intellect, in so far as it is knowing, must be true, so far as it has the likeness of the thing known."[30] As a knowing power, the form proper to the intellect is a likeness to the form proper to the thing known.

At this point the similarity between intellect as knowing power and a sense as a perceiving power comes in. Sight, for instance, has the likeness of a visible thing, but sight does not know the degree of likeness that exists between the thing seen and the image that sight has formed of it. "But the intellect can know its own conformity with the intelligible thing."[31] We note that according to ancient and medieval philosophy, there is a close similarity between perceiving a thing by our senses and knowing it by our intellect. The general principles of science become "intelligible" if we know them by our intellect, our inner eye. St. Thomas Aquinas referred to this insight as his first criterion of truth. It is, for example, intelligible that perfect beings would move in perfect orbits.

110   The Humanistic Background of Science

It is obvious that we cannot derive such statements by science in the modern sense, as we will present in Part II. By checking the observable results of such general principles we can never prove beyond a doubt that such principles are true, while by "knowing" in the Thomistic sense, by equating the form of our intellect with the form of the thing, we can prove them with certainty. Statements that are obtained in this way that cannot be obtained by the methods used in modern science will be called "metaphysical statements" in the traditional sense of this word. They are not based upon the specialized method of science, and their truth is therefore independent of the advance of science.

11. Metaphysics, Common Sense, and the Inner Eye

If we eventually trace the statements of science back to metaphysical principles, science is no longer dehumanized; its principles are known by equating the human intellect to the form proper to the thing to be known. We can learn from a contemporary Thomist in what sense the statements of metaphysics are the results of "seeing with the inner eye" or "using the intellect as a knowing power." Jacques Maritain, a strong advocate of traditional metaphysics, writes the following:

> Metaphysics . . . is a type of intellectual visualization in which not only are the conditions of singularity wanting but all reference to the perceptions of the outer senses or to a probable construction by the imaginative intuition. The object is seen and defined by the intellect only in reference to the intelligibility of being itself, in other words in a purely intelligible and immaterial fashion.[32]

It is instructive to note that by using the expression "intellectual visualization," Maritain describes metaphysical knowledge as somehow similar to visual perception, or seeing with the eye of the intellect.

For another example, the French philosopher Henri Bergson was certainly one of the strongest advocates of metaphysics at the turn of the century. He distinguished metaphysics from science by claiming that science describes, for instance, the physical world "from the outside"; it investigates the connections between the quantities which define the position and velocities of bodies with respect to other bodies, the temperature with respect to other temperatures, and so on. Metaphysics, on the other hand, attempts to penetrate further and describe the physical world "from the inside." What is meant by this can be seen clearly if we examine how Bergson applied the distinction between "outside" and "inside" to the motion of physical bodies. He regarded this application as a particularly lucid example of his conception of science and metaphysics. Toward the beginning of his book, he wrote:

> Consider . . . the movement of an object in space. . . . My expression of it will vary with the system of axes, or the points of reference to which I relate it; that is, with the symbols by which I translate it. For this . . . reason I call such motion *relative*: [describing motion in this way] I am placed outside the object itself. But when I speak of an *absolute* movement, I am attributing to the moving object an interior and, so to speak, states of mind.[33]

Physical science knows only relative and absolute motion as a metaphysical concept. We cannot find out by experiment and the connected logical arguments whether a physical body has felt "moving" in its mind or not. If someone ascribes to a body an "absolute" motion, he does what the man does who says at sunrise that the sun is "really moving." We regard the sunrise as analogous to a phenomenon that is familiar to us from childhood: the motion of a ball launched in a room. Then we learned to say that "the ball is moving" and not "the room is moving." In this same way we speak about the sunrise. In metaphysics we say that "the sun is absolutely moving," but what we actually mean is that, in the situation of daily life we regard as analogous to the sunrise, the traditional role of our everyday language suggests that we say the "sun is moving." If we use the language of metaphysics we would say, "Our intellect knows by equating its own interior form with the interior of the rising sun that the sun is moving." But if we describe the actual behavior of an observer, we notice that with a certain effort the observer can see that the sun is moving upward just as well as notice that the sun is resting and the horizon sinking. In each of these alternatives man envisages different situations of his daily life and uses them as analogies to the sunrise. We can see from this example that what we called "philosophical interpretations of science" and "metaphysical insight" applies practically to one and the same class of statements. However, if we speak of philosophical interpretations, we imply that they are the result of our desire for a humanization of science and that an individual interpretation cannot be called "true" or even "probable" in the same way as a statement of science proper can be called. We call the sun "moving" or not according to how we want to round out the general world picture. Should a picture be drawn in the "Aristotelian style" or in the "Newtonian style"?

12. The Nature of Metaphysical Statements

What Bergson actually meant by metaphysical statements about the "absolute motions" of a body can easily be seen from his explication of "absolute motion." He elaborated as follows:

> I also imply that I am in sympathy with those states, and that I insert myself in them by an effort of imagination. Then, according as the object is moving or stationary, according as it adopts one movement or another, what I experience will vary. And what I experience will depend neither on the point of view I may take up in regard to the object, since I am inside the object itself, nor on the symbols by which I may translate the motion, since I have rejected all translations in order to possess the original. In short, I shall no longer grasp the movement from without, remaining where I am, but from where it is, from within, as it is in itself. I shall possess an absolute.[34]

It is certainly not easy to imagine how I shall feel when I attempt to insert myself into moving bodies and "sympathize" with them; but one thing is clear—the feeling cannot be very different from the way I feel in a moving ship, plane, or some more extravagant vehicle. Hence this feeling can only be a state of mind that is familiar from everyday experience and that what Bergson calls "knowing the body from within, as it is in itself" is actually an imagined analogy taken from the commonsense experience of an average human being. What we do in following Bergson's advice is to interpret commonsense statements about motion by commonsense statements about the feelings of a passenger in a vehicle.

Scientifically minded thinkers have always felt that these "philosophical interpretations" are not necessary or even helpful for science proper or for predicting future events and handling physical devices. They are helpful for giving a picture of the universe in a language that is close to the language that has since our childhood seemed to most conveniently describe the world of our daily experience. If, however, one believes that the statements by which he interprets science philosophically are as certain as our immediate sense observations—that they are the result of "knowing with an intellect" or "seeing with the inner eye"—we say that he makes a "metaphysical statement." To say, for instance, "It is certain that the sun is at rest" is a metaphysical statement; but to say that one can interpret the universe by introducing a sun that is absolutely at rest is a "philosophical interpretation." It describes the astronomical phenomena by a picture taken from our daily life experience.

In our commonsense language we can distinguish between absolute and relative motions because in every practical situation a certain system of reference is taken for granted. In scientific language every motion is relative; but in commonsense language we can speak about absolute motion, although we must be aware that we introduce in this way a subjective human element into the description of motion. This can be helpful for some purposes: for instance, in correlating astronomy with human behavior. However, such statements can become confusing if we regard them as "true" in the same way in which we regard scientific statements as true. While the truth of scientific statements is tested by comparing the logical consequences of a statement with observations, the metaphysical statement would be checked in a similar way if we compare its logical consequence with the observations of our inner eye.

13. The Inner Eye and Intuition

Ancient and medieval theory held that metaphysical statements are not essentially different from scientific statements, except that sense observations are replaced by observations of the inner eye. There is, however, one great difference: agreement among men can be achieved concerning the data of immediate sense observations but not about the "data" of our inner eye. Reference is often made to what is known by immediate sense experience as the "given"; but rarely is what is known by direct application of our intellect referred to as "given." There is one expression, however, that lumps together both types of knowing—the word "intuition."

Some authors use the word in the sense of "direct sense impressions." Filmer S. C. Northrop says, for instance, "A concept by intuition is one which denotes, and the complete meaning of which is given, by something which is immediately apprehended."[35] The way Northrop uses the term "intuition" becomes particularly clear if one compares sentences in which he makes use of this term. He writes: "Positivism . . . is the thesis that there are only concepts by intuition."[36] On the other hand, according to Northrop, a metaphysical theory is one maintaining that there are also other concepts apart from the concepts by intuition.

However, by "intuition," other authors mean the metaphysical insight—the seeing with the inner eye. Bertrand Russell describes the period of science in which sense observation was minimized in comparison with conceptual thinking by stressing the point that "thought is superior to sense, intuition to observation."[37]

This equivocal use of the term "intuition" has certainly contributed to the confusion reigning among authors who examine the relationship between science and philosophy. In many cases the equivocal character of "intuition" has been used to give statements about the relationship between science and philosophy a particular flavor. We may quote, for instance, what Edouard Le Roy writes about dissatisfaction with modern science: "There are with us also disinterested desires that strive for satisfaction. Later on these tendencies are dissociated and cultivated separately from each other. *They would become Science and Philosophy.* The first one systematic organization; the latter living on as rich intuition."[38]

Chapter 3

Metaphysical Interpretations of Science

1. The Founder of Pragmatism on Science and Philosophy

The departure from common sense has been one of the main reasons why dissatisfaction with science has arisen and why many feel that it has become "dehumanized." It is interesting to examine the opinion of those who were equally well versed in science and philosophy on the issue of the role of common sense and philosophy.

There is a widely held opinion that philosophy has always attempted to stay on the level of common sense, while for science an important abiding tendency in its growth and evolution has been that it is drifting increasingly farther away from common sense.

There were two great turning points in the history of science, the transition from medieval to modern science around 1600, characterized by the names of Galileo and Newton and, around 1900 the transition from Newtonian science to twentieth-century science characterized by the Theory of Relativity (Einstein) and the Quantum Theory (Max Planck, Niels Bohr, Werner Heisenberg, and others). Each of these turning points was also a strong departure from what was regarded as common sense during that period. Around the year 1900, the second great turning point in the history of science and philosophy, one of the most prominent thinkers was Charles S. Peirce.

In his classification of the sciences he suggests a very revealing distinction between philosophy and the special sciences, such as physics, biology, and so on. For class I, Peirce proposes mathematics. Class II is philosophy, which deals with positive truth, indeed, yet is content with observations such as those in range of everyone's normal experience and for the most part in every waking hour of one's life. These observations escape the untrained eye precisely because they permeate our whole lives, just as someone who never takes off blue spectacles soon ceases to see the blue tinge. Obviously, therefore, no microscope or sensitive film would be useful in this class (i.e., philosophy). The observation is an observation in a peculiar yet perfectly legitimate sense.

Peirce explicitly states that advances in the special sciences do not contribute much to the advance of philosophy. "If philosophy glances now and then at the results of special sciences, it is only as a sort of condiment to excite its own proper observation."[1] If, for example, we learn from twentieth-century science that a very large quadrangle cannot have four right angles, that fact lies far beyond our commonsense experience; but this broad generalization may serve to stimulate us to investigate more carefully the commonsense concept of a quadrangle.

In class III, according to Peirce, are "the special sciences, depending upon special observations, which travel or other exploration, or some assistance to the senses, either instrumental or given by training together with unusual diligence, has put within the power of its students."[2]

2. Peirce's Conception of Philosophy

According to Peirce, in the formulation of every science there is some measure of philosophy because we have to start every special investigation using only commonsense concepts and commonsense laws. Peirce writes that

> philosophy, whose business it is to find out all that can be found out from those universal experiences which confront every man in every waking hour of his life, must necessarily have its application in every other science. For be this science of philosophy that is founded on those universal phenomena as small as you please, as long as it amounts to anything at all, it is evident that every special science ought to take that little into account before it begins to work with its microscope, or telescope, or whatever special means of ascertaining truth it may be provided with.[3]

Peirce divides philosophy into "Phenomenology, the ultimate analysis of all experience," "normative sciences, which investigate what ought to be," and finally metaphysics.⁴ The attitude of metaphysics toward the universe is, according to him, nearly that of the special sciences. However, it is distinguished from sciences like physics or psychology by the fact that it confines itself to those parts of physics or psychology that can be established "without special means of observation. But these are very peculiar parts, extremely unlike the rest."⁵ This is what separates metaphysics from physics.

Peirce gives an instructive example of what constitutes both a scientific and a philosophical solution to a question. Since the period of Epicurean philosophy in ancient Greece, one of the most important theories about the structure of matter is "atomism," according to which matter consists of small solid lumps with a large empty space in between. But there has been, at least since Descartes, another theory according to which the whole world space is filled with a subtle liquid; there is no empty space, and where the existence of solid particles (atoms) had been assumed, there are actually vortices of liquid. If we wish to decide scientifically which theory is preferable, we can try to derive logical conclusions from each of the two theories that could be checked by sense observations. If the two theories lead to different sense observations, we can by direct observation decide whether matter is divided into small lumps or fills space continuously.

From the philosophical point of view, a decision cannot be reached unless we "look at the alternative with our inner eye and see which is the true one." This means, in other words, that we can try to find out which alternative, atoms or vortices, discontinuity or continuity, is nearer to our commonsense experience and nearer to the experience of our everyday life. Peirce writes that "the question . . . must evidently depend upon what we ought to conclude from everyday, unspecialized observations."⁶

This means that we have to decide whether atoms in a void or vortices of fluid are more reconcilable to our commonsense experience. This recommends the same methods as those by which Francis Bacon decided in favor of the Ptolemaic system rather than the Copernican.

3. Metaphysics Nearer to Common Sense than Science

From Peirce we learn a very important point: metaphysics is not based on the results of modern science. It does not need, for instance, twentieth-century theories, as it can draw all its conclusions from the theory that presents our everyday life experience in the most practical way, and this theory is the doctrine of common sense. We find that even the most fervent advocates of metaphysical insight, provided that they have good judgment about scientific method, will agree that metaphysics is not based upon science but upon common sense. We may quote the judgment of Jacque Maritain, one of the most prominent scholars and thinkers following the philosophy of St. Thomas. He writes the following:

> The knowledge of common sense is a natural spontaneous growth, the product, so to speak, of rational instinct, and has not yet attained the level of science. . . . Nevertheless, this infra-scientific knowledge is more universal than that of various particular sciences. . . . It possesses a certain metaphysical value in as much as it attains the same object as metaphysics attains in a different fashion. Common sense is therefore . . . a rough sketch of metaphysics, a vigorous and unreflective sketch drawn by the natural spontaneous instinct of reason. This is why common sense attains a certain though unscientific knowledge of God, human personality, free will, and so on.[7]

While metaphysics is based directly on common sense, the path leading from commonsense experience to the principles of quantum theory or relativity is long and arduous. All concepts of metaphysics bear a resemblance to the concepts we use to describe our everyday experiences and that we call commonsense concepts, while the main concepts of theoretical physics are very remote. To understand this we can take an example from geometry. The expression of "straight line" used in non-Euclidean geometry, or of "mass" used in relativity theory, or of "momentum" in quantum theory can be connected with observational concepts like "red," "warm," or "round," only by a long chain of rather difficult arguments. We shall clarify the situation when we discuss the "metaphysical interpretations" of particular physical theories.

However, at this point we shall recall how a French logician and philosopher, Edmond Goblot, described this particular relationship between science and metaphysics, which is very different from the way in which it has been traditionally envisioned by scientists as well as philosophers who have not given sufficient thought to the behavior of scientists and the actual behavior of philosophers. Goblot wrote, "Physicists and Metaphysicians arrive at principles of physics which are very different from each other."[8] Here again appears the dichotomy described by St. Thomas. The principles of the physicists are formulated in terms that allow one to deduce, by means of operational definitions, statements about sense observations. The nature of the terms used is irrelevant to physicists. The metaphysician prefers formulations that have a direct meaning in the domain of commonsense experience, or that can be interpreted by close analogy in that domain. Goblot continued:

> Physicists start from concepts by means of which one can construct all other concepts; they arrive at these principles at the end of long research work and of profound analysis. The metaphysicians start from the simplest beings and the most palpable facts, taking their first elements from most common experience.[9]

Like Peirce, he strongly stressed the point that metaphysics is not based on science in its recent form but on commonsense experience, or more pointedly, on the system of statements by which our daily life experiences are presented in the most practical way. Metaphysics consists in a general view of our common experience or, when it is concerned with advanced science, it does not offer any view on the principles of contemporary science but rather on the analogies to these principles in the domain of commonsense experience. From these considerations it follows that metaphysics is much nearer to the concrete facts of our experience than physics, which makes use of very abstract concepts like "the straight line" of non-Euclidean geometry, "the particle without trajectory" of quantum theory, and so on. Goblot continued:

> It is theoretical physics which is really abstract: it is metaphysics that allows a piece of empiricism. It may seem that metaphysics, which flatters itself that it proceeds *a priori*, deserves the reproach of empiricism while physics, an experimental science, is shown to be perfectly rational.[10]

Goblot saw the empirical nature of metaphysical knowledge in the fact that all metaphysical prepositions, although verbally referring to "knowing by the inner eye," actually refer to phenomena in our commonsense experience, while the principles of quantum theory cannot be interpreted as direct descriptions of our daily experience. As we shall later see, one can give "metaphysical interpretations of quantum theory," which attempt to find in our commonsense experiences some analogues to the principles of quantum theory.

Goblot also recognized clearly that the "seeing with the intellect" claimed by metaphysics is actually a disguised appeal to commonsense experience. He said:

> Despite its rationalist and aprioristic tendencies, metaphysics in its search for the real cannot attain it without an appeal, at least a disguised one, to experience, since only experience can teach us what is.[11]

From all these considerations, it can clearly be seen that the "man in the street," whose thoughts are fully restricted to common sense, will be more inclined to accept the "intuitions" of metaphysics than the scientist. Henry Vincent Gill, an advocate of Thomistic philosophy, writes,

> The fact that a few "specialists" call in question some intuition generally accepted by men does not present a valid reason for doubting its truth. The specialist is, indeed, perhaps the one whose view or first principles should be taken most cautiously. Before anyone becomes a specialist he must recognize the first principle on which specialist knowledge has been based.[12]

The author means to prove that before going to work, the scientist has to accept the "first principle" as advanced by the metaphysician because such principles are based upon common sense, and no science can develop without accepting the doctrine of common sense. The argument seems to confuse two different ways in which common sense is connected with science. As we have seen, again and again, all experimental observations are described in the language of common sense, and hence all science is based on common sense. However, these descriptions are certainly metaphysical statements. The metaphysical interpretations, on the other hand, also arise when we look within the domain of common sense for analogues to the general principles of science. For instance, we may interpret Newton's gravitation as a tendency of masses to attract each other, or the law of inertia as a passivity, an inert character of masses that will not change their state without being forced to. This "metaphysical interpretation" of Newtonian mechanics adds nothing to the conclusion, which is drawn from Newton's laws. Hence, the work of the scientist is by no means based upon these metaphysical interpretations. They are neither conclusions drawn from commonsense statements nor inductions or generalizations based on commonsense statements but rather pictures in the language of common sense whose purpose is counteracting the dehumanization of science. Gill argues, "The very fact that the scientist denies the need of accepting first principles would seem to be a confession of want of confidence in his own assertion."[13] This statement by an advocate of Thomistic metaphysics is instructive because it reveals the misunderstandings that have arisen and persisted between scientists and philosophers. They have their basis in the lack of a common language that covers the entire domain of science and philosophy, as well as physics and metaphysics.

This book attempts to find this common language by defining as clearly as possible the place of common sense, on one side, and of scientific principles on the other. Then we can clearly see that the "first principles" of metaphysics are commonsense illustrations of the general principles of science but are by no means the basis upon which science itself is built. In the first two chapters we have already learned to underscore the fact that metaphysical statements are not generalizations of scientific statements and, therefore, metaphysics is not a kind of super science or superior type of knowledge. On the other hand, it would be a misunderstanding to say that the metaphysical statements of science do not serve any useful purpose. They "humanize" the principles of science: they counteract the dissatisfaction of which we spoke previously (e.g., in Part I, chap. 2, §1–3) and contribute in this way to the well-being of the scientist and the "consumer" of science. This point has been recognized by many advocates of traditional metaphysics.

4. The Purpose of Metaphysical Interpretation

One of the most prominent contemporary advocates of Thomistic metaphysics, Jacques Maritain, writes,

> It is . . . true that metaphysics brings no harvest to the yield of experimental science. . . . Its heuristic value, as the phrase

> goes, is entirely nil. . . . This universe in which metaphysics issues . . . is not intelligible by its dianoetic or experimental means . . . it is not connatural to our powers of knowledge, it is intelligible only by analogy.[14]

In order to emphasize his principal point, that metaphysics has a purpose that is fundamentally different from the purpose of science, Maritain concedes to the enemies of metaphysics the point that "metaphysics has no heuristic value whatever." This is obviously exaggerated. By "humanizing" the principles of science, by linking their content with other activities of the human mind, interest in scientific research is stimulated. It is certainly true that the metaphysical interpretations do not produce any conclusions about observable facts which had been unknown about them, but they help integrate science into the whole of human enterprise.

Alfred North Whitehead, who more than anyone else in our period combines mastery of science and philosophy, gives much attention to this point in his book *Adventures of Ideas*.[15] He reveals the intellectual situation that could arise in our culture if science were presented and taught from the scientific aspect. This would mean restricting the presentation of science to the logico-empirical aspect; one would present only the system of axioms, the logical conclusions drawn from this system, and the operational definitions of the occurring terms (see, e.g., Part I, chap. 1, §4). This could exclude all information about the historical context in which that system of axioms had developed. It would exclude the social, educational, and religious aims that system had been supposed to serve. Speaking about geometry, if one means by science not just geometry but also psychology, sociology, and so on, this restriction to the purely scientific aspect is not actually serving the cause of science as a whole.

Whitehead writes,

> The intimate timidity of professionalized scholarship circumscribes reason by reducing its topics to triviality, for example, to bare sensa and tautologies. . . . The world will again sink into the boredom of a drab detail of rational thought, unless we retain in the sky some reflection of light from the sun of Hellenism.[16]

Whitehead stresses, as had Maritain (but in a subtler way), the fact that metaphysics or philosophy in its state of separation from science cannot be directly helpful to the special sciences. He writes: "A philosophic system, viewed as an attempt to coördinate all such intuitions, is rarely of any direct importance for particular sciences."[17] He describes with vigor and ardor, however, the actual use of this philosophy for the scientist. "Even from the view of the special science," he continues, "philosophy systems, with their ambitious aims at full comprehensiveness, are not useless. They are the way in which the human spirit cultivates its deeper intuitions."[18]

He describes in a psychologically lucid way the actual effect of these systems on the scientist: "Such systems give life and motion to detached thoughts. Apart from these efforts at coördination, detached thoughts would flash out in idle moments, illuminate a passing phase of reflection, and would then perish and be forgotten."[19] Philosophy attempts to give an interpretation of science that alleviates the dissatisfaction produced by the "dehumanization" that has been so deplored by those interested in individual and political education. The ancient and medieval philosophers, as well as their contemporary followers, attempted to provide the satisfaction that science had been unable to provide. These philosophers dug deeper than science had, penetrating by way of the "eyes of the intellect" into the ultimate causes of things and their final explanations.

5. Metaphysics as Science

Maritain described this hope in his book *The Range of Reason*:

> There must be a science, a knowledge, where the intellect . . . may engage *in the inside task, within* the workings of knowledge and where it may develop freely its most profound aspirations, the aspiration of the intellect as intellect. Such knowledge directly concerns the being of things intelligibly grasped, it is philosophical and metaphysical knowledge.[20]

He clearly describes how the scientific method, by proceeding from the commonsense aspect of things to a system of mathematical symbols, prevents us from approaching the inner being of things. This argument is correct only if what the metaphysician calls the inner "being of things" involves an analogy to some domain of commonsense experience. By formulating general principles outside the domain of common sense, Maritain asserts, "Science works . . . against the grain of the natural tendencies of the intellect."[21] He astutely explains why philosophical schools that take their cue from scientific knowledge like "positivism, old and new, and Kantianism, do not understand that metaphysics and philosophy are authentically *sciences*, that is to say, fields of knowledge capable of certitude which is demonstrable, universal and necessary."[22] The reason for this lack of understanding is that "they do not understand that the intellect *sees*. . . . In the eyes of the Kantians and Positivists, the senses alone are intuitive, the intellect serving only to connect and to unify."[23]

And here Maritain certainly makes a salient point. According to the typical procedure of modern science, every principle has to be verified by checking its consequences with sense observations. A metaphysical statement, according to Thomists like Maritain, may also be verified by checking its consequences through, one might say, intellectual observations. Maritain occasionally calls this procedure of checking "the mystery of abstractive intuition" or "the mystery of analogical intellection."[24] The last expression hints somehow that the mystery turns out to be the description via an analogy in the domain of commonsense experience.

Whatever one may say in praise of "seeing with the intellect" as "looking from the inside instead of from the outside," the problem arises in all these cases: how can we describe the results of our "inside track"? Then it becomes obvious that these results are actually analogies from the domain of our daily experience and are formulated in commonsense language.

Certainly, these metaphysical interpretations are given in a language that is much less abstract and less general than the language of science and much closer to the language we use in everyday life. There is, as a matter of fact, no language in which we could formulate our knowledge that we acquire by the act of "knowing with the intellect." One well-known Thomist philosopher, Gerald B. Phelan writes,

> When the intellect of man is confronted with *being as being*, it is dazzled, and can only hope to see it by gazing upon its analogical reflection or participation in the things that are.[25]

Prominent advocates of Thomistic metaphysics are well aware that "seeing with the intellect" yields only "analogical reflections" of real being; by "analogical reflections," we mean, practically, an analogy in the domain of commonsense experience. This use of the term *analogy* originated in theology from the custom of speaking of "God's will" or "God's finger" without giving to these expressions their literal sense; if we speak, for instance, of "God's will producing God's deeds," this sentence is an analogical reflection of a sentence about human will producing human deeds. We recall also how Maritain characterized metaphysical knowledge; as we quoted at the beginning of this section, he stressed that the universe is intelligible to metaphysical inquiry by no direct logical or empirical procedure, but what can actively be done is to point out analogies within the domain of commonsense experience.

6. The Laws of Physics and Their Metaphysical Interpretation

In modern physics we understand "physical law" through a system of relations between symbols (the axioms), including the operational definitions of the symbols as well as of the relations. This system logically implies statements about observable facts. The axiomatic system and the operational definitions imply, in this way, relations between observable facts. This system as a whole is now called "the laws of physics.' We also speak, of course, of a "law of gravitation," "law of inertia," and so on. Strictly speaking, these "laws" as isolated statements do not imply any statement about observable facts, but they are part of the structure that we call "the laws of physics."

If we enumerate these laws, we are certainly using a language that describes these laws as pieces of legislation, as parts of a code of laws. This way of speaking originates in the analogy between the universe and a human society. As the behavior of any human society is regulated by laws, the behavior of the physical universe is regulated by "physical laws." The abandonment of this analogy by modern science is one of the chief reasons for the dissatisfaction with modern science we described previously (Part I, chap. 2, §1).

The replacement of the "laws of nature" by a formal system of axioms is the central issue in the complaints about "dehumanization." In traditional metaphysics a major effort has also been made to prove by "metaphysical insight . . . by penetrating the physical world with the eyes of the intellect," that the uniformities in the world of our sense observations are due to "genuine laws" in the sense of this ancient analogy between the universe and a human society. We shall discuss these "proofs" later on in the chapter on "Thomistic Philosophy" (Part

II, chap. 7). We shall see that what has been called "results of metaphysical intuition" can in most cases also be interpreted as the result of efforts to find analogies with the world of commonsense experience. The search for these analogies manifests itself with great clarity in the efforts to give "metaphysical interpretations" of the concept of "physical law" or more generally "law of nature."

We may take as a starting point the alternative interpretations discussed by Whitehead. According to him, at the present time there are four prevalent main "doctrines" concerning the laws of nature: (1) the doctrine of laws as immanent; (2) the doctrine of laws as imposed; (3) the doctrine of laws as observed order of succession; and (4) the doctrine of laws as conventional interpretations.[26]

It is easy to see that none of these "doctrines" can be confirmed scientifically. We cannot derive from them facts that could be checked by observation, and they would differ from each other according to which of the four doctrines that served as our fundamental hypothesis. If we look at "causality" from the scientific aspect, the law of causality asserts the existence of laws in nature. The existence of law means the possibility of building up formal structures of a simple type from which we can derive the phenomena that are actually observed. A physical law is certainly not an "observed order of succession" as the third doctrine asserts. The laws of physics, the recurrent series of events, certainly do not refer directly to observation. No "observed order of succession" is given to us by our experience.

If we say flatly that a law of nature is a uniformity in the succession of observed facts, we do not correctly present the scientific concept of physical law but an analogy to some fact of our everyday life. If we consider, for instance, a familiar experience like the regular succession of days and nights, or of the seasons of planting and harvest, we can describe this type of commonsense experience correctly as an "observed order of succession." But if we should claim that these events describe the actual characteristics of physical laws, we would be very wrong to say that they deal with "observed succession." If we characterize the laws of nature in this way, we advance a metaphysical interpretation of this concept. We interpret the very general concern of "physical law" based on a small range of phenomena when we interpret it as successions of observed events. Whitehead called this interpretation of physical laws the "great Positivist doctrine. . . . It tells us to keep to things observed, and to describe them as simply as we can. This is all we can know. Laws are statements of observed facts."[27] However, as we have just pointed out, this "positivist" doctrine is actually a metaphysical interpretation. In the twentieth century, the term *positivism* has been used to denote the "scientific aspect." In order to avoid confusion, the expression "logical positivism" has been introduced. This point will be elaborated later (in Part II), when we will present those lines of philosophical interpretation that have been influential in our own century.

If we now examine the idea of a law being either "immanent" or "imposed," it is obvious that these two concepts do not belong to the scientific aspect but are two metaphysical interpretations. Going by the doctrine that law is immanent means, according to Whitehead, "that the order of nature expresses the characters of the real things that the physical objects behave according to their real nature."[28] As we learned in discussing causality (Part I, chap. 2), the "real nature of things" cannot be observed, and what really happens is that we observe order in our observation and set up mathematical patterns by which this order can be described. Since David Hume we know that by saying "the laws are immanent" nothing is added to the result of observation and logical deductions. What is added is an analogy with commonsense experience. The physical changes in material bodies are compared with changes in human individuals and human society, which have their sources in the inner constitution, with the physiological and psychological structure of the human being participating in these changes. The commonsense language says that man is directed by his "mind" or by his "interior."

On the other hand, the doctrine that "laws are imposed" means that the world consists of four parts. Outside the region where physical laws are valid, there are regions that are not subject to these laws but have imposed these regularities upon our physical world. This means, in particular, an analogy to daily life situations in which a lawgiver imposes laws upon a population. To simplify this distinction, let us consider the orbit of the sun around the earth. We can say either that the sun is driven by a god or spirit inside the sun or perhaps identical with the sun, or that it is driven (as assumed in the Bible) by a god who is outside the physical universe. In the first case, we say that the law of motion is immanent; in the second, it is imposed. It is obvious that both cases involve metaphysical interpretations that are actually illustrations of scientific statements by comparison with familiar types of human behavior. We see from these considerations that the purely scientific aspect of physical law has been regarded as a part of the general "dehumanizing" trend in modern science. All the metaphysical interpretations—that the laws are immanent, imposed, and observed rules of succession—they all have in common the fact that they bring a human element into the concept of law and attempt to make science more palatable to the nonscientist who is, above all, interested in the effect of science upon human conduct.

7. How Scientists Have Interpreted Their Own Theories

Quite a few scientists have claimed that metaphysical interpretations of science are actually "misinterpretations" given by authors who have no real understanding of the intricate physical theories of our twentieth century, such as relativity and quantum theory. However, quite frequently, such interpretations were given by the very authors who had discovered or, to be exact, produced these new theories. It is perhaps instructive to look at the example of Heisenberg, who is one of the most prominent authors of quantum theory and who, in particular, is the "discoverer" or maybe "inventor" of the "uncertainty principle."

According to the theory, the "position of a moving particle" is not a property the particle has by itself. Only within a specific environment a "particle with a position" can be described. The same thing is true for the "velocity of a particle." Expressions like "this particle has this and this position" or "this particle has this and this velocity" are not statements about isolated particles but statements about readings on measuring instruments. Such readings can only be made unambiguously if the arrangements of the observational instruments with respect to physical systems of reference are precisely described.

Heisenberg explains this state of affairs occasionally in a more humanized or personalized way by saying that physical science does not describe the particle by itself but rather the act of observing and measuring the particle. In a more general way, he writes: "*The object of research is no longer nature itself, but man's investigation of nature.*"[29] He continues this argument by stressing that the picture of nature no longer means what it used to mean in the era of Newtonian science. Today, according to him, science no longer tries to give a picture of nature which exists independent of man, it gives merely a "*picture of our relationship with nature.*"[30]

Heisenberg writes bluntly: "*The scientific world-view has ceased to be a scientific view in the true sense of the word.*"[31] This way of speaking seems to set a green light for unscientific and antiscientific worldviews. Actually this and similar formulations have been employed for the justification of spiritualistic views and for the support of traditional interpretations of religion. If we take these formulations as expressions in our language of daily life, we can, according to Heisenberg, proceed to the statement that "*for the first time in history modern man on this earth now confronts himself alone.*"[32] Heisenberg links this statement to man's struggle with the dangers that surround him. While previously man had been threatened by savage animals, diseases, hunger, and violent forces of nature, in our age man is threatened by those who claim the right to the goods of the world. In this way the relation of uncertainty in physics leads to a statement about the position of man in the physical universe.

It is, of course, debatable whether there is really a logical and empirical chain of argument that leads from the relation of uncertainty in physics to statements about the struggle of man with the world's dangers. It may be that this result can only be achieved by formulating scientific laws (like the uncertainty relation) first in the language of our daily life, through a kind of commonsense analogy, and to draw conclusions within this commonsense language. In this way we would not achieve an actual scientific result but only a popularized presentation of science. What we call "popularized science" is essentially a presentation of science by an analogy formulated in the language of our daily life. There is no doubt that such popularization of science is helpful for the understanding of science for the layman and beginner, but it is dangerous to draw philosophical results from these analogical formulations. We shall see later (Part II, chap. 7) that this metaphysical thinking was characteristic of medieval scholastic philosophy and strongly antagonistic to scientific thinking.

Chapter 4

The Sociology of Metaphysical Interpretations

1. Can Science Be "Purged" of Philosophy?

If we consider the great variety of arbitrariness of philosophical interpretations examined in the previous chapters, we can understand the desire often expressed by scientists to present and teach science in a way that adheres to the "strictly scientific aspect" and avoids all philosophical or metaphysical interpretation. In the parlance of traditional metaphysical systems, this would mean restricting oneself to the registration and ordering of observed phenomena without attempting to investigate the ultimate reality behind the phenomena. In the language used in this book, this restriction would mean presenting physical science by elucidating its formal systems and operational definitions, without any addition of commonsense analogies. As we have repeatedly stressed, the formal systems and operational definitions are not uniquely determined by the "purely scientific" or "logico-empirical" requirements of agreement with experience and logical consistency. Requirements like "simplicity," "economy," and "agreement with common sense," must be added.

The decision, however, as to which theory is simpler or more in agreement with common sense depends upon the historical and social situation in which the theory is constructed.[1] In order to satisfactorily present the reasons for recommending scientific theory, we must therefore present the historical and social context and point out how, under these circumstances, certain standards of simplicity and common sense have been accepted. We must discuss, moreover, interrelations between the "purely scientific" (logico-empirical) and the "sociological" criteria. If we have done all this, however, we have actually presented the type of argument often emphasized in books on the "philosophy of science" and, specifically, the present book.[2] What would happen, however, if we should "purge" science of this "philosophy of science"? We would present only systems of axioms and operational definitions, without any argument as to how these systems should be chosen among all the possibilities compatible with logical consistency and experimental confirmation. In this case, every choice will obviously be accused of being arbitrary. If we enumerate, however, all possibilities and refrain from making a decision, we will be branded "skeptical," "relativistic," or "agnostic."

Let us assume that scientists take the risk of being called "dogmatic" or "skeptical," provided they keep their science free from philosophical contamination. Will they achieve their purpose? If a physical theory has existed over a long period and has proved helpful, its axioms would be regarded as commonsense statements. A convenient commonsense interpretation would establish itself by tradition, and there would be less searching for "philosophical interpretations." Such a situation existed when Newtonian physics was recognized as the final universal truth. But when, after 1900, Newtonian physics had to be modified in many respects, the rift between the axiomatic systems of physics and their commonsense interpretation, which had faded away during the unchallenged rule of Newtonian mechanics, reappeared in relativistic and subatomic physics. Those scientists who attempted to "purge" physics of philosophy maintained that the whole of physics consists in the mathematical formalism of relativity or quantum theory. However, the human mind—and we must remember that scientists are human, too—longs for understanding through analogies with commonsense experience. Since these very axioms in the period dominated by Newtonian axioms had been identified with commonsense "truth," the new physics seemed to be "obscure" (i.e., contradictory to common sense).

According to the "philosophy of science" presented in this book and in many writings of contemporary authors, this happens to every new theory. If we do not understand precisely the relation between axiomatic systems and commonsense interpretations, we have a strong urge to fill the rift by some familiar analogy to commonsense experience or, if we use the parlance of chapters 1 and 3, by some familiar metaphysical interpretations originating in ancient and medieval systems of thought. They have left deep imprints on our education of children; they act specifically through our traditional teaching of language, folklore, and religion; they have introduced superficial concepts of "matter" and "mind," of "reality," and "appearance," of "freedom" and "law." They have manifested themselves in the metaphysical interpretations of twentieth-century physics, for example, in calling the statements about measurements of length in relativity theory statements about "apparent" and not about "real" length. In the same way, "space" and "time" by themselves were regarded as "apparent" entities, while only a four-dimensional combination of both was recognized as a "real" entity.[3]

2. Science and Chance Philosophies

Once again, the great British philosopher and scientist Alfred North Whitehead characterizes this situation as follows:

> It is legitimate (as a practical counsel in the management of a short life) to abstain from the criticism of scientific foundations so long as the superstructure 'works.' But to neglect philosophy when engaged in the reformation of ideas is to assume the absolute correctness of the chance philosophic prejudices imbibed from a nurse or a schoolmaster or current modes of expression.[4]

The salient point is the remark that science in its "pure state" leaves a vacuum filled in practice by the "chance philosophies" acquired by scientists during their childhood.[5] If we investigate the degree to which scientists imbibe obsolete "chance philosophies," we find that they swallow them more easily, the more they have been trained in "purged science" presented as a collection of recipes without analysis of the reasons why general rules have been accepted.

Engineering students receive training that results from being handed down readymade rules more so than students of theoretical science, who are more frequently inclined to investigate the reasons why a new theory is accepted or rejected. We notice, therefore, that engineers are more inclined to accept "chance philosophies" as implications of their science than are workers in the field of theoretical physics. Engineers have often been enthusiastic adherents of the view that "science has confirmed materialism," has "refuted materialism," or has "led to the belief in 'absolute truth' or in 'agnosticism'" and other popular beliefs, while scientists trained in the methods of theoretical physics will take to a "philosophy of science" that investigates the circumstances under which theories are accepted or rejected. If, for example, a theory is preferred because it helps to fight "materialism," the student who has been trained in a critical "philosophy of science" will understand that this theory easily lends itself to a "commonsense interpretation" that may support a desirable way of life; conversely, the average graduate of an engineering school will believe that the theory itself (such as relativity theory) implies a refutation of materialism.

Since the philosophies infiltrating the vacuum left by a meticulously purged science are, as Whitehead said, determined by the "prejudices of nurses and schoolmasters," these "chance philosophies" will support the ways of life that seem desirable to these groups. Speaking in terms of observable facts, these ways are determined either by old tribal traditions or by order of the authorities who control the schools. These interpretations often infiltrate and merge with science to such a degree that many students get the impression that "pure science" or "science proper" supports the ways of life that are desirable to "nurses" and "schoolmasters." There are, of course, always groups that oppose the ways of life that are supported by these "chance philosophies." They are opposed either to the tribal customs or to the policy of the present authorities. Such groups have frequently attempted to purge science of the chance philosophies by denying them the status of truth and pointing out that they are merely interpretations by commonsense analogies.

Such "purges" of science can be carried out for different purposes. One can attempt to stress the difference between valid statements of science and their philosophical interpretations as much as possible; then science consists of the axiomatic system, its operational meaning, and the argument by which the automatic infiltration of chance philosophy can be checked. Every philosophical interpretation is then introduced as arbitrary as far as "scientific truth" is concerned, while the influence of these interpretations on human behavior is discussed precisely. This attitude toward science is often called "positivistic." But the purge of chance infiltration can also be used as a preliminary step toward a deliberate interpretation that supports a cherished way of life. This attitude can be called "metaphysical" because it tends toward regarding the commonsense analogies as truth, revealed by our intellectual or inner eye. Those groups who sympathize with the ways of life supported by the infiltrated chance philosophies would not feel the desirability of any purge but would be prepared to admit the chance philosophy of "nurse and schoolmaster" as legitimate implications of "science proper."

From these considerations we can easily understand why agencies who feel the duty and the right to guide human conduct will be concerned with philosophical interpretations of science. Since the technological applications of science are based completely on the "purely scientific aspect," the "philosophical interpretations" can be influenced by agencies like governments, churches, and educational institutions without seriously impairing the technological usefulness of science. It is also clear what attitude these authorities will assume in order to bring about the interpretations they consider desirable. They would in some cases force scientists to refrain from any interpretation and to keep to a "completely purified" science. In this way a vacuum is kept open into which interpretations can be poured—interpretations that seem desirable for purposes of moral education. In other cases authorities who believe themselves to be in a powerful enough position will force scientists to permit a "merger" of "desirable" interpretations with science proper and to refrain from a strict distinction between statements of science proper (logical and empirical statements) and their philosophical interpretations.

The attitude of scientists can be construed in a similar way. In order to avoid any encouragement for a way of life that they may not cherish, they would adhere strictly to the "purely scientific aspect"; this positivistic attitude would also be the scientist's protection against any suspicion of supporting an "undesirable way of life." If, however, scientists feel that they are in a strong position and are not obliged to continually prove their agreement with secular or church authorities, they will attempt to "purge" science of "chance philosophies"; they will, moreover, claim the right to advocate philosophical interpretations that support the way of life that seems desirable to them.

3. The Attitudes of Scientists and Authorities

In order to verify these statements about the attitudes of scientists and educational authorities, we have to know how the concept of "purged" science has actually been described by both groups. Again, pure science consists in a formal (axiomatic) system of relations accompanied by operational definitions. If we emphasize the formal system, we would say that "science consists in a set of formal relations which are free creations of the human mind." If we put more emphasis on the operational definitions by which symbols and relations are connected with sense observations, we would say that "science consists in relations between sense observations." Both descriptions are expressions of the "positivistic attitude"; the first follows the line of Henri Poincaré, the second adheres to the conception of Ernst Mach.

Since around 1920 an objection of the Communist Party in Soviet Russia against the theory of relativity has been that as a product of the bourgeois class decomposition, the theory is infiltrated by idealistic interpretations and thus furnishes support for counterrevolutionary ideas:[6] it is pure idealism and not reconcilable with materialism. Despite the adverse effect of his theory on human behavior, Einstein was elected by the Russian Academy in 1923 as an honorary member because of the merits of his theory—if purged of philosophical infiltrations. Such infiltrations have been seen and criticized by Soviet authors even in geometry and Newtonian mechanics. We have previously pointed out (Part I, chap. 1, §4) that geometry and mechanics consist of a formal (axiomatic) system and operational definitions. A Russian author writes,

> The idealist who works in axiomatics will not take any responsibility regarding the axioms he assumes. He cannot prove them logically and does not recognize any other kind of proof. Dialectical materialism insists upon the necessity to prove the axioms themselves. No logical proof is required, of course. They have to be proved by practice, by means of the materialistic conception of experience. This conception gives to the scientist a much greater responsibility than the mere requirement that the axioms have to be convenient, independent and consistent.[7]

According to the author, official Soviet philosophy disapproves "setting up arbitrary axioms which are not dependent upon experience."

As another example of Soviet antagonism to the axiomatic method we may point out a criticism directed by Newski against Einstein's famous paper "Geometry and Experience" (1921).[8] He accuses Einstein of explicitly being responsible for the infiltration of idealistic philosophy into his presentation of geometry and of having encouraged the idealistic interpretation of his theory. He elaborates that the theory of relativity would never have arisen without the recognition of geometry as an axiomatic system, the conclusions of which we later check by experience.

The infiltration of philosophical interpretation into science is seen in the "recognition that science is a free creation of the human mind which cannot tell us anything about the objects of physical reality." As a matter of fact, the structure of a physical theory contains within itself, according to the official Soviet philosophy, a great many "infiltrations." Mitin's textbook on dialectical materialism, which has been used at Russian universities, points out that what is regarded in "bourgeois countries" as 'science proper" is actually heavily infiltrated by "Machist" or "anti-materialistic" philosophy, which has served to support anti-Communist political doctrine. The textbook describes this positivistic conception of science as follows:

> Science has nothing to do with the representation of objective things and their relations, but with physico-mathematical signs, which are free inventions, with symbols which are assigned to subjective sense-impressions. From combining these signs by means of mathematical equations, they derive new combinations which they denote by new signs. The work of science consists, according to those Machists, in the construction of that system of signs.[9]

This "positivistic" or "Machist" conception of science supports, according to Soviet doctrine, the political and moral goals of the bourgeois class; it describes the world as a rigid frame instead of a rapidly moving flux of changes. The text continues:

> This reactionary play with abstract mathematical toys, according to Machists, has to take the place of the development of a living, objective knowledge of the immensely complex reality. Here it becomes particularly clear how subjective idealism goes hand in hand with mechanism.[10]

In Soviet philosophy, "bourgeois science" is often accused of supporting idealism because positivists regard science as a system of statements about mental facts or sense impressions. But positivistic philosophy is also accused of supporting "mechanism" because of the "rigidity" inherent in every axiomatic system.

The Russian physicist Vavilov wrote in 1939 that there are three types of infiltration into physics that are detrimental to a desirable way of life: "mechanical materialism, idealism of all possible shades, and philosophical indifferentism."[11] By "indifferentism" he meant obviously a complete purge of philosophy that would produce a vacuum that could be filled by harmful infiltrations. The prominent Russian physicist Abraham Fjodorovich Ioffe attempted to separate "physics proper," which is not dependent upon the social class of the physicist from the "philosophical interpretations" that have always served to support a certain way of life. He wrote the following:

> When physicists (like Bohr and Heisenberg) go in for philosophical generalizations, their philosophy is sometimes the effect of the social conditions under which they live and of the social task which they carry out, consciously or unconsciously. Heisenberg's scientific theory is materialistic; he intends to give the best approximation to physical reality that can be reached today.[12]

His philosophical interpretation, however, may according to Soviet doctrine reflect an idealistic or mechanistic creed that supports the sociopolitical goals of the bourgeois class. To avoid misunderstandings we must remember that the term "materialism" in Soviet philosophy is not to be understood in the sense of "mechanism" but rather, perfunctorily speaking, in the sense of "physical realism."

4. The Battle of Worldviews

We can obtain a better understanding of the human (sociopolitical) role played by the philosophical interpretation of science if we consider the situation in different countries under different conditions. When the Nazi government in Germany was established in 1933, the political authorities attempted, on the one hand, to purge "academic science" of all infiltrations that had been absorbed under bourgeois liberalism and, on the other hand, to clear the path for new philosophical interpretations that would support the political line of the new government. The "Führer" of German university teachers once said the following in an address:

The Sociology of Metaphysical Interpretations 139

> Our science is free and will remain free . . . unless it would attempt to destroy again our domestic unity, which we have achieved by tough fighting. There are commitments which no man, may he be scientist or politician, can ignore: the commitment to one's biological and historical heritage. . . . There is no science that is completely objective and without any presuppositions.[13]

While in this address the restriction to "science proper" is declared incompatible with loyalty to the new government, Professor Heinrich Kunstmann's[14] address stressed that what has been presented as "objective science," as "science proper," in the science curriculum at German universities before the Nazi regime was actually infiltrated by the materialistic and positivistic philosophy that had dominated that period of "liberalism" and "democracy." He said, "Scientific research is not threatened today. However, the Third Reich has turned against the positivistic-materialistic philosophy that has been implanted into science like a suffocating crust."[15] He directs the attention of the students to the fact that their most beloved and admired teachers will deliver that obnoxious philosophy hidden within what looks like "objective science." Kunstmann said,

> Don't allow yourself to be confused by what is called the irrefutable power of scientific facts, if your natural sense tells you that your professor does not teach you science but indoctrinates you with a Weltanschauung (ultimate world view).[16]

In this address we clearly see the interaction between the "scientific aspect" and "commonsense analogies." It reminds us also of Lenin's oft-quoted warning that "one should not trust a bit to professors, who did excellent work in physics or chemistry, when they speak about philosophy."[17]

We learn from these examples that there has frequently been cooperation between the "positivistic" and "metaphysical" attitudes toward the purge of philosophy from science. As a first step, science was reduced to a formal system or to relations between sense data; then, as a second step, the vacuum produced by this purge was filled by a metaphysical interpretation or a commonsense analogy that served to support a desirable way of life. We can trace back this cooperation to the beginnings of modern science in the sixteenth and seventeenth centuries. When Copernicus prepared to publish his new (heliocentric) world system, he felt some apprehension that this system would easily entail philosophical interpretations that could be criticized by the advocates of the dominant philosophical and religious creeds. In 1541 Copernicus received a letter from the Lutheran theologian Andreas Osiander (of Nuremberg) in which can be read as follows:

> As for my part, I have always felt about hypotheses that they are not articles of faith but bases of calculation. Even if they are fake, it does not matter much provided that they describe the observed phenomena correctly. [. . .] It would, therefore, be an excellent thing for you to play up a little this point in your preface. For you would appease in this way Aristotelians and theologians, the opposition of whom you fear.[18]

In medieval parlance, science purged of all philosophy was generally described as a system of hypotheses that "saves the appearances or phenomena" without claiming to contain "truth" in the philosophical sense.

By a strange series of coincidences, the fundamental book of Copernicus, *On the Revolution of the Celestial Bodies*, was published in 1543 with a preface written by the same Osiander. This Lutheran theologian obviously attempted to help Copernicus in appeasing the Roman Church. He wrote in the preface: "The hypotheses of this work are not necessarily true or even probable. Only one thing matters. They must lead by computation to results that are in agreement with the observed phenomena."[19] In a similar way, the Roman Church favored, in the sixteenth century, a presentation of astronomy as a formal system that can successfully account for observed facts but that does not contain the "truth." Cardinal Bellarmine,[20] the leading figure in the "Holy Office" (Inquisition), wrote the following in a private letter to a friend of Galileo:

> I think that you and Galileo would act more prudently if you presented your opinion as a hypothesis and not as an absolute truth. To assert that the earth is really moving is a very dangerous thing, because it would irritate the philosophers and theologians. [. . .] To prove that the hypothesis of the immobility of the sun and the moving of the earth saves the

appearances is not at all the same thing as to demonstrate the reality of the movement of the earth.[21]

5. Purging Physics and Metaphysics

We can clearly see that the Copernican system as a strictly scientific theory was not irritating to the people who judged science as an instrument to influence human conduct directly. This irritation was only produced when metaphysical terms like "real" or "reality of the movement" were introduced into science. Therefore, the church strongly advocated a purge of philosophical interpretations from astronomy, of terms like "reality," which are actually commonsense analogies. As a more recent example of this attitude, we may quote some attempts to reduce Einstein's theory of relativity to a merely formal system. In his book on electrodynamics, Alfred O'Rahilly writes that

> the only scientifically relevant and effective portion of a book on physics is its quantitative formulae, what we have called its algebra. The remainder, the 'discourse,' often consists nowadays of irrelevant talk about moving clocks, moving rods, imaginary observers, travelling twins, men in rockets. The *science* of physics is concerned with the physical laboratory on earth, the quantitative results obtained therein and their qualitative operational background. [. . .] A clear distinction must be drawn between a physicist's algebra and his discourse.[22]

There is certainly much truth in this statement. We must consider, however, that an author's "algebra" contains his formal system only, while the "discourse" contains not only the "philosophical interpretations" but also the operational meaning of the symbols. In a review of O'Rahilly's book, the Reverend P. J. McLaughlin writes: "All attempts to attribute any *physical* theory to Einstein must be abandoned. Einstein's theory can be accepted only as a piece of pure mathematics."[23] This would mean purging science not only of philosophy but also of its operational meaning. O'Rahilly says explicitly that he is not opposed to metaphysical interpretations of science, but it has to be the right metaphysics.

According to him, physics has not contributed one iota toward modifying or clarifying our ideas of matter and electricity. Physicists are only bluffing when they pretend otherwise. He does not deny that there are problems of metaphysics and epistemology. He denies only that physics can shed light on them.

The necessity to "purge" science of philosophy is stressed in order to provide space for a philosophical interpretation that supports a desirable way of life. O'Rahilly is certainly an advocate of Thomistic metaphysics. Therefore, it will be instructive see how "purged" physics can be interpreted, according to Thomistic doctrines, by a cooperation of the "positivistic" and the "metaphysical" attitudes. We take as an example the treatment given to the principle of inertia by a well-known contemporary advocate of Thomistic philosophy: the Reverend Reginald Garrigon-Lagrange. The author asked the French physicist Pierre Duhem to outline his attitude toward the principle of inertia, knowing that Duhem had attempted a radical "positivistic purge" of physics but did not reject a metaphysical filling of the vacuum left by this purge. Duhem wrote to him:

> Dear Father: [. . .] (1) *We shall never have the right to affirm categorically any one of the principles of the mechanical and physical theory, that is true.* (2) *We are not allowed to affirm of any one of the principles on which the mechanical and physical theory rests, that it is false, so long as there has been no discovery of phenomena that disagree with the consequences of the deduction of which this principle constitutes one of the premises.* What I just said applies particularly to the principle of inertia. *The physicist has no right to say it is certainly true*; but still less has he the right to say it is false, since we have so far met with no phenomenon (if we leave out of consideration the circumstances in which the *free will* of man intervenes) that compels us to construe a physical theory from which this principle would be excluded.[24]

Duhem makes the point that "as a physicist" he cannot say more than what is said in the previous lines. But he emphasizes that "as a philosopher" he could raise further objections. He could ask, for example, whether the law of inertia makes the affirmation that there is, in all bodies in motion, a certain reality, an *impetus*, or whether these propositions apply to other beings endowed with free will. According to Duhem, these are problems that the physicist leaves to free discussion of the metaphysician.

Based on this argument from a prominent physicist, our author purges the law of inertia of every admixture of "chance" philosophical interpretation and gives the green light to the systematic interpretation by professionally trained metaphysicians. There is, however, according to Duhem, one restriction on the interpretation of the metaphysicians: it must preserve the objective and pure character of the scientific principles (e.g., the law of inertia) obtained by the "purge." He concludes his letter: "There is only one case which would induce the physicist to be opposed to this liberty of the metaphysician. It is that in which the metaphysician would formulate a proposition directly contradicting the phenomena."[25] Duhem agrees that the metaphysician has the "liberty" of introducing any statement into the vacuum left by the purge, provided that these statements do not alter the phenomena predicted by purged science.

Our author makes use of this liberty by asserting that a body hurled into empty space would stop moving unless divine power imparted to it a motion that has the effect of keeping the body moving eternally. "How," he writes, "without exceeding the limits of his science, can the physicist maintain that the *divine motion is not necessary*" to keep a body moving?[26] In this way the author can reject Newton's principle of inertia and stick to the Aristotelian principle according to which a body cannot keep its speed without being pushed continually by an external cause. Aristotelian physics can be retained without any possible refutation by observed phenomena.

6. Science and Reality

We learned in the previous section that the "irritation" produced by the Copernican system was due only to those advocates who claimed that the earth is "really" moving, while the Holy Office in Rome did not raise any objection to the assertion that the mobility of the earth is a useful hypothesis to "save the appearances." Although, from a strictly scientific aspect (logical consistency and agreement with observations) there is hardly any difference between these two presentations of the

Copernican theory, there was, in fact, a great difference in their direct effects on human behavior. This difference was obviously connected with the effect of the word "real," or the "meaning" of the word "real."

Our best starting point is perhaps to investigate how the words "real" and "reality" have been used in physical science. The advocates of the Copernican system called the mobility of the earth "real" because this hypothesis led to a more convenient description of the observed facts than the geocentric hypothesis. In this sense "real" is defined within the realm of "logico-empirical" discourse: the more a theory is confirmed by logico-empirical evidence, the better it describes physical reality. However, the opponents of Copernicus maintained that his heliocentric theory might very well be confirmed from the scientific point of view, but they denied that the mobility of the sun was "real." In this denial the Holy Office in Rome agreed with the anti-Aristotelian Francis Bacon. In what sense did these anti-Copernicans use the word "real"? They denied reality to the motion of the earth because the statement "the earth is really moving" would be contradictory to statements of our commonsense parlance (according to which the earth is at rest) and, moreover, to the philosophy that all our religions and moral doctrines were based on at the time.

As a matter of fact, these two reasons for denying the reality of the earth's motion are clearly connected because all rules for human conduct can only be formulated in a commonsense language. In other words, the statement "the earth is really at rest" served a higher purpose than the statement "the motion of the earth is real." The first achieved an improvement in astronomy, while the second was an improvement of human conduct. To prefer the "reality of the earth's movement" would mean to prefer to advance in astronomy rather than to advance in moral conduct.[27] Hence to apply the term "real" means to make a statement of preference or, as is often said, to pronounce a value judgment. If we investigate the controversies that have arisen in physical science about how to use the term "reality," we shall see clearly that the problem has been in every case to pronounce a judgment as to which of two formulations is more useful. The higher usefulness may consist in a higher heuristic value (helpful in discovering new laws), in higher simplicity, or in higher agreement with common sense or with cherished philosophies.

There have been old controversies in science about "physical reality." Is matter "real" and "force" only a word (a construct)? Or, on the contrary, is "force" a reality and "matter" only an abstraction? Is "electric charge" a physical reality and the electromagnetic field only an abstraction, or the other way around? The same controversy arose again between "atomism" and "energetics" and between "de Broglie waves" and "particles." The conception of physical reality prevalent among scientists who have not put much effort into the construction of a critical philosophy of science is lucidly presented in the writings of the great German physicist Max Planck. His writing brilliantly reflects the infiltration of chance philosophies that have their origin in what Whitehead calls the teachings of "nurses and schoolmasters," or, more generally speaking, in the cultural environment the scientist was brought up in.

7. Max Planck and the Real World

According to Planck, a physical theory presents three worlds: the world of sense data, the world of physical quantities, and the "real world." By "physical quantities" he denotes terms like "mass," "force," and "electric charge," which are about the same as what we have called the "symbols occurring in a formal or axiomatic system." To be precise, the sense data and the "physical quantities or symbols" are connected by operational definitions. Newton's mechanics, for example, consist of formal laws of motion and the operational definitions by which these laws can be translated into statements about sense data. As we have learned, all results that are confirmable by physical experiments and observations can be derived from the formal system and the operational definitions. This instrument, which is used to derive observable facts, is called a "physical theory" and consists of two types of terms: sense data and "physical quantities." The observable facts, as far as they are covered by the theory, can be derived from it as logical conclusions.

Now what is the role of the "real world" in the theory? To be sure, it does not contribute anything to the derivation of observable facts. Statements about the real world of mechanics may, for example, assert that "matter" is a real thing but that "force" is only a verbal expression, or vice versa. No observable results of the theory are influenced by what we say about "reality." Hence, these statements play the same role as what we have termed "philosophical or metaphysical interpretations." Planck himself calls the restriction to the two worlds of sense data and physical quantities the "positivistic attitude"—in contrast to the "metaphysical attitude," which adds to the theory the "third world" of physical reality.

In the terminology of this book, the positivistic attitude attempts to present science as an instrument for the derivation of observable facts from general principles, while the metaphysical attitude attempts also an understanding of the principles by analogies taken from experience on the commonsense level.

I discussed Planck's views in a paper that appeared in 1935, and I stressed the point that from the scientific aspect, it does not matter which objects—"matter," "energy," and so on—are called "real."[28] "To call a thing 'real,'" I wrote, "amounts merely to give to it an honorary title" and I called Planck's "third world" (the real world) the "Third Reich," alluding to the Nazi government that had been established at that time in Germany. The expressions "honorary title" and "Third Reich" hinted, rather whimsically, that by calling a thing "real" one attributed to this thing a value for a certain purpose. Hence, a statement about the "real world" attempts to influence human conduct; it has an ethical or political aim.

8. Meanings and Examples of "Real"

Later, I was encouraged in my view by the writings of John C. Ducasse.[29] He points out that by calling a thing "real," one does not advance a scientific hypothesis about the thing. A criterion of reality, he wrote

> does not formulate a hypothesis and is therefore not susceptible of being proved, disproved, or assigned a probability; but rather formulates simply the *criterion of interestingness* which we use or propose to use at a given time in appraising any given thing as interesting or uninteresting to us.[30]

In a paper published in 1945, Ducasse writes bluntly: "To be real is to be relevant to the purpose or interest which rules at the time."[31] If we put the question, often asked by philosophers, of whether, for example, this piece of wood we call a "table" is "in reality" an agglomeration of electrons and nucleons between which there is empty space, we can only answer: it depends on what results we want to derive from the statements in which the world "table" occurs. If we are looking only for results needed in our daily life, it is reasonable to treat the table as a massive piece of wood. If, however, we are looking for statements about subtle physical properties of the table, such as its penetrability by X-rays or alpha particles, we must draw our conclusions from the assertion that the table is an agglomeration of electric charges. Which description of the table is nearer to the "real" table depends upon which type of conclusion we are most interested in. If our aim is to fit our statements into the language of our daily life experience, we will say that the table is "really" a massive piece of wood and the agglomeration of charges a "construct." If, however, we are looking for the principles of "science proper," we would say that the table "really" consists of electric charges separated by empty space. By giving to a specific description the title of "reality," we express our preference for the aims that can be reached by this description.

The investigation of electromagnetic phenomena by men like Faraday and Maxwell had shown that the best way to derive them was to start from principles that contained physical properties of the space between charges (e.g., field strength and dielectric constants) while the older theories (e.g., Coulomb's law) contained only the electric charges. These new types of theories (e.g., Maxwell's field equations) were characterized by the assertion that the "electromagnetic field" between the charges is "real" while the charges are merely constructs, which means precisely that the field is now more valuable for the formulation of the principles than the charges. We learned from the theory of relativity that only the four-dimensional space-time continuum is "real," while three-dimensional space is only an appearance. This means, of course, that the four-dimensional presentation of relativity is of great value. It has a high degree of mathematical simplicity and beauty; therefore, it is very helpful for the construction of new theories that embrace more facts than Einstein's original theory. The most brilliant example is Einstein's theory of gravitation and general relativity.

When the quest for physical reality was applied to subatomic physics, the problem arose of whether the wave function (the ψ-function) or the "particles" is "real." We know that both answers have their advocates.[32] This is understandable if we keep in mind that the problem of "reality" is the question of whether the particles or the ψ-function is more valuable. This means, precisely speaking, asking which is more valuable for a certain purpose. The ψ-function is more valuable because it helps to build an analogy with the commonsense concept of causality, while the particles support an analogy with the motion of the material bodies (stones, men, etc.) occurring in our daily life. In both cases "real" means "valuable for a certain purpose," and in each case the purpose is to stress some analogy with commonsense experience.

9. Sociological Role of "Reality"

By observing the behavior of those who speak about science, experts and laymen alike, we notice that often very heated debates develop about whether, for example, "forces" are "real" or whether they are only "auxiliary concepts" that are helpful for the formulation of laws that describe and predict the motions of masses (material bodies), which are the only "real objects" that occur in mechanics. What is the cause of this heat? It can hardly come from a difference of opinion about whether "force" or "matter" is nearer to our commonsense picture of the world. We have learned from a great many examples that metaphysical interpretations of science have been the basis of rules for desirable human conduct and, therefore, the basis of ethical, religious, and political doctrines. Since the struggle between rival doctrines has frequently been accompanied by highly emotional behavior, the debates between rival interpretations of science have frequently become quite emotional, too.

We can easily see that in these emotionally loaded controversies concerning the interpretations of science, the most controversial assertion is usually about the "reality" behind a certain verbal expression. If we maintain that "force" is a "reality," we claim that there are "real objects" that are "immaterial," while the denial of the "reality of forces" implies that only material objects are real. Very frequently in our cultural tradition, high estimation of material things has been linked to undesirable conduct, while belief in immaterial entities has been praised as leading to highly desirable behavior. Hence the belief in the "reality of forces" has frequently been preferred, for social or moral reasons, to the belief that only "masses" are the "real objects" of mechanics. If we look back to the Copernican conflict, we notice again that for the Roman Catholic Church, the objection was the assertion that the motion of the sun was "real," while no authority rejected the mobility of the sun as a "hypothesis" or basis of calculation.

One of the deepest rifts in the history of philosophy has been the rift between "realism" and "nominalism." The supporters of the former maintained that universal concepts like "horse as genus" denote "real objects," while, according to nominalism, only individual horses are "real." Here again, "realists" have fought for the recognition that immaterial things, like the general concept of horse, are "real objects." This recognition would contribute to the defeat of "materialism," according to which only material objects are "real." On the other hand, nominalists who denied the "reality" of universal concepts supported "materialism" and undermined the belief in immaterial things.

We can easily understand that the concept of "reality" has played an immense role in the attempt to derive the rules for desirable human conduct from general principles. This becomes obvious when we look at how the term "reality" is used in the social sciences. Let us consider three fundamental concepts: individual man, nation (in the political or racial sense), and mankind as a whole. Some have maintained that only individual human beings are "real things," while nation and mankind are not real things but rather 'constructs." Other authors, however, make the point that "nation" is a 'reality," while "mankind" is merely a "word" that denotes a construct.[33] To understand the "operational meaning" of such statements about reality, we have only to find out what observable conclusions are drawn by people who make such statements.

It would be wrong to maintain that these statements have no operational meaning and are thus "meaningless." There is no doubt that the authors of these statements have tried to foster the belief that desirable human conduct consists in putting one of these concepts at the center of human interest. To maintain that only the individual is "real" supports anarchism; to declare nation as the only "reality" supports nationalism; and to regard mankind as the only "reality" leads to internationalism. In order to fully understand the practical consequences of these metaphysical statements about "reality," we may quote an expert in practical politics: Joseph Goebbels, minister of propaganda in Hitler's cabinet. When he was still a simple writer without a cabinet position, he liked the language of Kantian philosophy and spoke of "ultimate reality" as the "thing in itself." He wrote,

> Mankind is no thing in itself and the individual is no thing in itself, either. The thing in itself is the nation. . . . Mankind is something imagined only, not a fact. . . . The materialist regards the nation only as a means to an end and will not recognize it as a substance by itself. . . . Mankind is for him the end. . . . Therefore the materialist is necessarily a democrat.[34]

In a similar way, a leader of the Slavophile Movement in Russia, Nikolai I. Danilevskii[35] pointed out that "mankind" is a product of "abstract generalizations," while "nation" is a "culturo-historical type" and a product of "intuition and personal experience." Such statements, whether they come from German or Russian sources, cannot be checked by sense observations. But they have a clear political purpose: to degrade love for mankind and to extol nationalism. Frequently we find that indoctrination of a certain value system such as nationalism takes the form of asserting the "reality" of some concepts that are regarded by others (by "bad" people) as merely abstract constructs.

We mentioned at the beginning of this section the great value that has been attributed to the belief in the "reality" of physical forces. In the official philosophy of the party that ruled Germany between 1933 and 1945, the belief in the "reality of physical forces" was declared a characteristic of "superior races," while the attempt to deny this reality was regarded as a subversive action of racially inferior groups. In the periodical *Journal for General Science*,[36] the mouthpiece for the official philosophy of science, we can read the following:

> The "concept of force" has been introduced by Aryan scientists in order to formulate the causal laws for the changes of velocities. This concept has obviously its roots in the living experience of human labor, the creative work of the craftsman which again has been the essential content of the life of Aryan man. The world-picture that has originated in this way reflects in its foundations and in its whole structure the characteristics of the Aryan type of man. . . . Therefore, it can in all its details be grasped by intuition because it is in agreement with the form of our internal and visualization. This agreement produces satisfaction and happiness in every kindred mind.

From this interpretation of the concept of "physical force," we see clearly that according to the author a certain racial type will produce a commonsense analogy by which it will interpret the concept of "force" that has been used in modern physics. It is, according to the author, the habit of "inferior races" to abstain from these commonsense analogies and to regard the concept of "force" merely as a symbol interpreted by operational definitions. This means that these "inferior races" stick to what we have called "science proper"; they regard science as a system of logical or mathematical relations between symbols from which the observable facts are derived by logical conclusions and operational definitions. Our author describes the difference between the superior and the inferior types as follows:

> It is the characteristic property of the great Aryan researchers to surrender to nature with their whole soul and mind. In contrast to it, the Jewish character attempts to grasp nature by cold intellect alone; it refrains from any intuition; it does not look for the possibility of a visualization that may be fitting to our (the Aryan) mind; but it replaces intuition by purely logical and mathematical formulae.

In a similar way, official Soviet philosophy has denounced scientists who formulated the task of science as "mathematical description of observable phenomena" because desirable human conduct would be better supported by formulating the goal of science as the best possible description of "objective reality." In the article "Aether" in the *Large Soviet Encyclopedia*,[37] the author blames the theory of relativity for having dropped, along with the "aether," the search for objective reality. The author writes:

> The special theory of relativity takes refuge in a mathematical description; it abandons the investigation of the question about the medium in which the electromagnetic events take place. Along with this question, it abandons altogether the question of the objective reality of the physical phenomena; this means that this theory accepts essentially the views of Mach.[38]

In Soviet philosophy the advocates of the opinion that physical science consists in a system of mathematical formulae have often been called "formalists" and linked with idealistic philosophy because mathematical formulae express mental, not physical, facts.

10. "Reality" in Soviet Philosophy

In the article 'Theory of Relativity" in the *Large Soviet Encyclopedia*, the author recognizes as merits of this theory the abandonment of Newton's absolute time and space and its "setting up of laws governing the connection between space-time and the distribution of moving masses."[39] But the author also objects to some elements in Einstein's theory:

> However, we have to underline that this breakdown of the old conceptions of space and time has been exploited by the reactionary Bourgeois philosophy, in particular by Machism and similar subjective idealistic schools. To a certain degree, some physicists have participated in this interpretation, among them Einstein himself.[40]

We remember that the Copernican system was condemned by the Roman Church because it regarded the motion of the earth as "real." Since, according to the theory of relativity, the motion of the earth is as much "real" (and unreal) as the motion of the sun, this theory is accused by the Soviet authorities of giving support to reactionary doctrines of the Roman Catholic Church in its fight against the "true," the Copernican, system. The *Encyclopedia* continues:

> Some formalists-physicists have claimed that the Ptolemean geocentric world system is as "real" as the Copernican system. It cannot be physically distinguished (if we accept Einstein's theory of gravitation) whether a body rotates relative to the fixed stars or the other way about. This is extreme relativism as advocated by Mach and other relativists.[41]

To deny the attribute "real" to the Copernican system would mean, according to Soviet philosophy, a derogatory attitude toward this system and, therefore, toward the advance of science and human culture in general. The Soviets accuse Mach of supporting such a derogatory attitude by regarding science as a "description of phenomena," instead of an "insight into reality."

The author of the article continues: "Mach's assertion of equal rights for the Copernican and Ptolemean systems of reference is one of the attempts to 'refute' the materialistic theory of knowledge."[42] This theory is, according to Soviet philosophy, the "picture theory of knowledge," according to which our knowledge resembles objective reality as a photograph resembles the original. This theory is similar to the Aristotelian and Thomistic doctrines, which teach that knowing is "seeing with the inner eye."

"According to Mach," our author continues, "any knowledge is a pure description and has only to meet the criteria of 'convenience' and 'economy.'"[43] But when he tries to give arguments proving why the Copernican system is the "real" one, he cannot find any reasons except those of convenience and economy. This means that the "real" system of reference is more convenient for the description of fact. We must, of course, consider that "convenient" always means "convenient for a certain purpose." This purpose may also be to support the cause of what one believes to be cultural and political progress. This is easy to recognize if we remember that this author writes about how to recognize the "real" system of reference practically. According to him, the role of the "real" system of reference becomes clearer if we consider the historical importance of the different systems. Only the heliocentric system has allowed us to ask the question about the origin of the solar system. If we analyze this argument carefully, we find again that the author calls those concepts "real" that are relevant for achieving purposes he considers desirable.

The official Soviet philosophy denounces "Machism" because it refuses to attribute the predicate "real" to some concepts and hence refuses support to the practical goals that are regarded as desirable by the ruling group. In the present century, the view that "science is a system of formulae by which observable phenomena can be predicted or produced" has become so familiar among scientists that it occurs frequently in the presentations of "science proper" in textbooks, lectures, and personal conversations. In Soviet discourse such presentations are taxed by the derogatory term "bourgeois science" and apprehension is frequently expressed that the "reactionary philosophy" of Machism or relativism might infiltrate Soviet science through textbooks on science (e.g., the theory of relativity), which pretend to be objective and without philosophic admixture of any sort. The "positivist" doctrine that science can be presented without philosophical interpretation is declared a reactionary bourgeois doctrine because it allows harmful philosophies to be smuggled into the Soviet orbit under the guise of "pure science."

In the article on "Philosophy" in the *Large Soviet Encyclopedia*, we read:

> The mechanists claim that philosophy can be reduced to a synopsis of the results of contemporary science. Philosophy was declared a scholastic survival and the thesis of bourgeois positivism was upheld that science by itself is philosophy. This attitude leads necessarily to a fetishization of bourgeois science

and a flirtation with the idealistic tendencies of contemporary science. Forgetting that the results of empirical science can be systematized only on the basis of the method of materialist dialectics, which is itself the result of the historical development of science and philosophy, mechanists capitulated to bourgeois science.[44]

As we have seen, the relation between philosophy and science is often marked by political and moral preferences. In order to obtain a better picture of this relation, the next chapter will investigate further the relations between science, society, and politics.

Chapter 5

Philosophy of Science and Political Ideology

1. Sociology of Knowledge

We have learned in what way and to what degree philosophical and metaphysical interpretations of science have been influenced by moral, political, and educational purposes. The question now is whether scientific theories themselves have also been similarly influenced. The answer is certainly "yes," since no sharp line can be drawn between theories themselves and their philosophical interpretations. This will be clear later from the presentation of "theory building" (chapter 6). When the metaphysical view of a "true theory" was prevalent, the influence of social factors on a theory could only be a distortion of truth. Hence, it is understandable that the investigation of these influences began with derogatory judgments occasionally launched against some scientific theories and their authors. The metaphysical theory of knowledge has maintained that every set of propositions—and hence every scientific theory—is either valid or invalid and that the ultimate criterion of validity is an act of "intellectual vision," a "seeing with the inner eye," an "act of intuition or visualization," or whatever one may call it. If an author were not motivated by his "insight" to assert the validity of a theory but rather by his social situation or by some psychological motives, he was accused of having failed in his duty to search for objective truth; his behavior was seen as a sign of moral turpitude. His motives were frequently characterized as "selfish" because the announcement of the "objective truth" would have been disadvantageous to the author because of his life situation.

From their theory of "historical materialism" as an "objective truth," Marx and Engels[1] derived the prediction that a revolution will arise in which the working (proletarian) class will take over the government. They claimed that people connected with the capitalist (bourgeois) class would deny this "objective truth" because it would not be in their interest to create or support belief in this coming victory of the revolutionary class. According to Marxist doctrine, bourgeois scientists have been hampered in the search for truth by subconscious drives caused by their life situations. By these instinctive pressures the "objective" search for truth was diverted into channels of wishful thinking. Since the "objective truth" was regarded as favorable to the proletariat, this class could consistently search for "objective truth"; it was able to produce "science proper" in the narrowest sense of the word, while the "bourgeois" class would always deviate from strictly scientific research because of their life situations. One of the most prominent theoreticians of Marxism, Georg Lukács, writes the following in his book *History and Class Conscience*:

> For the proletariat truth is a weapon that brings victory the more ruthlessly it is applied. By this fact we understand the fury of desperation by which the bourgeoisie combats historical materialism; when the bourgeois is forced to accept ideologically this doctrine it is lost. But in this way we can understand that for the proletariat and *only* for the proletariat a correct understanding of the nature of society is a powerful weapon of primary importance; it is perhaps the weapon that will carry the final decision.[2]

While the proletariat and its advocates are able to proceed methodically in the search for "truth," the bourgeoisie is prevented by selfish motives from a strictly scientific method of investigation. However, bourgeois thinkers do not realize in their consciousness that they combat objective truth for selfish motives. They replace the selfish motives, in their conscious thinking, with a type of argument that refutes the proletarian doctrine: not by using scientific method but by introducing wishful thinking in the form of scientific reasoning.

This type of argument is called in the Marxist language "ideology." As Engels wrote in 1893:

> Ideology is a process accomplished by the so-called thinker consciously indeed but with a false consciousness. The real

motives impelling him remain unknown to him otherwise it would not be an ideological process at all. Hence, he imagines false or apparent motives.³

Robert K. Merton writes that according to Marx, ideology is "an *unwitting, unconscious* expression of 'real motives,' these being in turn construed in terms of the objective interests of social classes."⁴ Somehow anticipating Freud's psychoanalysis, the Marxist doctrine assumed that the selfish motives of bourgeois thinkers are repressed by a moral censor into the subconscious. In order to satisfy this moral censor, the ideological argument that becomes conscious is always unselfish and frequently religious. Max Weber, in his famous book on the sociology of religion, attempted to show that Calvinism has frequently been a disguised plea for a capitalist economy. The "freedom of the individual soul" has been used as an ideological argument in favor of "free enterprise."⁵

2. The General Sense of Ideology

If ideology is a distortion of truth, scientific argument should be used to "debunk" ideological argument. In this respect, the Marxist doctrine runs parallel to the pragmatic theory of truth. If we ask for the "operational meaning" of a statement that is formulated in very abstract terms, we challenge the author of the statement to "show cause" as to why we should not regard it as meaningless. The method of "operational definition" we explained by examples taken from geometry, Newtonian mechanics, as well as relativistic and subatomic mechanics in the chapter on "science of science" (Part I, chap. 1). In the working of that method, we can often note a "debunking" process. In some ways the theory of relativity "debunks" the use of concepts like "mass of a body" or even "length of a body" in the sense in which they had been used in traditional physics. We have learned that we can keep on using these old concepts, but this is not practical from a purely scientific point of view. We have learned that the use of the old concepts allows in some cases for a metaphysical interpretation that is believed to support a desirable way of life. In such cases, the discourse in which the old concepts are used usually plays the role of an "ideological discourse" in the Marxist sense. An instructive example is the suggestion to refrain from adopting Newton's law of inertia in order to keep physics in closer agreement with Aristotelian and Thomist philosophy, which some regard as supporting traditional ethics and religion.

If we use the terms "ideology" and "ideological discourse" in this way, we depart, of course, from the way in which these terms were used by Marx and Engels. For them, ideology was a means to prevent man from using scientific methods and from discovering in this way a "special truth"—the coming proletarian victory. For them, as Merton writes,

> The theory of ideology is primarily concerned with discrediting an adversary, *à tout prix*, and is but remotely concerned with reaching valid articulated knowledge of the subject-matter in hand. It is polemical, aiming to dissipate rival points of view.[6]

If we now say that presentations of physical phenomena, like relativity theory or quantum theory, are not accepted, and obsolete theories are retained because of some ideological argument, we obviously use "ideological" in a more general sense than the Marxists did. We say that the new theories are much more satisfactory from the purely scientific point of view, but the old theories are retained because they are regarded by some ideology as supportive of a desirable way of life. If we look at the examples given in this book, we note that different kinds of ideology—in some cases Nazi ideology, in some cases traditional idealist philosophy or traditional religion—rejected theories that were scientifically satisfactory.

3. Mannheim, Ideology, and Sociology of Knowledge

The concept of ideology in this more general sense was introduced by Karl Mannheim.[7] He called the Marxist concept the "special concept of ideology" because it was restricted to bourgeois ideology; the existence of others, such as Communist ideology, was not admitted. Instead of this special concept, Mannheim introduced the "general concept of ideology" by applying it to ancient Greece in the period of Plato. We could say that Plato in his work *Laws* made use of an "organismic ideology," which served to prevent natural scientists from assuming that celestial bodies consisted of the same material as the earth. Followers of Plato would say that there is a "materialist ideology" that distracted the scientist from the "natural" assumption that the sun and stars are essentially different from the earth. This "materialist ideology" supports the establishment of a unitarian pattern that fits some political and ethical purposes. If we speak in this way, the "general" concept of ideology does not discriminate between parties or groups that fight one another. Ideology becomes "symmetrical"; it is related to all groups in the same way. As Merton characterizes the difference, "In the special formulation, only our adversaries' thought is regarded as wholly a function of their social position; in the general, the thought of all groups, our own included, is so regarded."[8]

Mannheim himself wrote that

> with the emergence of the general formulation of the total conception ideology, the simple theory of ideology develops into the sociology of knowledge. What was once the intellectual armament of a party is transformed into a method of research in social and intellectual history generally.[9]

But even if we do not discriminate between ideologies and grant them all equal rights, an investigation of ideologies would be an investigation of the actual and potential distortions of truth. One can investigate by what social factors truth has actually been distorted, a problem in social and intellectual history; but one can also systematically investigate the ways in which truth can be distorted by the influence of the social situation to which the investigator is exposed. Both types of investigation are rewarding for sociological research. In both cases, social facts appear as the subconscious roots of ideologies that oppose and stall the advance of scientific knowledge. Up to this point, sociology of knowledge is a kind of pathology of knowledge because it treats exclusively distortions of truth. Since, according to the ancient and medieval theory of knowledge, our knowledge is a picture of the real object, and knowledge is produced by the object in our intellect. Hence, any factor *outside* the object of our investigation could only distort the picture. Merton points out correctly that sociology of knowledge cannot transcend the limits of a "pathology of knowledge" unless it starts by investigating the social factors that are involved in making "true statements," in building up "true theories." Merton writes:

> The "Copernican Revolution" in this area of inquiry consisted of the hypothesis that not only error or illusion or

> unauthenticated belief but also the discovery of truth was socially (historically) conditioned. . . . The sociology of knowledge comes into being with the signal hypothesis that even truths were to be held socially accountable, were to be related to the historical society in which they emerged.[10]

When Mannheim developed this idea of a "sociology of knowledge," he was well aware of the basic difficulty: that an influence of social factors upon the formation of a "truthful" man was incompatible with the idealistic or metaphysical theory of knowledge that dominated German philosophy and science. Mannheim pointed out that according to this theory, the validity of a statement is independent of the properties of the author who makes the statement; social factors certainly determine the life situation of the author, but they cannot influence the validity of his statements. We may agree for the moment that the validity of a statement does not depend upon the person who has uttered it. However, there is no doubt that the process by which men produce statements or systems of propositions is a part of the social process by which the human race develops. The production of scientific systems is, therefore, just as much a topic of sociological studies as the production of grain or motorcars.[11]

4. Forms of Social Influence

There are two main types of problems that have been investigated in sociological studies of science. The first is obvious. If the solution to a scientific problem is helpful in making human life more pleasant, working toward the solution will be supported by social powers that are, in some way, interested in producing or increasing this kind of pleasure. The word "pleasure" is used here in the most general sense. Consider, for example, that nuclear reactions are helpful in the production of atomic bombs. Those powers who believe that the use of atomic bombs can bring victory in a war will support the study of nuclear reactions because they believe that victory will bring some pleasure—or at least the avoidance of displeasure. Through this support from powerful social groups, the investigation of some problems is certainly promoted or at least made possible. This influence existed when medieval kings supported research in astrology and alchemy, and it exists today when government and industry support research that is (or is supposed to be) relevant to military and industrial success. There has been much research in this field, but it has little to do with the philosophy of science, and we shall not discuss it in this book.[12]

Systematized knowledge is also influenced in another way by social factors; this type of influence is, as we shall see, much more closely connected with the philosophy of science. As a matter of fact, this connection between social factors and systems of knowledge is brought about largely by the philosophical and metaphysical interpretations of these systems. When Marx and Engels launched their attacks against the economic and sociological doctrines of the period, they called them "bourgeois science" and accused them of distorting scientific truth by ideological argument. The typical reply to Marx and Engels was, of course, that all scientific research starts from "stubborn" facts that are gleaned by experiment and observation; they cannot be influenced by the wishes of any social class, bourgeois or proletarian. On the basis of these facts, science proceeds by the scientific method of inductions to generalizations, which are also beyond the reach of any social influence.

5. Facts and Interpretation

Georg Lukács defends the Marxist view by analyzing the conception of "fact." He writes:

> Of course, every cognition of reality starts from facts. However, one has to ask which datum and in which methodical connections does it deserve to be taken into consideration as a fact that is relevant for this cognition? Narrow-minded Empiricism denies that facts do not become facts without a methodical treatment that depends upon the purpose that we are to achieve by cognition. In this way it is overlooked that the simplest enumeration of 'facts' without comment is practically an 'interpretation:' the facts are caught by a theory, by a method; they are separated from their original living environment and are fitted into the context of a theory.[13]

Given by Lukács in 1922, this analysis of the relation between facts and theories is not a mere interpretation of Marxism by its own conceptual system, for it has absorbed the analysis of science given around 1900 by men like Henri Poincaré, Pierre Duhem, Abel Rey, and other representatives of what Rey called the "new positivism." What has been called "old positivism" described scientific research as a collection of objective facts and an inductive inference of law from these facts. The "new positivism," however, pointed out that one cannot establish laws between "brute" facts. The scientist has to make use of the brute facts as raw material to construct concepts that are fit to be connected by laws. This is an achievement of the creative imagination, as Poincaré has insisted. In the language of scientific psychology this means that these concepts and laws are part of the individual behavior of the scientist and, therefore, dependent upon all factors that determine that scientist's behavior. This means especially that the "conceptual scheme" of science, as James Bryant Conant prefers to call it, is clearly dependent upon social factors.

To keep the historical record correct, we must admit that this influence of theories upon the choosing of relevant facts was not unknown to "old positivists" like Auguste Comte and certainly Ernst Mach. There is no doubt, however, that it was emphasized strongly in the writings of Poincaré and Duhem and gained even greater strength from the creators of twentieth-century physics like Albert Einstein and Niels Bohr.

While in the natural sciences these ideas did not develop until the end of the nineteenth century, they had been vaguely accepted in the social sciences for a long time. It has been obvious that the results of social science depend greatly upon the social situation of the author. It has also been easy to realize that conceptual schemes used by different authors depend upon their social status. Hence, the Marxists, as well as the followers of Mannheim, have directed attention to this conditioning of the conceptual schemes of social science. Mannheim gives an example from which we can see how the basic concepts used in social and political science are not just reflections of brute facts but are conditioned by the life situation of the author:

> So, for example, early nineteenth century German conservatism . . . and contemporary conservatism, too, for that matter, tend to use morphological categories which do not break up the concrete totality of data of experience, but seek rather to preserve it in all its uniqueness.[14]

By "morphological categories," Mannheim means groups of observational data that are recognized by commonsense language as "natural units." In other words, conservatism is inclined to discuss political and social problems by using as much as possible, even on higher levels of abstraction, terms from commonsense language; such words are "state," "nation," "leader," and "loyalty" to a religious denomination. Mannheim continues:

> As opposed to the morphological approach, the analytical approach characteristic of the parties of the left broke down every concrete totality, in order to arrive at smaller, more general, units which might then be recombined through the category of causality or functional integration.[15]

"State" and "nation" will not be treated as natural wholes but will be described by combining recurrent traits of human behavior in various ways. The political conservative will describe human life and history as consisting of larger agglomerations like "state," "nation," "national spirit," "loyalty to his faith," which may reappear at any point in human history. Adherents of radical liberalism have been inclined to build up human history by using elementary traits of behavior, like conditioned reflexes, memory, etc. By using this description one can be fairly certain that the basic concepts (e.g., conditioned reflex) will be relevant at every point in human history, while it can by no means be taken for granted that concepts like "state" or "nation" will be helpful for presenting an evolution of mankind in the distant future. We note, however that the Marxist leftists have rejected the use of general traits of behavior and have preferred, like the conservatives, to use "morphological wholes" like "classes" and "internal contradictions" as basic concepts.

6. Sociology of Science

From this example, we can easily understand why different social and political groups—conservatives, liberals, and Marxists—build up different systems of social science by choosing appropriate basic concepts and how the results of these systems were determined by the social status and social aims of their authors. It has frequently been said that the situation is quite different in the mathematical and physical sciences.

Among the theoretical advocates of Marxism, there is a complaint that social scientists attempt to ape the methods of natural science and have, in this way, produced the illusion that an "objective social science" is possible. According to the Marxist complaint, "bourgeois science" has been identified with that objective science, and the bourgeois scientists have denied that the social situation of the author has a bearing upon the result of his research. Lukács, the advocate of "orthodox Marxism," writes that the opponents of this doctrine

> take their cue from the method of natural science, from the way in which this science is able to find "pure facts" by observation, abstraction, experiment, etc. and to establish their connections. This ideal of cognition they compare favorably with the arbitrary constructions of the Dialectical method.[16]

The author, certainly one of the scientifically minded advocates of Marxism, describes the "methods of natural science" precisely as they were conceived before the end of the nineteenth century. He does not consider the ideas of the "new positivism," not to mention pragmatism, operationalism, and logical positivism, which will be discussed later. In this way he can prove that natural science furnishes objective results that are not dependent upon the life situation of the scientist.[17] Even Mannheim, who introduced the sociology of knowledge, has always been uncertain whether this method can be applied to science, whether one can also speak of the "sociology of science." Robert K. Merton, who has, more than any contemporary sociologist in the United States, worked toward building up a sociology of science, remarked that Marxists strongly tend to separate the social from the natural sciences and to declare the results of natural science independent of the social situation of the scientist. Marx writes the following:

> The distinction should always be made between the material transformation of the economic conditions of production which can be determined with the precision of natural science, and the legal, political, religious, aesthetic, and philosophic—in short, ideological forms in which men become conscious of this conflict and fight it out.[18]

As we have learned previously, the main tenet of Marxism is to assert that the use of strictly scientific methods leads to the prediction of proletarian revolution and the ascent of the proletariat to power. Any other result is due to distortion by bourgeois ideology. By "strictly scientific method" one can hardly mean anything but the methods successfully used by natural scientists. Therefore, Marx was eager to emphasize that political economy works with the precision of natural science. It is obvious that this assertion can only make sense if we assume that natural science gives precise results, provided one follows the "scientific method." From this argument one could easily conclude that Marx gave full endorsement to the results of exact science and regarded them as independent of the social situation of the scientist. In order to get a more concrete understanding of this attitude, it is instructive to look at the actual conditions in the scientific life of the Soviet Union, where teaching and research are organized according to the Marxist doctrines. It is easy to note that the results of the exact sciences established in the international community of scientists have by no means been accepted and taken for granted by the Communist Party and the government of the Soviet Union.

In order to characterize the official attitude, we quote again from Mitin's textbook. The author explicitly rejects the view that what is taught in the books and schools of the Western world under the title of "exact science" is identical with the "exact science" that Marx proclaimed to be objective and impartial. While there have been authors in the USSR who claimed that Western science is essentially objective and impartial—only superficially tinged with a bourgeois ideology that manifests itself as an idealistic philosophy and appearing occasionally in the introductions of textbooks and anniversary speeches—Mitin objects. He writes that

> idealism is not a surface feature of bourgeois science. As a matter of fact, no reactionary idealistic philosophy compels a blameless and classless science to become a servant of the ruling classes. To assume such an influence would mean to assume that only philosophy is a class science while the exact sciences are by themselves classless sciences which can, however, be exploited for the interests of one or another class. From such a conception that is cherished, in particular, by "our" mechanists follows an uncritical bow to science, their alignment to "exact science," and their fight for the "liberation" of science from the philosophy of dialectical materialism which is to divert science from the path of truth. In a class

society every science is, by its very essence, a class science. To follow "science" blindly and uncritically is nothing but to move to the position of bourgeois science.[19]

This clearly means that according to Marxist doctrine, only the proletariat and its advocates can produce objective and unbiased science while what is produced by the bourgeois scientists under the name of "objective and exact science" is actually tainted by bourgeois ideology and has become bourgeois science. The belief in scientific method as warranting the achievement of objective truth seems to the Marxist a very dangerous belief. It could be applied to research in social science done by bourgeois investigators. Hence, the use of scientific methods in social studies, praised by Karl Marx himself, has been disparaged by "orthodox Marxists" in our own period. In his book quoted above, Lukács writes:

> When the ideal of scientific cognition is applied to nature it serves the advancement of science, but if it is applied to the evolution of society it becomes an ideological weapon of the bourgeoisie. It is a question of survival for it to regard its order of production as a result of the eternal laws of nature, determined to eternal existence.[20]

7. Social Class and Social Situation

Marxism's aversion to the general applicability of the scientific method can be traced back to the following argument: as physical science is (and will always be) built on the basic concepts of mass, force, velocity, and so on, political economy is based upon the basic concepts of supply and demand, value of money, free competition, etc. If these basic concepts are valid forever, the capitalist theory of economy can never be replaced by a new one. The way to Marxist theory is barred from the beginning. The methods of "exact science" are rejected because of their rigidity.

When Mannheim and others replaced the Marxist theory of special ideology with the theory of general ideology and the sociology of knowledge, the role of the exact sciences and their method also led to great difficulties. As we learned previously, the social situation of the author determines the result of their research: in other words, to every ideology belongs a different system of knowledge. According to the traditional metaphysical theory of knowledge, which Mannheim calls the "idealistic" one, the validity of a proposition depends only upon which relation between the basic concepts is studied. It would be absurd to claim that the life situation of the scientist should have any bearing on the validity of any relation between mass, energy, force, etc. Therefore, Mannheim complained that the theory of knowledge has been formed after the example of the exact sciences instead of taking the cue from other types of knowledge (e.g., the social sciences). According to these considerations, it becomes hard to explain the meaning of the "validity of a statement."

However, Mannheim points out that we can find, in twentieth-century science, some argument that would support the conception of the sociology of knowledge. According to the theory of relativity, the "absolute" speed of a body cannot be determined by any experiment; but we obtain a definite value relative to a determined system of reference which is, of course, arbitrary. Just as we have to reject the answer to the question of what is the real or objective speed of a body, "we must reject," writes Mannheim, "the notion that there is a 'sphere of truth in itself' as a disruptive and unjustifiable hypothesis."[21] The statement that "there is an absolute velocity of a body, but we cannot observe it" is analogous to the statement that there is a truth independent of the life situation of the author but that we can never find out what that truth is. According to the original Marxist doctrine, a statement is valid if it is proved according to scientific methods by a scientist whose social interest is identical with the interest of the proletariat, while "bourgeois scientists" may distort the truth by bringing into the argument the ideology of their class, such as religious beliefs. However, if we argue according to Mannheim's sociology of knowledge, a social scientist reaches a result that depends upon his life situation. The results of different scientists differ from each other, and we must find out which of these results are "valid."

Instead of invoking the results of scientists who sympathize with the proletariat, Mannheim invokes as the arbiter of validity a group of men who do not belong to any definite social class. He makes use of Alfred Weber's view that the "intelligentsia" is a "socially unattached" group; he also speaks of the "freely floating intelligentsia." For Mannheim, the results obtained by this intelligentsia are the valid results. They consider all possible ideologies that cancel each other out in the intellectual's mind. For the Marxist, the ultimate criterion of validity has been the decision of the proletariat; for the Mannheim school it is the decision of the intelligentsia. Both criteria flagrantly contradict the criterion of validity taken for granted by the "metaphysical" theory of knowledge. Mannheim recognizes this very well. He writes that it is an axiom of the currently prevailing "idealistic" theory of knowledge that the "genesis of a proposition is under all circumstances irrelevant to truth. . . . It is regarded as impregnable and is the most immediate obstacle to the unbiased utilization of the findings of the sociology of knowledge."[22]

8. The Solution to the Puzzle

Neither the German school of *Wissenssoziologie* (sociology of knowledge)[23] nor the Marxist school dropped the metaphysical criterion of validity, but the Marxists call the "materialistic criteria" what Mannheim has called "idealistic criteria." Mannheim has repeatedly attempted to reconcile idealistic epistemology with the sociology of science and has stated that eventually a consistent sociology of knowledge would call for "revision of the thesis that the genesis of a proposition is under all circumstances irrelevant to truth." Neither the Marxists nor Mannheim, however, have made a real attempt to work out a new theory of knowledge. To use their own terminology, they were deterred by "ideological reasons." However, as Merton remarks,[24] the revolution in the theory of knowledge that Mannheim ascribes to the sociology of knowledge, in its bold outlines, has long been familiar to the American mind. Merton refers, of course, to the writings of the pragmatist school, including Charles S. Peirce, William James, John Dewey, and George Mead.

As a matter of fact, the departure from the metaphysical theory of knowledge started even earlier in the "positive philosophy" of Auguste Comte (1830).[25] He wrote flatly that there is no theory without experience and no facts without a theory. This means that he denied the existence of "brute" facts given to us by nature before we conceived a theory. As we mentioned previously, this point was stressed and elaborated on by Henri Poincaré and the school of "new positivism." What is immediately given to us can be described in commonsense language either as a continuum of sense impressions or as an agglomerate of "things," if we use this word as it is used in commonsense language when we say that a "table" or an "insect" are things. The actual rise of science has shown that there are no simple laws between "sense impressions" or between "things." We have to construct concepts like mass, mass point, acceleration, and so on in order to find simple laws like Newton's by which these concepts or constructs can be connected.[26]

As previously mentioned, and as will be discussed later in more depth, similar lines were pursued by logical positivism, pragmatism, and operationalism. Now if we take our actual problem—the solution of the puzzle posed by the sociology of knowledge[27]—there is no essential difference between the positivistic, pragmatic, and operational theories of knowledge. The puzzle had arisen partly from the vague meaning of the word "knowledge." One may mean by it "knowing that an animal with a long neck exists and is called a giraffe," but one may also mean "knowing that man has descended from inanimate matter," or "knowing that the earth is actually rotating." In order to clarify what is meant by sociology of knowledge, "knowledge" in this context can be understood as "scientific theories." And by "validity of theories" we mean "fitness of the theory to serve some definite purpose." According to the pragmatic "science of sciences" to be elaborated in Part II, theories are always conceived for certain purposes. "Validity of a theory" is only meaningful if we make this expression more specific and speak of "validity for a certain purpose." A theory can be valid for a technological purpose but invalid in the pursuit of a desirable moral life. If we drop the metaphysical concept of validity as agreement between the theory and objective reality, it is no longer true that the validity of a theory is dependent only upon the words by which it is formulated. The validity certainly depends also upon the author because he determines the purpose for which he advanced the theory.

Chapter 6

Sociology of Science and the Search for a Democratic Metaphysics

1. Validation and Theory Building

We mentioned previously that an impact of the social environment upon science certainly exists if we single out the technological purpose of scientific theories. Social and economic powers favor scientific research that will produce such technological progress as they consider desirable. As we mentioned previously (Part I, chap. 5, §4), this influence of social factors upon technology is not discussed in this book because it has little to do with philosophy of science. We learned, however, in our treatment of theory building, that scientific theories have more than just technological purposes; they have always been used to direct human conduct along desirable lines.

From Plato to Einstein, as we have seen, scientific theories have been interpreted philosophically and, accordingly, used for various moral, religious, and political purposes. If we include these interpretations in the theories, it is evident that social powers influence the acceptance of scientific theories. If we try to sharply distinguish between the scientific theories in the proper sense of the word and their philosophical interpretations, it is easy to see that this is impossible if we wish to include very general theories like determinism in physics, spontaneous generation in biology, the structure and origin of the universe in cosmology, and so on. This again amounts to the question of whether science can be "purged" of philosophy, which we have already discussed. There is, in any theory, a part that is not determined by technological criteria but by requirements of economy or beauty. The axiom that the validity of a theory does not depend upon the life situation of the author is wrong in the pragmatic theory of knowledge. The validity cannot be judged without knowing a theory's purpose, and its purpose cannot be known without knowing the life situation of the author.

The revision of that axiom, required by Mannheim, has actually taken place. In the pragmatic theory of knowledge, or science of science, the dependence of theories upon the life situation of their authors is indispensable for explaining how a certain theory was constructed. We have to discuss the effects of the different criteria of validity according to the intended purpose. Eventually, we obtain not one unique theory but a choice among several, to be exact, among an infinite number of theories. We can present the reality of this choice in two different ways: we can say that the choice is arbitrary, in which case we describe the situation in the language of "conventionalism" as suggested by Henri Poincaré and others. The conventional choice is, however, only logically arbitrary; this means that no decision can be achieved by logical conclusions and physical experiments. If we include, however, not only physics and mathematics but also the fields of psychology and sociology, we can derive reasons for preferring determinism to indeterminism or the Copernican system to the Ptolemaic system. The consistent investigation of theory formation in the physical sciences leads to the result that without considering the social situation of the theory builder, no unambiguous criteria for the validity of a theory can be established. From the point of view of the pragmatic theory of knowledge, the puzzling questions posed by the sociology of knowledge do not even arise: the influence of social factors upon the validity of scientific (even physical) theories is a necessary part of pragmatic epistemology.

2. Science as a Compromise between Technology and Political Philosophy

If we carefully consider how the validity of a scientific theory is judged, according to the pragmatic theory of knowledge, we notice that in the case of very general theories the answer to whether a theory is valid cannot be "yes" or "no." A compromise has to be worked out as to what weight is given to each different purpose of a theory, particularly to the technological and the moral or political purposes. Achieving agreement about a compromise often has the character of negotiations toward a compromise in political action; it often looks like the search for a "happy formula" that may solve a diplomatic conflict. In order to understand this point clearly, it will be instructive to revisit some concrete cases in which such compromises were established.

As a first example, we return to the compromise proposed to Galileo in the matter of the Copernican theory of the planetary system. Both parties agreed that from the astronomical (technological) point of view, the Copernican (heliocentric) system was acceptable. While Galileo wanted to call it "true" or "valid" for this reason, the representatives of the Church pointed out that from the religious and moral point of view the "mobility" of the earth raised grave difficulties; it implied complications in the interpretation of the Bible and conflicted with the generally accepted philosophy that was also the generally accepted basis of religion. On April 4, 1615, Cardinal Bellarmine wrote a letter to Paolo Antonio Foscarini, a Carmelite monk and a good friend of Galileo. From the letter it is evident that the church did not press for a "yes" or "no" but for compromise acceptable to both parties. Cardinal Bellarmine wrote:

> It seems to me that you and Galileo should content yourselves with speaking not positively but tentatively, as I have always believed Copernicus did himself; the supposition that the movement of the earth and the immobility of the sun accounts for appearances seems to me legitimate, is harmless, and suffices for a mathematician. . . . Even if there were a positive proof for the immobility of the sun and the movement of the earth it would be necessary to proceed with the greatest prudence in the explanation of the sacred books which would seem opposed to the fact. But I cannot believe in such a proof until it has been shown to me; for it is one thing to save appearances and another to destroy them, and in case of doubt the Holy Scriptures should not be questioned.[1]

The letter sounds indeed like proposals made—say, in the United Nations—to settle a conflict via diplomatic formula.

As a second example, we may look at the formula that has been proposed in the Soviet Union to establish agreement between physics and political ideology with regard to the theory of relativity. While the official philosophy of dialectical materialism requires belief in the objective reality of the external world—and in the objective truth of physical theories altogether—the theory of relativity has frequently been interpreted as denying this belief. At the Jubilee Session of the Academy of Science of the USSR, Mark Borisovich Mitin spoke on the subject "Twenty-five Years of Philosophy in the USSR":

> As the result of the tremendous work that our philosophers and physicists have carried out, as a result of . . . the battle of ideological principles, it may now be said that our philosophical conclusions concerning the theory of relativity have been firmly established.[2]

The physical content in its purely technological interpretation can be fully accepted, but from it

> springs *neither* the rejection of the existence of an objective world *nor* the rejection of the objective concept of nature. *The theory of relativity does not deny that time and space, matter and movement are absolute in the sense of their objective existence outside human consciousness.* . . . No 'point of view' of an observer, no 'system of reckoning' . . . has any more power to destroy the objective fact of natural processes.

These formulations establish the way in which the term "objective" has to be used. On the other hand, the term "relative" can be used in the following formulations:

> The theory of relativity established only the *relativity of the results* of measuring time and space by observers who are moving *relatively to one another.* . . . Time and space are indivisible from the moving body and must be regarded relative to the movement. In this respect time and space are relative.

What is really achieved by this formula is ensuring that the commonsense term "objective" has a legitimate place in the general principles of physical theory.

As a third example we may describe the compromise on the theory of relativity in Germany under Hitler. Such a compromise was necessary because relativity theory, as well as quantum theory, had introduced a language that was much more "sophisticated" than the commonsense language, and that led to results that looked absurd when expressed in this language. To state that the speed of light was the same in different systems of reference, or that the position and velocity of a particle cannot exist together, seemed to make it possible to deny all statements expressed in commonsense language. Since the government held it necessary for the well-being of the German people to impose rules that everybody had to believe and follow without the slightest doubt, it was impossible to tolerate theories that showed how gaps in the validity of seemingly self-evident rules could be pointed out and even confirmed. These dangers associated with the new theories were usually denoted by the derogatory names "relativism" and "sophistications." On the other hand, it was clear that these new theories were the scientific basis for new weapons that were necessary to vanquish the enemies of the German Reich. Just as in the Soviet Union, a compromise had to be prepared in order to preserve the scientific basis of the atomic bomb and, at the same time, to separate "relativism" from science.[3] The German government looked for a way to combine the blessings of the atomic bomb with the elimination of relativism and sophisticated conclusions.

On November 1, 2, and 3, 1942, the leadership of the University Teachers convoked a "physicists' encampment" in Seefeld (in the Austrian state of Tyrol). About thirty highly reputable physicists were invited to discuss with the "leadership" representing the ruling party a compromise between the technological requirement and the political or moral requirements for the validity of contemporary physical theories. The compromise was recorded in a protocol, a copy of which was brought to the United States by Professor Samuel Goudsmit. The content of the compromise was, of course, as in analogous cases, the acceptance of the theory of relativity as far as its technological consequences were concerned. It was regarded as legitimate to draw technologically valuable conclusions from the principles. But drawing logical conclusions "for their own sake" in order to investigate the logical consistency of the principles was flatly rejected. The meeting unanimously passed a resolution to reject "an elaborate investigation of certain subtleties and paradoxes of a typically Jewish character which are presented as the essential content of the special theory of relativity." Moreover, the encamped group rejected the "philosophical interpretations of the physical theory of relativity that have been given by Einstein and his followers."[4] This means, practically, that physical science was to be tolerated only as a technological device, while philosophical or even logical investigation was reserved for the leadership of the political party.

This fully agrees with the well-known advice of Lenin that "one should not trust a bit, even to the most prominent and deserving physicist or chemist, when he starts speaking about philosophy."[5] It can easily be confirmed, by quoting statements of the highest authorities in the German government of that period, that the main goal was to reduce science to a purely technological instrument in the hands of political authorities and to reduce scientists to machines directed by politicians. At an inspection of the government institution for physics and technology in 1934, Prussian Minister of Education Bernhard Rust said, "National Socialism is not hostile to science; it is only hostile to theories."[6] Hermann Goering, minister of the air force, said the following in the same year:

> We esteem and honor science, but it must not be pursued for its own sake and degenerate into intellectual arrogance. Our scientists are now faced by a fertile field of research. They should concern themselves with how we can produce in Germany some raw material that has had to be imported previously from abroad.[7]

From this speech we learn the determination of the political leaders to restrict science to its technological task and to reserve all philosophical interpretations to the politicians. This interpretation had to completely agree with the party line. Precisely this adjustment was demanded in a speech delivered by German Minister of Justice Hans Frank. He said, "I am in favor of full freedom of teaching and thinking. . . . Since science serves truth, it has to serve National Socialism."[8] These views are, again, in accord with Lenin's doctrine that science has to adjust itself to the party line.

3. The Scientific Conscience

This compulsory compromise is obvious in totalitarian countries.[9] However, the tendency toward such a compromise is visible under any form of government, even though refusal to align oneself with the compromise does not always mean "liquidation." Michael Polanyi writes: "The propositions embodied in natural science are not derived by any definite rule from the data of experience."[10] As we have seen, this holds particularly for general propositions that do not seem to belong either to a science or to its philosophical interpretation. Polanyi stresses that there will always be a conflict between conceivable general theories, which can only be solved by prescribing an order of predilection for the different purposes of these theories. He says:

> The conflict can be resolved only through a judicial decision by a third party standing above the contestant. The third party is . . . his scientific conscience. . . . This indicates the presence of a moral element in the foundation of science.[11]

If we analyze this concept of "scientific conscience" according to contemporary depth psychology, it is clear that such a conscience cannot develop without a community that cultivated the belief in scientific method and has proved its usefulness by its success in advancing our knowledge.

This "community of scientists" has, to a high degree, established itself in universities and research institutes. Polanyi asks under what practical circumstances this community can become an authority and a guide of scientific conscience. He asks for a "practical art" by which the spirit of the scientific community can be cultivated and transmitted. An essential part of this, Polanyi writes, is

> the art of free discussion, transmitted by a tradition of civic liberties and embodied in the institutions of democracy. This art, this tradition, these institutions will be discovered in the purest form in countries like Britain, America, Holland, Switzerland, where they were first and most efficiently established.[12]

Whatever the practical manifestations of "scientific conscience" may be, when we want to define it in operational terms it boils down to the social pressure of a group. Polanyi writes, "A child growing up in a modern community will be forced to abandon the magical outlook to which it is primarily inclined and to adopt instead a naturalistic view of everyday life."[13]

Considering all these points, we must understand that in shaping the compromise leading to the selection of a scientific theory, the authority of the "scientific community" or, in other words, of the "authorized scientific method," will play an important part. In an address to a group of Jesuit scientists, the Reverend Father Walsh, head of the Biology Department of Boston College, compared this authority of the "scientific method" to the authority of the ruling party in the Soviet Union:

> One is using the party principle as the sole criterion of truth, and the other is using the scientific method just as dogmatically as the only source of human knowledge. . . . We must remember that American and British scientists can be just as wrong as Soviet statism.[14]

In terms of what we have been saying in this book, this means that in judging the validity of a scientific theory we must not omit any factor that participates in forming a compromise. For no single factor, not even the "scientific method," can warrant validity. The effect on human conduct, the moral aspect, has to be considered with the same weight as the technological aspect.

How the different aspects are to be weighed cannot be derived from any logic of science, but only from our knowledge of human behavior. If we want to express ourselves in a slightly flippant manner, we might say that decisions about the choice of a theory are based upon sociological propositions or, if we prefer, they ultimately involve moral decisions. If this is so, we must not be surprised that every political, religious, or moral creed is connected with a specific philosophical or even metaphysical interpretation of science that serves to give to the creed in question the blessing of science. It is well known that totalitarian systems have been eager and serious in their attempts to set up philosophical interpretations of science that belong to the creeds of the ruling parties and are an article of indoctrination in all schools, from elementary schools to universities. In this book, and particularly in the present chapter, are numerous examples of this kind of "official" interpretation.

4. Philosophical Interpretations and Democracy

The great propaganda value of this connection between politics and science has been a matter of grave concern to followers and advocates of a democratic outlook. Repeatedly the wish has been voiced for a philosophical interpretation of science that would support the democratic way of life.¹⁵ Many advocates of democracy have even suggested making this democratic philosophy an article of indoctrination in schools and a creed that every good democrat must believe. The official Soviet doctrine holds that the political division of our world corresponds to an "Eastern division" and a "Western division" in the philosophical interpretation of science. As we shall learn in more depth later (Part II, chap. 8), the official doctrine of the Communist Party and the government of the USSR describes these two philosophies as "materialism" and "idealism."

Many statesmen and philosophers in the Western world have accepted this doctrine and have proposed to declare "idealism" and "antimaterialism" as the official philosophy of the Western democracies. A clear-cut example of the intentions of prominent Western authors to provide a philosophical background for their political ideals was given by Jerome Frank. He starts from the consideration that contemporary physics cannot, from the purely scientific point of view, decide whether our world is ultimately determinist or indeterminist. Both philosophical interpretations are possible. "Since," argues Frank, "from the purely scientific (technological) angle, the choice between determinism and indeterminism is arbitrary, we can choose the theory that is preferable from the moral point of view." He goes on, "We can choose the theory which is a better support for the American way of life. This is obviously the hypothesis of 'indeterminism' because it allows for freedom, the essential trait of American political philosophy."¹⁶

Perhaps the most instructive example is Charles Malik, a statesman as well as philosopher. An Easterner by birth and a Westerner by learning, he represented Lebanon at the United Nations and took courses at Harvard under Whitehead. He writes, "Because man is a rational being, the evil of his own doing always has its origin in an error of his mind."¹⁷ According to the official Soviet doctrine, every departure from the party line has its origin in an error in philosophy, a departure from the authorized interpretation of dialectical materialism. Malik's statement practically translates this doctrine into the language of scholastic philosophy. He continues: 'I shall now list eight basic errors committed by the metaphysics of Communism, and contrast them in each case with the truth of the Western positive tradition."¹⁸ We shall quote four of the eight "contrasts" that closely relate to the philosophical interpretations of scientific theories.

1. Communism: The ultimate reality is through and through matter.

   Western Tradition: The truth is that besides matter and utterly irreducible to it, there is an independent and superior reality, namely, mind and spirit.

2. Communism: The proper attribute of reality is change and strife.

   West: The truth is that there is a changeless and stable order of existence, on which the mind can really rest.

3. Communism: There is no objective and external truth.

   West: The truth is that such a truth exists, and that only by humbly seeking it and finding it can we achieve genuine understanding and real peace.

6. Communism: That so far as the nature of things is concerned, only the tradition of Democritus, Lucretius, Feuerbach, and Marx is right.

   West: The truth is that this materialistic tradition is thoroughly absorbed by the more concrete positive tradition from Plato and Aristotle to Hegel and Whitehead.[19]

This example shows the great difficulty of producing a "democratic metaphysics" that could match totalitarian metaphysics. The first three items are typical metaphysical interpretations of our experience or, for that matter, of our systematized knowledge called science. Both sides of the iron curtain attempt to characterize the world of our observations by propositions couched in commonsense terms. The same propositions should also be fit to guide human conduct. Since in a democracy no philosophical system is enforced by the authority of the government, it should be no surprise that the "Western positive tradition" described in Malik's "contrasts" is not generally in agreement with or even similar to the philosophy that is actually accepted in the Western democracies. The entire philosophy of empiricism and pragmatism, so instrumental in giving philosophical form to the ideas of democracy among the English-speaking nations, sharply contrasts with Malik's "Western tradition." As a matter of fact, Malik describes the contrast between two philosophical systems:

dialectical materialism and scholastic metaphysics. But only a part of the Western world ascribes to this "Western tradition"; and it is certainly not the part responsible for the advance of science. This becomes clear in Malik's sixth contrast. Western science has certainly been influenced more by Democritus and Lucretius than by Aristotle, and in Communist philosophy of science Hegel certainly plays a greater role than he does in the formulations of Western science.

Perhaps the most stimulating effort made in this country to correlate a certain political philosophy and a certain philosophy of science was presented by F. S. C. Northrop. He starts from the fact that what he calls the "normative propositions of the social sciences cannot be proved nor even confirmed by an investigation of the social facts."[20] "Normative statements" advise people on how to conduct their lives; they are "statements about values" or "appraisals" or "predilections.' Even if we knew the laws connecting the social facts, Northrop argues, we would not learn from them what a "good society" or "good conduct" is. Yet the system of values accepted by a certain nation or party is rooted in a philosophy that is connected to the scientific theories accepted by that nation or party.

According to Northrop, these philosophies "are always regarded by the people who hold them as called for by the scientific knowledge which they take into account."[21] In the language of the present book, these philosophies are, to be exact, philosophical interpretations of scientific theories. They are "called for" or "determined" by the results of science and determine, in turn, the value system of the same period and the same group.

Northrop emphasized the close connection between the normative theory of social science and the factual theory of natural sciences:

> By making primitive concepts and postulates identical in both the normative theory of social science and the factual theory of natural science . . . one obtains normative social theory which can be verified, since the deductively formulated theory of natural science is scientifically verifiable.[22]

For example, Northrop also analyzes the difference between the value system of Western democracy and the Soviet state. He attempts to reach the underlying differences in philosophy, each of which is closely connected to a certain conception of science: the "true" science corresponds to the "good" value system. In this way, Northrop finds the philosophies at the roots of Western democracy and Russian Communism, respectively. But his results are very different from Malik's.

According to Northrop, Western democracy is based upon the empirical philosophy of Locke, while Russian Communism is based upon a philosophy much closer to Aristotle than Locke. In terms of the scientific roots of these philosophies, Locke's is based upon Newton's physics, while Aristotle's is based upon his own organismic physics. The soul or personality in Locke's moral and political philosophy is identical with the person of the observer in Newton's physics. The only connection between different persons is achieved by the sense impressions of these persons.[23] In Aristotelian or Thomistic philosophy, however, the individual is a part of a hierarchical order. The nature of the individual person is connected with the nature of others, with the nature of higher beings like angels, and, ultimately, with God. Respectively, these conceptions of the human person are based upon the conceptions of nature prevalent in the periods of Aristotle, St. Thomas, and Locke. For Northrop, it is

> the lack of any intrinsic social relations between persons which is the Lockean modern foundation of the political economy expressed in the American Declaration of Independence to the effect that there is no basis for government, no normative social theory, apart from a social convention.[24]

According to Northrop, therefore, the most articulate formulation of a democratic state, the Declaration of Independence, is based on Locke's philosophy that, in turn, is based upon Newton's physics.

This connection between political philosophy and philosophy of science is also clear in the analogous connection between Aristotle's physics, Thomistic philosophy, and the political philosophy of authoritarian states. According to Northrop, "In the Aristotelian theory of a person, a person is organically related to other people and must have this relation expressed in society, if society is to give expression to his own individuality." He continues, "Consequently, a good state is one in which a man cannot be himself except insofar as he operates through an organized social church and through an organized political government."[25] According to Northrop, therefore, Marxist political theory is much nearer to Aristotelian than to Lockean philosophy. Northrop stressed Marx's theory that the human individual has no meaning apart from this status within the organic structure of society. The Communist state is based philosophically upon Aristotle and of course, Hegel.

It is easy to see from these passages how differently Northrop sees the coordination between political philosophy and philosophy of science from the way Malik sees it. Although a connection seems to exist, the precise coordination does not seem to follow unambiguously from the actual state of science. While there is a strong belief in our Western world in some connection between political ideology and the philosophy of science, there is a great difference of opinion as to what specifically this coordination may be. We do not even know with certainty on which side of the iron curtain such a great philosophical system as Aristotle's belongs. There is no doubt that in all periods of intellectual history, philosophical interpretations of science have been analogies to daily life experiences formulated in commonsense language. There is no doubt either that these interpretations have been used as philosophical support for desirable human conduct and that this has been possible because this advice is formulated in the same commonsense language of philosophical interpretations of science.

Northrop points cut that every philosophy of science can be tested by scientific methods and, therefore, by implication political philosophy can also be tested by scientific methods. If and when we become convinced that Newton's physics is scientifically preferable to Aristotle's physics, we know also that Locke's philosophy is, at this point in time, preferable to Thomistic philosophy and that by implication Western democracy is preferable to the totalitarian state. There is no doubt that this way of arguing is attractive to those who like to be guided by scientific argument; but there is some inherent weakness in this argument. As we have learned in several cases, a scientific theory does not imply an unambiguous scientific interpretation. The theory of relativity, for example, has been used to support idealistic, materialistic, and positivist philosophies. The choice of the interpretation is determined by the political philosophy (the "normative system" in Northrop's language) of the group concerned.

Hence the statement that by confirming a scientific theory by experiment, I confirm simultaneously a philosophy of science and by implication that a political philosophy implies a vicious circle because the philosophy of science itself is determined by a political philosophy. If we hold fast to the scientific way of arguing, we will probably notice that the connection between philosophy of science and political philosophy cannot be established satisfactorily by analyzing their language. It is necessary to study social facts directly and to realize that building up science belongs to these social facts as much as building roads and houses.

5. The Physical and the Socio-cosmic Universe

One of the most ancient parts of science is a human picture of the physical universe, one that always bears some analogy to human society. The gods operate the sun and stars, oceans and mountains, just as human rulers direct their officials and subjects. Once the physical universe is established as a "cosmic state," man attempts to make states on earth as similar as possible to the cosmic state. The closer to this ideal a state comes, the more it is regarded as a "good" state. We remember that the geocentric and the heliocentric systems were regarded as two different "constitutions" of a cosmic state, and they implied two different ideals for human states. If we look at the facts in this way, it becomes clear that "normative principles" of social philosophy are not necessary. The only "normative" factor occurs in this human urge to imitate the cosmic state. Every philosophy of science has interpreted the laws of science as a "cosmic state" and in this way has created an ideal for human states. This is probably the historical and sociological root of the well-known fact that philosophical interpretations of science continue to be used as foundations of moral, religious, or political philosophies.

The notion that our pictures of the physical universe are not based upon intellectual research but influenced by our moral and political ideas has been strongly upheld and lucidly presented by John Dewey. He suggests

> as a reasonable hypothesis the idea that philosophy originated not out of intellectual material, but out of social and emotional material. . . . If anyone will commence without mental reservation to study the history of philosophy not as an isolated thing . . . ; if one will connect the story of philosophy with a study of anthropology, primitive life, the history of religion, literature and social institutions . . . the history of philosophy will take on a new significance. What is lost from the standpoint of would-be science is regained from the standpoint of humanity. . . . When it is acknowledged that under the disguise of dealing with ultimate reality, philosophy

has been occupied with the precious values embedded in social traditions . . . it will be seen that the task of future philosophy is to clarify men's ideas as to the social and moral strifes of their own day.[26]

For an instructive and simple example of how the picture of the physical universe has been used for guiding the conduct of man toward desirable goals, we need only look into the first book of the Scripture (Genesis). We are so familiar with this narrative that we do not always realize how closely physical hypothesis and moral guidance are interwoven. No one would doubt that the most characteristic feature in our picture of the physical world is the existence of physical laws or, in other words, the law of causality or uniformity. Yet if we carefully read the story of the Great Flood in the Bible, we learn that these laws were established to reward good behavior and can be revoked at any time if men do not behave well.

These considerations lead to the conception of a general analogy between the physical universe and a "good" human society. Ernst Topitsch named this conception the "socio-cosmic universe" and discussion of this will be taken up later (e.g., in Part II, chap. 8, §3).[27] The relation between physical and moral laws in the Bible will be more elaborately discussed when we present the conscious and unconscious reverberations of ancient and medieval philosophies throughout modern science and philosophy.

PART II

Chapter 7

Scholastic Philosophy and Thomism

1. The Meanings of Rational and Intelligible

Many authors emphasize that modern science is less "rational" than ancient and medieval science. Whitehead states flatly in his book *Science and the Modern World* that modern science is based on faith, while medieval science was based on reason.[1] This means, obviously, that modern science takes for granted laws of nature, such as Newton's laws of motion, as "hard facts" which cannot be "proved," while medieval Thomists attempted to derive physical laws from "intelligible" facts. The prominent American historian Carl Becker[2] argues that modern science has become less and less rational, and that twentieth-century science should be bluntly called "irrational."

As a matter of fact, however, Becker and many "humanists" use the words "logical" and "illogical" in ways that do not agree with how words are used in science. A system of statements would be called "illogical" in science if and only if it contained a logical contradiction. But this is certainly not the case in modern science any more than in medieval science. What Becker calls "logical" is what Aristotle called "intrinsically intelligible" or what is often called simply "understandable." A statement or system is "intelligible" if one can derive it logically from principles that are self-evident. In this sense, every "stubborn fact" that we take just as it comes is "illogical."

In *The Heavenly City*, Becker characterizes the spirit of modern science:

> Facts are primary and what chiefly concerns us; they are stubborn and irreducible and we cannot get around them.

They may be in accord with reason, let us hope they are; but whether they are so or not is only a question of fact to be determined like any other.[3]

To say that "a series of facts is in accord with reason" means only that they can be logically derived from intelligible principles; it has nothing to do with the logical consistency of these facts. "This subtle shift in the point of view," Becker continues, "was perhaps the most important event in the intellectual history of modern times, but its implications were not at once understood."[4]

The "subtle shift" was, of course, the decline of belief in the possibility of deriving observable facts from intelligible principles or, in other words, the decline of belief in a rational world picture. Becker writes:

In the course of the nineteenth century this optimistic outlook became overcast. The marriage of fact and reason, of science and the universal laws of nature, proved to be somewhat irksome, and in the twentieth century it was, not without distress, altogether dissolved. . . . If logic presumes to protest in the name of law, they (the scientists) know now how to square it, so that it complaisantly looks the other way while they go on with illicit enterprises—with the business, for example . . . of teaching 'the wave theory of light on Monday, Wednesday, and Friday, and the quantum theory on Tuesday, Thursday, and Saturday.'[5]

This is certainly an incorrect presentation of the twentieth-century theory of light, and it would be interesting to know what light happens to be on Sunday. Becker presents the history of science as if omissions in logic increase more and more in modern science and, while the scientist works, "logic has to look the other way."

What has actually happened in the history of science is that science has become less and less rational, and this means, again, that the departures of twentieth-century theories from commonsense experience became greater and greater. In the twentieth-century theory of light there is no departure from logic, but conspicuous departures from the ways by which our daily life has been properly described. As we discussed previously, the increasing departure from "logical"—or rather, "rational"—theories in the physical sciences marks a departure of physical theories from formulations that can be expressed in commonsense language. However, the human mind is not satisfied with this trend, and wants, as we discussed (in Part I, chap. 1), to return to a rational picture of the physical world. To these "rational pictures" belongs, in particular, the picture given by traditional Christian and Jewish religion and all kinds of teleological philosophies that explain physical phenomena by analogies with events that happen in an artisan's shop as the result of the artisan's purpose and blueprint.

2. The Role of Philosophical Schools

This inclination the human mind has toward rational pictures has resulted in "philosophical systems;" they are, practically, rational pictures of the universe that have been worked out systematically and elaborately. As we have learned previously (Part I, chap. 1, §8), people who embrace the same rational picture of the physical universe will also share many moral and political ideas. Hence, a certain comradeship or friendship develops and a "school," in the philosophical sense, emerges. These schools have had wide cultural and social influence; so knowing and analyzing the main philosophical schools adds much to our understanding of historical periods, particularly our own. In order to understand contemporary philosophy of science, it will be helpful to know the philosophical schools of our period that pay attention to philosophy of science. In order not to diverge from our main topic, we shall discuss only those schools that, on the one hand, place philosophy of science at the center of their efforts and, on the other, are believed to be pertinent—for good or evil—by political, social, and religious organizations.

According to these criteria, we restrict ourselves to three schools or systems: the Thomistic philosophy of the Roman Catholic Church; dialectical materialism, which is the official philosophy of the Communist Party and of the Soviet state; and positivism or pragmatism, which has been the main philosophical stream in twentieth-century science and has decisively influenced lines of thought in the Western democracies, particularly in Anglo-Saxon and Scandinavian countries. A fourth school of philosophy that fulfills our requirements would be "idealism" of the Kantian or Hegelian variety. While this has played a tremendous role in the understanding of intellectual and social life in Germany and the countries that have been under German influence, we shall restrict ourselves to those schools whose doctrines have shaped the international interpretation of contemporary science. Occasionally, however, we shall discuss some of idealism's effects upon the interpretation of science.

3. Science and "Thomism"

If it is generally true that philosophical interpretations of science originate in a longing for a rational understanding, this is particularly true of the Thomistic School. We can obtain a lively understanding of this attitude by reading passages from *Recollections of Seventy Years*, by Cardinal William Henry O'Connell.[6] The late cardinal had a strong interest in philosophical interpretations of twentieth-century science and was gravely concerned about the materialistic implications he believed could be derived from the theory of relativity were it were presented without a proper philosophical interpretation—an interpretation he believed could only be provided by Thomistic philosophy. Speaking about Pope Leo XIII, who was responsible for a mighty revival of Thomistic philosophy that preceded the rise of twentieth-century physics, the Cardinal wrote the following:

> He saw very clearly that while great discoveries had been made in the world of physical laws and material science, the conclusions drawn from the premises, though superficially attractive, were logically false and destructive of genuine Catholic philosophy.[7]

It is particularly instructive to read how Leo XIII, in his seminary in Perugia, characterized Thomistic philosophy in contrast to the way of thinking that was customary among physicists. He led his students, as Cardinal O'Connell expresses it,

> to the teaching of the great master of thought (Thomas Aquinas) of which the church had such reason to be proud, but which the world of physical science was striving to bury beneath the disorderly mass of mere speculation—the vogue at the universities of that time.[8]

This refers to the time after 1880, and if Pope Leo were alive today he would scarcely believe that the vogue at the universities is any less inclined to speculation in physics.[9] The cardinal writes that "the somewhat flamboyant mentality of the scientists of that day had little reverence for the exact intellectual processes of the Thomistic system."[10] Although

Thomistic philosophy has had a conspicuous revival, one would hardly say that the mentality of the average scientist today is "less flamboyant" or has "more reverence" for the syllogistic system of St. Thomas. What the Thomistic interpretation has pledged to offer the scientist is, as we have learned, "less speculation and more exact syllogistic processes."[11] A prominent Thomistic philosopher, James A. McWilliams, writes: "Perhaps the greatest mission of our metaphysics today is to justify itself in the eyes of the scientists and to show them that it is what they need in order to rationalize their various scientific disciplines."[12]

The main feature in the Thomistic interpretation is the elaboration of analogies with the phenomena observed in living organisms, particularly in human beings. Edward Francis Caldin writes:

> Thomist philosophy . . . pays greatest attention to intellectual beings. . . . Non-rational living organisms occupy a subordinate position and dead matter is considered as interesting largely because it is capable of being actualised by some higher form. More specifically, Thomist philosophy places God at its apex. . . . It treats of man as an intellectual being . . developing towards God; it treats of other living organisms is analogous to man in respect of life, but distinct in respect of . . . intellect.[13]

Thomistic philosophy believes that "physics," in the modern sense, a theory of the motion of "dead matter" that is not living and intelligent, is impossible. Caldin continues:

> Because of this attention to living and intelligent beings (Thomistic philosophy) perceives the scale of perfection among essences, the *analogy of being* in what might be called its *'vertical' dimension*, which is overlooked by those who restrict the object of their philosophising to the dead world.[14]

This attitude of contemporary Thomism to physical science is lucidly described by the Reverend Joseph T. Clarke, S. J. who is thoroughly familiar with contemporary scientific methods:

> The same physical reality of a common, shared experience forms the material object or subject matter of both science

and philosophy. But their respective formal objects, or the precise points of view under which each study examines an identical subject matter, are distinctly different.[15]

Consider as an example the motion of a falling stone. The scientist asks, according to Father Clarke, *what* is the motion of the stone? The answer describes the stone's position as a function of time and derives this motion from simple laws, like Newton's laws of motion. If we know all this, we know *what* the motion is. The answer is given in terms of centimeters, grams, seconds, etc. After the scientist answers his question, the philosopher asks a further question: what *is* motion? And how can I describe motion by using intelligible properties instead of measurable properties? By intelligible properties, we mean expressions that occur in our intelligible, self-evident propositions about objective reality.

There are, according to Clarke, "alternative vocabularies, each of which is irrefutably valid, irreducibly different, but not irreconcilably opposed, rather mutually complementary."[16] This suggests that the relation of scientific description to philosophical description of the same condition may be analogous to complementary descriptions of an atomic object, according to Bohr's principle of complementarity. In order to understand how these complementary descriptions are related, we must, above all, know what properties constitute a philosophical description of a situation. Clarke gives the following example: from the scientific point of view, we may "describe an atom or an organization as a configuration of electrons or as a colony of cells in thermodynamic equilibrium," but from the philosophical point of view one may describe the same atom or the same organization "as a substance, composed of two substantial coprinciples, known as prime matter and substantial form in the technical jargon of philosophy."[17] Both descriptions are equally true, Clarke says, equally true in the sense in which it is equally true that a Mr. Smith is a Democrat, a wheat farmer, and a Baptist all at once. As physical science has a theory of matter according to which an atom consists of nucleons and electrons, so Thomistic philosophy has a theory of matter according to which the same atom consists of prime matter and substantial form.

4. The Thomistic Theory of Matter

The assertion that every atom consists of prime matter and substantial form has its origin in an application of commonsense experience to physical and biological science. Greek philosophers, like Plato and Aristotle, compared objects that we find in nature, such as stones, plants, and animals, with objects made by human artisans. If a sculptor produces a cube in marble, we can easily distinguish between the material, marble, and the form he gives to it, a cube. The marble is passive, it serves only as the material on which the sculptor works; but the cubic form is active, since it presents a goal that the sculptor attempts to achieve. If we apply a more subtle analysis to this conception of "marble" as passive material, we easily find that it is not completely passive. Its parts are held together by forces of cohesion that act upon them. Hence Aristotle described a piece of marble as consisting of "prime matter" or "primary matter," which is completely "passive" and "inert," and a "substantial form" to which belongs not only the form or shape in the geometrical sense but also all "forces" that act upon the marble—not only cohesion but also repulsion (impenetrability), gravity, electromagnetic field, and so on.

In his book *Modern Thomistic Philosophy*, Richard Percival Philipps[18] presents the theory of matter in three propositions: (1) There is in bodies a substantial material principle and a substantial formal principle; (2) both these principles are incomplete substances; and (3) the material principle has the same relation to the formal principle as potentiality to actuality. Philipps explains these last two terms with the following example: when we say "John can read," John has the "potentiality" of reading; when, however, we say "John is reading," he has the actuality. To explain the terms "material and formal principle," Philipps offers a popular example: "The flour, raisins, eggs, etc." the author writes

> are the material principles of the plum pudding, but these have to be combined in the proper proportions, mixed and boiled in a certain way in order to obtain a plum pudding. The result of such combination, mixing, etc., is the formal principle of the plum pudding, making it . . . different from all other kinds of pudding.[19]

Such popular examples are in some respects misleading. The "prime matter" of Thomistic philosophy is not the same concept as that of "matter" in everyday language and elementary science. The elements—earth, fire, water, etc.—are all the same prime matter in different forms. "Prime matter" means what is common to all kinds of matter in the ordinary sense of this term. If a change takes place in which one kind of matter is transformed into a different kind, we speak of "substantial change." The existence of such changes, for St. Thomas and his school, proved the coexistence of matter and form in every material body because in every case the form changed while the matter remained the same. For medieval philosophy, the conversion of wood into ashes or into fire was a substantial change. Every death converted living matter into inorganic matter. In every material body, prime matter and substantial form are not only conceptually distinguishable, but according to a theorem of Thomistic philosophy, "even in inorganic bodies prime matter is physically distinct from substantial form."[20] This was evident from the existence of substantial changes where actually the form changed while prime matter remained unchanged.[21]

The rise of modern science brought about a new conception of combustion. According to the new chemistry, the same types of matter existed in the ashes and the fire and in the fuel and the air. No substantial change took place and there was no reason to assume that wood and ashes contained the same prime matter under different forms. Along with the advance of the physical and biological sciences, it became obvious that it would be difficult to tell in each individual case whether what was taking place was a "substantial change" or just a change in the positions of matter. If hydrogen and oxygen combine to form the compound water, the Thomistic philosophers have not all agreed as to whether this is a "substantial change" or if the forms of the constituents continue to exist in the compound. Colligan, in his book on *Cosmology*,[22] writes that the transmutation of chemical elements and the formation of organic compounds from their constituents are certainly substantial changes but that the formation of an inorganic (nonliving) compound, like water, from its components is "probably" only a substantial change.

As a matter of fact, to speak in the language of science, the concept of "substantial change" has no operational meaning. The situation became still more complicated when isotopes and nuclear reactions were discovered. There has been much discussion among Thomists as to whether these transmutations are substantial changes or not. Fortunately, for the social significance of Thomistic philosophy, it is irrelevant whether or not "substantial changes" really exist. What matters is only whether one can prove that "prime matter" and "form" are physically distinct from one another in every material body.

5. The Social Significance of Thomistic Philosophy

If Mr. Smith is a Democrat, a wheat farmer, and a Baptist all at once, it depends upon the circumstances whether we prefer to describe him by one of those words. When we canvass votes, it is important to know whether or not he is a Democrat; but if we intend to provide food for India, it is important to know whether or not he is a wheat farmer. Each description is equally true. As we learned recently in section 3 from Clarke, it is equally true that our material world consists of atoms and organizations of atoms and that it consists of prime matter and substantial form. If we are interested in making an atomic bomb, it is practical to describe matter by atoms and their nuclear structure, and this description is certainly of high social significance. At the same time, however, as Clarke writes,

> But to know the essential structure of an atomic substance (i.e. the description of matter and form) is to know the basic formula of the irresistible evidence for the existence of God. And that knowledge is, I submit, also of colossal social significance.[23]

Clarke compares the technological usefulness of nuclear physics with the moral usefulness of the Thomistic philosophy. Each is an example of social usefulness or significance.

In some respects, this Thomistic view of science and philosophy is no different from the pragmatic view that we presented in the chapter on the sociology of philosophical interpretations of science (see Part I, chap. 4). According to this pragmatic view of science, one distinguishes between the technological and the moral purposes of a theory. The difference between the two consists mainly in how the moral "truth" is used. In order to better understand this point, consider how the Thomistic description of nature is used to derive statements about God, the angels, and other concepts highly valuable for supporting rules of desirable human conduct. Perhaps the most important argument in this line is the Thomistic proof for the existence of God, based upon the Thomistic theory of motion. "Motion" in the Aristotelian and Thomistic terminology means every type of change, from hot to cold, from red to green, from life to death; it includes as a special case "local motion," which coincides with our common concept of motion. The characteristic definition of motion in this philosophy is "the passage from the potential to the actual," or "the actualization of the potential." However, a body cannot perform this passage by itself because it requires the introduction of a form into prime matter that is apt to receive it. Since prime matter is completely passive and inert, it contains only the potentiality of change; the passage to the actual must occur, as Étienne Gilson writes, "under the impulse of an act that is already realized" or, in other words, motion originates in "an imperfect act that is being completed."[24]

From this we conclude that "whatever is in motion is moved by another." This proposition, introduced by Aristotle, is the starting point in a Thomistic proof for the existence of God. We find it in St. Thomas's *Summa Contra Gentiles* where, drawing heavily on Aristotle, Aquinas writes the following:

> Whatever is in motion is moved by another: and it is clear to the sense that something, the sun for instance, is in motion. Therefore, it is set in motion by something else moving it. Now that which moves it is itself either moved or not. If it be not moved, then the point is proved that we must postulate an immovable mover: this we call God. If, however, it be moved, it is moved by another mover. Either, therefore, we must proceed to infinity, or we must come to an immovable mover. But it is not possible to proceed to infinity. Therefore, it is necessary to postulate an immovable mover.[25]

This mover cannot be a body because an immovable body cannot produce motion in other bodies. The "prime mover"—already introduced by Aristotle—is thus identified by St. Thomas with the biblical God.

It is important to understand that this and similar proofs in Thomistic philosophy are not based upon scientific analysis of our experience but on metaphysical analysis. Science eventually led to Galileo's and Newton's law of inertia, according to which a body that possesses a certain speed can move without being moved by another body. This law would invalidate the Thomistic proof for the existence of the prime mover. However, we must keep firmly in mind that the philosophical analysis of experience, in terms of matter and form, actuality and potentiality, is completely independent of the advances in science. Speaking in terms of physics, even if we accept the law of inertia, it is possible that "metaphysical analysis" in the Thomistic sense leads to the proposition that nothing can be moved unless it is moved by something else. Some Thomists interpret the law of inertia by assuming that a moving body does not keep its speed without action from something else but would lose its speed unless at every instant God gave a push to the body. By this and similar interpretations, one can accept Newton's laws and mechanics as technologically valuable, while sticking to the Aristotelian and Thomistic assumption that a body cannot move without getting movement from something else.

6. On Angels and Genuine Laws

Before we ask how the Thomistic proof for the existence of a prime mover and similar proofs are different from proofs occurring in science, we shall present other proofs given by Thomistic philosophy and regarded as socially useful. In order to give to the structure of the universe a more balanced character, for example, St. Thomas introduced between God and man the conception of "Angel." In his *Treatise on the Angels* he writes the following:

> We consider the distinction of corporeal and spiritual creatures: firstly, the purely spiritual creature which in Holy Scripture is called angel; secondly the creature wholly corporeal; thirdly the composite creature, corporeal and spiritual, which is man.[16]

Like everything in the world, angels consist of matter and form. To understand these two fundamental concepts, it is instructive to learn how St. Thomas applies them to the angels. The main problem is whether spiritual substance consists of both prime matter and form. If this were so, corporeal and spiritual substance would contain the same prime matter. St. Thomas, however, argues:

> But one glance is enough to show that there cannot be one matter of spiritual and corporeal things. For it is not possible that a spiritual and a corporal form should be received by the same part of matter; otherwise one and the same thing would be spiritual and corporeal. Since this is not possible, there would be one part of matter with a spiritual and one with

a corporeal form. Since matter is quantity, the spiritual part must also be divisible, which is not possible.[27]

Therefore, it is impossible that corporeal and spiritual things should have the same matter, and it has been proven that an angel has only "form" and no "prime matter."

We shall understand the Thomistic way of arguing better when we glance at how St. Thomas derives the properties of angels,[28] in particular the problem of "whether the motion of an angle be instantaneous." According to the laws of Aristotelian mechanics, the speed of a body is proportional to the "power of the mover." St. Thomas writes, "But the power of an angel moving himself exceeds beyond all proportions the power which moves a body. . . . If therefore a body is moved in time, an angel is moved in an instant."[29] To this argument, he answers,

> The speed of the angel's motion is not proportional to the quantity of his motion, but to the determination of his will. . . . The time of the angel's motion . . . will have no proportion to the time measured by the motion of corporeal things.[30]

In other words, the meanings of the words "power," "speed," and "time" applied to an angel are not the same as when they are applied to material bodies. Therefore, the laws of motion that are valid for bodies like launched stones cannot be applied to an angel that is moving "under his own power." The same holds for the concept of "being in a place." as St. Thomas writes:

> It is evident that "to be in a place" appertains quite differently to a body, to an angel, and to God. For a body is in a place in a circumscribing fashion, since it is measured by the place. An angel, however, is not there in a circumscribed fashion, since he is not measured by the place, but definitely, because he is in one place in such a manner that he is not in another. But God is neither circumscriptively nor definitely there, because he is everywhere.[31]

St. Thomas mentions an interesting objection involving the infinite speed of light in order to make it plausible that an angel could move with infinite speed. He replies that "illumination is the term of a movement; and is an alteration, not a local movement. . . . But the angel's movement is local, and, besides, it is not the term of movement; hence there is no comparison."[32] We note again that words like "place," time," "speed," etc. have different meanings when applied to a moving angel than to moving stones. Nevertheless, the word "place" in sentences referring to an angel and the word "place" in sentences referring to a material body are not homonyms or "equivocal" to each other. Two equivocal words like "rubber" as an elastic material and "rubber" as in a game of cards have completely different meanings; the word "rubber" is equivocal. But "place" referring to an angel and referring to a moving stone has two meanings with a certain range of meaning in common. The word "place" is not used equivocally, as is the word "rubber," but rather "analogically," as some scholastic philosophers might describe it. Some sentences formed from such "analogical" words make just as much sense when they are spoken about an angel as when referring to a stone (e.g., "If an angel moves from one place to another, it has a speed and requires time").

7. Thomism and Physical Laws

We shall now give a second example in which analogical ways of speaking are used to prove important propositions of Thomistic philosophy. One of the most important points in a philosophical interpretation of science is the conception of "physical law." We presented three conceptions of physical law that were discussed by Whitehead (Part I, chap. 3, §6). One of these conceptions was the "imported law": law is an order imposed on nature by a lawgiver.[33] Thomistic philosophy has attempted to prove that physical laws are actually imposed. It is very instructive to study the nature of the proof. We shall follow the presentation of a textbook used in colleges where Thomistic philosophy is a part of the prescribed curriculum.

The author starts from the definition that "physical law is the fixed inclination with which irrational beings are endowed for the regulation of natural events."[34] This definition contains the analogical term "inclination," which means verbally a state of mind and can obviously be attributed only to living beings. But our Thomistic definition speaks of the inclination of an "irrational being," which may even be a stone.

The "inclination of a stone" makes sense only if we take "inclination" analogically or, as Charles Dickens said, in a "Pickwickian" sense.[35] By observation and inductive inference, we find uniformities in nature but no inclinations, as David Hume strongly emphasized. Therefore, in the sense of our textbook definition, "*Uniformity* of activity is not by itself a law, it is the effect of a law."[36] In the usual language of scientists, mention is often made of Newton's laws or Ohm's law. But these statements or formulae refer to uniformities and not to "laws" in the sense of our present definition.

In order to avoid using the "law" in an equivocal sense, Thomistic philosophy speaks of "genuine laws" if one refers to laws in the sense of our definition. "Genuine Law," our text states, "includes in its concept a superior and a subject; a genuine law is imposed. For that reason we say that natural objects are endowed with certain tendencies."[37] Then, Thomistic philosophy claims to prove the following: "The activities of natural bodies are governed by Genuine Laws."[38] We shall attempt to present this proof in an abridged form. According to Thomistic philosophy, study of the uniformities of nature belongs to the "special sciences" like physics, biology, and so on. By means of the data of these sciences, the existence of genuine laws is not proved but neither is it excluded. Our textbook reads:

> If we advance beyond the point at which science stops, that is only because philosophy should have no justification if it contributed to our knowledge nothing more than is contributed by the restricted natural sciences.[39]

What Thomistic philosophy adds to the "restricted natural sciences" is proof that observed phenomena are regulated by laws that emanate from an intelligence and are imposed by a lawgiver. The proof starts with a generalization of our observation. We find the characteristic property that

> the activities of our material world are highly rational. They fill the mind of man with wonder at the intelligence displayed. . . . Since material beings have no intelligence of their own, it follows that the intelligence they display is not their own. The case is parallel to that of a machine which displays the intelligence of the maker.[40]

This proves that material bodies act according to a plan that is imposed by an intelligence from the outside. On the other hand, they act according to an internal necessity; therefore the plan is implanted in their very nature. "There it is imposed by the One Who has supreme charge of the material world,"[41] in other words, by the Creator of the physical world, by God. According to Thomistic philosophy this is the only way to understand matter. The theory leads necessarily to the result that matter is governed by genuine laws. If we add this argument to the proof for the existence of a prime mover, the belief in God is supported as well as the moral rules that traditionally have been connected with this belief. There is no doubt that as a philosophical interpretation of the scientific theory of matter this argument is of high social significance.

However, it is instructive to investigate the logical structure of this argument if it is regarded as proof that general laws regulate physical uniformities. It begins with the remark that our empirical world displays a high degree of rationality. This means, on the one hand, there are many uniformities that are governed by simple mathematical laws (e.g., Newton's law of gravitation) and, on the other, there are a great many phenomena that can be accounted for simply by ascribing to them a simple purpose: the physical phenomena in the human eye, for example, serves the purpose of seeing. Both kinds of "rational behavior" have one thing in common: it is possible to see in the general physical behavior of bodies similarities to the behavior of bodies in our everyday experience. The law of gravitation can be interpreted as a combination of our daily experience of push and pull; the physical phenomena in the eye can be interpreted as being similar to the work of an artisan who manufactures a certain gadget. In other words, the physical universe resembles the product of an intelligent craftsman.

Now the terms "intelligence" and "craftsman" are used in an analogical way. They mean something different from (but similar to) these words when applied to human beings. As an intelligent craftsman produces a gadget, a superior intelligence and a superior craftsman produces the physical universe, regarded as a machine or gadget in the analogical sense of the words. The Thomistic argument is based essentially on the use of the analogical way of speaking and the analogical way of drawing conclusions. We take for granted that what it is true for human intelligence and manmade gadgets is also true for craftsmen, intelligence, and gadgets on a higher level. The analogical character of the terms used is seen clearly from the fact that after it is shown that the world is made by its Creator, as a machine by its manufacturer, every effort is made to prove that the world isn't really a machine in the scientific sense of this word. The world is a machine only in the analogical or, as we previously mentioned, Pickwickian sense.

8. Analogical and Scientific Thinking

The prominent Thomistic philosopher Gerald B. Phelan writes, "The importance of analogy in the philosophy of St. Thomas literally cannot be overestimated."[42] As we saw in the previous examples, the most important arguments make use of analogical thinking: from the fact of our experience that a machine is always made by an engineer, and the fact that our universe is in some respects similar to a machine, we conclude that there is a Creator who is in some respects similar to an engineer who "made" the universe—where the word "made" is analogical to the identically sounding word in the sentence, "an engineer made a machine." Although "machine," "engineer," and "universe" are terms that we often use to describe our daily life experiences, the concept of the "Creator of the universe" is produced by our analogical argument.

Because of its great importance, we shall discuss this concept in more depth. Although a great deal has been written about it by philosophers, we can do it only in an abridged and perfunctory way. According to a frequently used definition, analogy is the mean between *univocity* and *equivocacy*. In the two statements "an oak is a tree" and "an elm is a tree," the word "tree" designates one and the same thing; it is used in a "univocal" way. But in the sentences "I climb a tree" and "At night I put my shoes on a tree," the term "tree" has different meanings. It is used in an "equivocal" way, since "tree" in the second sentence means a "shoe tree." But if we compare "the oak is a tree" and "Eve ate from the tree of knowledge," the term "tree" is used in an analogical way because a tree in the ordinary sense of the word could give no knowledge to Eve.

The *Catholic Encyclopedia* attempts to familiarize the reader with the concept of "analogy" by relating it to concepts from daily life. The author discusses as a simple example the "analogy of proportionality":

> Two objects may be related to each other not by a direct proportion but by means of another intermediary relation.

For instance, the numbers 6 and 4 are analogous in the sense that 6:3 = 4:2; six is the double of three and four is the double of two.[43]

Six and four are obviously different but analogous because they play the same role in the proportion 6:3 = 4:2. This simple example shows, of course, that by the numbers 2, 3, 4, and the "proportion of analogy," the number six can be defined. But this number, like the numbers 2, 3, and 4, can be understood in terms of everyday life concepts, while in Thomistic metaphysics concepts like "Creator of the Universe" are products of analogical thinking. They are entirely new concepts and cannot be defined on the basis of everyday language.

Phelan points out correctly that the method of analogy cannot be expressed by using only concepts that can be made clear by relating them to daily life experiences. He writes, "Those who have attempted to express it in clear and distinct ideas have sinned against intelligence; for, clear and distinct ideas banish mystery and bring death to metaphysics."[44] We can easily understand how analogical thinking brings mystery into metaphysics. To say that every material thing consists of prime matter and form is analogous to saying that the product of a sculptor consists of the material (e.g., marble, and the shape which his skill gives to it). The latter statement describes events of our daily life experience in commonsense language, in "clear and distinct ideas," to use the famous words of Descartes.[45] But the concepts of "prime matter' and "substantial form" are introduced by analogy with those commonsense concepts. Because they are not directly reducible to commonsense experience, we say that they contain a "mystery." In a similar way, we have the commonsense statement that the universe has a Creator.

In order to understand the relation of Thomistic philosophy to science, it is instructive to consider the similarity and the difference between analogical and scientific thinking. It is certainly true that the search for analogies has played an important role in the evolution of physical science. An example on a large scale and of great importance is "mechanistic" philosophy. The motions of medium-sized bodies, launched stones, cannonballs, etc., and even the motions of the planets around the sun can nowadays be described by rules that contain only terms of commonsense language, provided that we extend the concept of daily life experience so far as to contain the phenomena governed by simple applications of Newton's laws. Mechanistic philosophy maintains that optical, electromagnetic, and chemical phenomena are in fact governed by laws that are analogous to mechanical phenomena in the narrower sense of this word. We then obtain statements in which terms like "intensity of light" or "density of electromagnetic energy" occur.

If we applied only analogical thinking, these concepts would be defined only by analogical argument; "electric energy" would be a concept not clearly reducible to concepts in commonsense language. There would be a "mystery" in "electricity," as there is in "substantial form" and "Creator of the universe." Instead the scientist introduces "operational definitions" or "semantical rules" by which he defines new terms like "light intensity" or "electric energy" by describing the physical operations that measure these quantities. The result of every measurement is a reading on a scale and can always be described in ordinary language. "Intensity of light," for example, can be measured by reading a photometer and "electric energy" by reading a voltmeter. The concepts introduced by analogies are all connected with statements about daily life experience and defined by expressions in the commonsense language. Then, statements in which these expressions occur can be checked by experiments that can be described in commonsense language. All experiments about electricity, for example, can be described in the language of elementary mechanics.

The situation is quite different in metaphysics. It starts at the commonsense level, like science, but if by analogical thinking metaphysical concepts are reached, those concepts are not "connectable" with commonsense statements; there are no operational definitions of metaphysical concepts, and you cannot return to the commonsense level. There is no operational definition for "prime matter" or "Creator of the universe." As Jacques Maritain, one of the most prominent advocates of Thomistic philosophy, told a symposium in which the relation between science and metaphysics was under discussion, the main difference between science and metaphysics consists in the fact that science uses only univocal predicates, while metaphysics uses words in an analogical sense. He meant that many propositions with a clear meaning on the "lower level" of daily life experience are used to designate analogous propositions on a "higher level" where they can no longer be checked by observations formulated in commonsense language.[46]

This difference between science and metaphysics was well formulated by Ernst Mach in his paper "On the Principle of Comparison in Physics."[47] He pointed out that a great many discoveries in science have been made by comparisons or analogies. The propagation of heat and light, for example, was treated as an analogy to the motion of a fluid. Even today the nucleus of an atom is treated as analogous to a drop of fluid. Yet these more or less vague ideas are not scientific until formulated in such a way that, by means of operational definitions, conclusions can be drawn from them about possible observations that can be described in commonsense language. Mach wrote, "Not the dim, half-conscious *surmises* of the acute observer of nature . . . belong to science, but only that which they possess *clearly* enough to *communicate* to others."[48] But whatever the role of analogical thinking may have been in advancing science, this way of thinking has undoubtedly produced a picture of the physical world that for centuries has had a great impact on the conduct of human beings.

Chapter 8

The Physical Universe as a Symbol

1. The Moral Universe

While the picture of the physical world developed by physical science has radically changed since the times of Plato, Aristotle, and St. Thomas, the picture outlined by Thomistic philosophy has not changed. The foundations of the Thomistic doctrine have been and remain today as follows: first, the doctrine that every material body in the commonsense meaning of this word consists of "prime matter" and "substantial form" and second, the doctrine that the laws of nature are not only uniformities but "genuine laws" that are imposed by a lawgiver. In close connection with these doctrines is the general feature of Thomistic philosophy that was described earlier (Part II, chap. 7, §3) as "an analogy of being in the vertical dimension," according to which all physical phenomena are analogous to the phenomena exhibited by living organisms. This is also an analogy to commonsense experience, since the main constituents of this experience are the behavior of simple mechanisms and familiar rules about the behavior of our fellow men. All this amounts to the doctrine that our physical universe, which according to physical science can be understood by a search for causal or statistical physical laws, is from the philosopher's point of view operated for certain purposes "for common good," as Thomistic philosophers say. The universe is not a dead machine but a society of living beings that is somehow similar to human society and can be used as an ideal to show how human society should operate. This means that physical phenomena can be interpreted not only by the traditional laws of physics but also by moral laws.

This interpretation of the physical universe as a moral universe began in antiquity in the writings of Plato and was vigorously continued in medieval philosophy, with particular lucidity in the work of St. Thomas. It is easy to understand that in medieval poetry this moral interpretation of the physical universe was a favorite topic. The most famous example is Dante's *Divine Comedy*,[1] in which the physical universe consists of Hell, Purgatory, and Paradise and is the place where rewards and punishments are handed out according to the moral merits of human beings.

The great American philosopher George Santayana, in his essay on Dante,[2] provided a lucid and instructive presentation of this medieval universe. Santayana directs our attention to the fact that the mere distinction between "prime matter" and "substantial form" introduces a moral element into the physical world. Natural lines of cleavage were obliterated and, Santayana writes, "moral lines of cleavage" were substituted. "Nature was a compound of ideal purposes and inert matter . . . or . . . a compound of evil matter and perfect form. . . . Evil was identified with matter."[3] This doctrine has infiltrated a great many philosophical interpretations of science. Even in our twentieth century, many philosophers and philosophically inclined scientists have extolled Rutherford's theory of the atom because its largest part is empty, and only very tiny pieces of matter (nucleus and electrons) remain. Like medieval philosophers, these contemporary authors believe that by reducing the quantity of matter in the physical universe the world is purged of evil.

For Dante, Santayana writes, "moral distinctions . . . are displayed in the order of creation. The Creator himself was a poet producing allegories. The material world was a parable, which he built *out of space and ordered to be enacted*."[4] The idea of medieval philosophy, Christian as well as Jewish and Muslim, was that nature's purpose is to produce "the good" and that physical laws are symbols by which moral laws are revealed to man. In order to put this in the right perspective, we must not forget that this conception of physical laws originated not in medieval scholastics but far back in antiquity. Santayana pointed out that this moral interpretation of physical phenomena was presented by Plato in his dialogue *Phaedo*.[5] Socrates, the spokesman for Plato in the dialogue, complains about the way in which philosophers (in our language, scientists) claim to explain the features of our physical universe—whether the earth is round or flat, how the earth is supported in its place, and so on. The philosophers disappoint him because they try to explain everything by what we would call causal or physical laws, while he expected them to prove that the actual physical universe is "the best" in the moral sense of this word. According to Santayana, this Platonic philosophy that later became the basis of medieval thought insisted that "the world is a work of reason. It must be interpreted as we interpret the actions of a man, by its motives."[6]

How Plato rejected causal or physical explanation is shown in the *Phaedo*. Socrates said:

> I imagined that he (the philosopher Anaxagoras) would tell me whether the earth is flat or round, and whichever was true he would proceed . . . to show the nature of the best, and to show that this was the best; and if he said that the earth was in the centre [of the universe], he would further explain that this position was the best, and I should be satisfied with the explanation given, and not want other sort of cause. . . . What expectations I had formed and how grievously was I disappointed. As I proceeded, I found my philosopher altogether forsaking Reason . . . but having recourse to air, and ether and water and other eccentricities.[7]

It is illuminating that the Platonic Socrates calls physical causes "eccentricities" in contrast to explanation by purposes.

2. Physical Science in the Bible

We can, however, go much further back in time to find examples of the symbolic role of the physical universe. We are all accustomed to relax every seventh day of the week, and a great many of us even every sixth and seventh day. The root of the habit is the Old Testament's theory of how our universe originated. Hebrew cosmology has been interpreted through the ages as a rule of human conduct, and it has not outlived its validity today. Among liberal theologians, it has become habitual to say the Bible is not a textbook of physical science but a guide to moral behavior. But it would be more correct to say the Bible *is* a textbook of physical science that has lost a great deal of its technical value but retains its social value to a high degree.

From careful study of the book of Genesis, we see that even the law of causality itself was interpreted by Old Testament writers as an incentive for moral conduct. After the Great Flood, God says (Gen. 9:11) to Noah and his family, "I established my Covenant with you and the waters shall never again become a flood to destroy all flesh." And (Gen.8:22), "While the earth remains, seedtime and harvest, cold and heat, summer and winter, day and night shall not cease." God pledged after the Flood that the law of causality would remain valid eternally unless man behaved in such an immoral way that God would punish him by abolishing causality. God behaves like a constitutional monarch who proclaims a constitution and promises not to violate it—except in case of emergency.

The fact that in the Bible the law of causality is regarded as a reward for the good conduct of man becomes completely clear from passages in the Talmud,[8] a commentary on the Old Testament that contains the oral tradition of the Jewish people:

> Said Rabbi Yochanan: "When the Holy One, blessed be He, created the first man, He made him Lord about everything, the cow obeyed the plowman, and the furrow obeyed the plowman. Adam having sinned, they rose up against him; the cow no longer obeyed the plowman, nor did the furrow obey the plowman. With Noah they quieted down.

Note that the reaction of the cow to man, a law of animal psychology, has equal footing with the reaction of the furrow to the plow, a physical law. In the same way, the laws that governed the motion of liquids served moral purposes:

> Rabbi Simeon ben Lakish said: "Before Noah the waters used to rise twice a day, once at morning prayer and once at evening prayer, and they used to flood the dead in their graves. With Noah, the waters quieted down."

Instead of the ancient Hebrew world picture, consider Copernicus's picture of the physical world and what we read in his most important book: "As though seated on a royal throne, the sun governs the family of planets revolving around it."[9] The planetary system is described by analogy with the system of government in a monarchy. We must remember that Copernicus used this analogy to bolster his heliocentric system. "As a quality," he wrote, "immobility is deemed nobler and more divine than change and instability, which are therefore better suited to the earth than to the universe."[10]

3. The Physical Universe and Human Behavior

All these examples lead to the general conception of the physical universe as analogous to a human society by which the universe becomes a useful symbol for encouraging desirable ways of life. Ernst Topitsch named this conception the "socio-cosmic" universe. He collected from a great many sources examples of this conception not only among the Hebrews and Greeks but in China, India, and Mesopotamia as well. "The whole world," Topitsch summarizes these examples, "is conceived as a state, a city, or a well-ordered household, and the regularities of nature correspond to the rules that govern civil life."[11] From this conception it became possible to derive the natural or ideal structure of human society: the structure that corresponded to the structure of the physical universe.

By the same argument, the history of mankind was also regarded as determined by the motions of the stars. According to the original idea of astrology, the stars did not determine the life of individual persons but only historical events. Obviously the life of kings belonged in this sphere, but the sale of "horoscopes" to indiscriminate individuals belongs to a decadent phase of astrology. In his book on Aristotle,[12] Werner Jaeger stresses the close correspondence between Greek ideas in astronomy and their ideas about the moral order in the city-state. Topitsch summarizes Jaeger:

> Human laws should be obeyed, because and insofar as they were a part of the rules governing the cosmos, which became by this reflection the pattern of the "right order" in society, the metaphysical foundation of the city-state morality.[13]

It would be a great mistake, however, to believe that this symbolic use of the physical universe has been abandoned in our scientific century. When atomic physics abandoned strict determinism as the supreme law for the motions of the smallest particles, this was used to symbolize freedom of human actions. We need only look again into *Fate and Freedom* by Jerome Frank to see the moral conclusion drawn from the "nondeterministic" character of the physical world picture in the twentieth century.

For another example, we can consider the philosophy of dialectical materialism that became the official doctrine of all Soviet governments.

The most palpable of its principles is the transition from quantity into quality: "If a property increases quantitatively more and more, a point will come at which the property undergoes a change in quality." The most familiar example in physics is the heating of water. As its temperature increases, the water remains for a time qualitatively unchanged; that is, it remains water. But at the boiling point, water is converted into a body of changed quality: water vapor.

Physicists would not see in this account of boiling much help for advancing the theory of heat. But Marxist philosophers point out that it clearly shows the analogy between the physical universe and social behavior. If in a human society the accumulation of the means of production (capital, machinery, etc.) increases, the character of the society changes only quantitatively, the society remains qualitatively a society of private owners. But if we apply the physical theory in its dialectical presentation, we can expect that after great quantitative changes, after great accumulations of capital in a few private hands, a qualitative change will take the means of production out of private hands and will become community property. This prediction of the coming collectivization of private property gets its strength and reliability from laws governing the physical universe. If we look "understandingly" at physical phenomena like evaporation, we feel immediately, according to dialectical materialism, how properties like temperature increase without qualitative change, until the accumulated drive toward a qualitative change becomes so strong that it frees itself in a sudden jump. Thus the necessity of a social revolution manifests itself to everyone who seeks a real understanding of changes in the physical world.

If we analyze this conception of the physical universe as a symbol for rules of human behavior, we run into difficulties that are evident to anyone who has done some "hard thinking" in the logic or epistemology of science. Again, if the picture of the physical world serves as a symbolic source of rules for human behavior, the shortest way to present this rule is to say that the physical universe is an ideal human society; every real human society attempts to emulate this ideal, and this striving becomes the rule of its moral behavior. If we describe this situation as a whole, we first note that our picture of the physical universe is created by the human mind as an image of our actual human society. But second, once this image is firmly established, it becomes an example by which human society attempts to become an ideal society. This, however, is a clear example of circular thought: man creates the universe in the image of his own social organization and then shapes his organization according to the image that he himself has created. It is logically clear that from this procedure no feasible rule for human behavior can be derived.

Topitsch gave special attention to this circular reasoning within attempts to derive moral rules from concepts like "cosmic order," "law of the universe," "natural form of society," and so on. He writes:

> These procedures lead into purely analytical propositions which are either "eternal truths" because they are disguised definitions or "eternal problems" because they are self-contradictions. These empty formulae have, or course, neither factual nor normative content and therefore neither descriptive nor prescriptive meaning.[14]

Nevertheless, the universe as a symbol has historically played a great role in shaping human behavior. In the hour of decision there is always a specific picture of the universe in the mind of man, while the larger chain of reasoning that reveals its circular character is never present at the same time to the conscious department of his mind.

In the nineteenth and twentieth centuries, the age of science, many authors attempted to make "science" the only valid basis of human thought and to eliminate metaphysical speculation. These attempts culminated in the doctrines of positivism and pragmatism. However the "sociocosmic" world conception, the interpretation of the physical universe as an image of a perfect human society, has never been completely eradicated from the human mind. We shall see later on (Part II, chap. 11) how the Vienna Circle proved metaphysics to be logically meaningless; but it was not actually exterminated. Metaphysics remains alive in the sociocosmic conception of the universe.

4. Scholastic "Scientism" and Modern "Positivism"

It would be entirely wrong to believe that the philosophy of St. Thomas, now the dominating doctrine of the Roman Catholic Church, was the only philosophical doctrine advocated by "schoolmen" who attempted, as did St. Thomas, to build up a system close to Aristotle but also compatible with the teachings of the Catholic Church. Even during his lifetime, Aquinas struggled against a doctrine, later called "scientism," that led to a breakdown of scholastic philosophy before the end of the fourteenth century.

Aquinas proved by one and the same method laws of physical science and the properties of angels or devils. This unifying method was described in medieval philosophy as the method of "reason." Obviously, the use of reason does not mean in this context the use of logical argument: it means analogical thinking. It was already known in the Middle Ages that this method achieved no practical advance in science. Although little science was known at the time, there was practical astronomy, practical building, and so on, from which it was clear that every law about observable facts has to be checked by sense observation before it can be declared reliable. This means that there existed a fairly accurate version of what we today call scientific method. It was also known that a law proved by reason, by analogical thinking, was not helpful in practical astronomy nor in practical building. On this basis, the view developed that only a proof by "scientific method" and confirmation of sense observations is reliable and not a proof by "reason" or a result of analogical thinking. This view was later called "scientism" by scholastic philosophers.[15] This led to the unreliability of the Thomistic proof for the existence of God and the properties of the angels, as the concepts are not interpreted by operational definitions and not checked by sense observations.

The philosopher William of Ockham[16] declared flatly that no kind of reasoning can prove the existence or nonexistence of real objects. This can only be done by sense observation. He distinguished this "intuitive knowledge" from "abstractive knowledge" obtained by the application of "reason." This means, in our contemporary way of speaking, that only scientific methods can establish the validity of a proposition. Since the method by which St. Thomas proved the existence of God, genuine laws, and the angels and similar propositions is unscientific, his demonstrations are not convincing. Still, Ockham and other opponents of Thomistic rationalism did not conclude that these unproved propositions are actually wrong. Their social significance was so great that they had to be established by methods that did not rely on reason alone. Besides "reason," understood as the combination of logical and analogical thinking characteristic of Thomistic philosophy, medieval philosophy also recognized "revelation." "Revealed" knowledge was given to great prophets, especially to Moses and Christ, by direct act of God; however, anybody could occasionally be so "illuminated" and acquire knowledge that could not otherwise be acquired by reason or sense experience.

Ockham's followers and others who denied any possibility of knowledge not acquired by scientific method had to say that the existence of God and propositions of comparable significance are known to man by revelation. Knowledge by revelation is essentially the same as knowledge by experience because the belief that a great individual has received knowledge is based upon the confidence we have in him as a witness to the revelation. The belief in the doctrine revealed to Christ, for example, rests upon the trustworthiness of the whole chain of witnesses who connect our historical period with the period of Christ. In this way, all human knowledge was reduced to two kinds: knowledge by logical argumentation and knowledge by experience. Experience also took two forms: direct sense experience and the experience of divine revelation. Arguments by reasoning in the Thomistic sense were excluded and what remained was science and sacred theology: that is, knowledge of God by revelation (in contrast to natural theology, which was knowledge of God by reasoning.)[17]

5. Shifting the Problem to Revelation

Roger Bacon, a contemporary of St. Thomas, stressed the point that statements about real existence can only be proved by experience, which included for him direct "illumination" of our minds by God. Bacon stated that

> the reason why philosophical wisdom is reducible to theological wisdom is not only that God has illuminated men's minds for acquiring a knowledge of wisdom, but that they possess wisdom itself from Him, and that He has revealed it to them.[18]

John James Wellmuth characterizes Bacon by writing, "Having thus converted philosophy into theology, he left the field of human knowledge wide open for the sciences."[19] The trend toward scientism reached a climax in the early fourteenth century in Nicholas d'Autrecourt (died 1350). He attempted to limit the field of certainty as much as possible so as to make at least this narrow area secure. He recognized that certainty could come only through experimental knowledge and conclusions based on the principles of ideality and contradiction (deductive logic). He has frequently been called the "medieval Hume" because his opinion reminds one of the famous statements of Hume, printed in 1777:

> When we run over to libraries . . . and take in our hand any volume; of divinity or school metaphysics, for instance; let us ask, *Does it contain any abstract reasoning concerning quantity or number?* No. *Does it contain any experimental reasoning concerning matter of fact and existence?* No. Commit it to the flames: for it can contain nothing but sophistry and illusion.[20]

St. Thomas distinguished between philosophical science (based on reason) and sacred science (based on revelation). Although the existence of God could, in his system, be proved philosophically, the intervention of revelation was nonetheless necessary because the common man could not fully grasp Thomistic proofs. St. Thomas writes the following:

> The truth about God, such as reason could discover, would be known by few . . . and with the admixture of many errors. Whereas man's whole salvation, which is in God, depends upon the knowledge of this truth.[21]

St. Thomas also attempted to explain how one and the same proposition can be proved by reason and by revelation. He compares it, to use our contemporary way of speaking, with the proof of a geometrical theorem by mathematical conclusion and by physical measurement. The first may be difficult to grasp and, therefore, for the layman the second may be appropriate. His actual example is the proof that the earth is round, which can be proved mathematically, abstracting from matter, and by means of matter itself.

St. Thomas puts up a strong argument for the recognition that sacred science is not different in kind from philosophical science: that is, the relation between them is no different from the relation between two sciences, both of which are based on reason. He wrote:

> Sacred doctrine is a science. . . . There are two kinds of sciences. There are some which proceed from a principle known by the natural light of the intelligence, such as arithmetic

and geometry and the like. There are some which proceed from principles known by the light of a higher science: thus the science of perspective proceeds from principles established by geometry, and music from principles established by arithmetic. . . . Hence, just as the musician accepts on authority the principles taught him by the mathematician, so sacred science is established on principles revealed by God.[22]

The brilliant American philosopher and writer Max Otto remarked:

Duns Scotus, who was born probably in the year St. Thomas died, found it impossible for a critical intelligence to harmonize the deliverance of faith and reason. Church doctrines remained inviolable with him, but they were no longer included, together with what was known about the world of nature, in one rational system. The famous pupil of Duns Scotus, William of Occam, carried this tendency farther. He separated the two realms as by a chasm. Theological and philosophical truths were regarded by him as mutually contradictory.[23]

This shifting of problems from philosophy into theology has continued among scientists for a long time and even now plays a considerable role among contemporary scientists. In order to keep science aloof from controversial problems about God, spirits, and immortality, many working scientists have been glad to find a solution to their personal conflicts, claiming that those problems are excluded from scientific thought and from approach by reason. They can only be solved by taking advice by revelation.

Scientific research has been carried out without interference from religious authorities, but scientists have refrained from touching upon these delicate problems, surrendering them to revelation. Even the keenest and most critical scientist may remain an uncritical believer in a field that does not belong to science. Otto writes:

No doubt this theory of dual truth was then and thereafter in some cases a disguise, conscious or unconscious, for a one-sided espousal now of religion and now of philosophy. What it was in a specific instance it is impossible to decide without reading too much between the lines.[24]

For some, shifting the problem to revelation was just a polite way of avoiding a decision that might have been socially dangerous, or at least inconvenient, while others wished, by this shift, to support philosophical propositions that could not be solved by Thomistic argument. Beyond doubt, as a consequence of this doctrine of "double truth," scientists who were particular and fastidious in accepting argument from reason had to appeal more to revelation than scientists who were prepared to swallow questionable philosophical arguments. Skepticism toward reasoning has frequently bolstered belief in revelation.

In all these respects, the founder of modern positivism, David Hume, resembled the late medieval scholastics like Nicholas d'Autrecourt. "All the objects of human reason," wrote Hume, "or enquiry may naturally be divided into two kinds, to wit, *Relations of Ideas* and *Matters of Fact*." To the first category belongs the "relation between figures [like] *that three times five is equal to the half of thirty*."[25] Like Ockham or d'Autrecourt, Hume emphasized that statements about the existence of God or spirits do not belong in either of these categories. But he also pointed out that we have to use analogical thinking in order to "prove" such statements by natural theology, or, in other words, by reasoning. Hume writes:

> The whole of Natural Theology, as some people seem to maintain, resolves itself into one simple, though somewhat ambiguous, at least undefined proposition, That the cause or causes of order in the universe probably bear some remote analogy to human intelligence.[26]

In this respect, not much has changed from the eighteenth century of David Hume to the twentieth century of Albert Einstein. If we look at a few simple mathematical equations that govern the complex phenomena in our universe, psychologically we have the same feeling as when we look at events masterminded by a great human genius. Looking at the universe, we get, according to Einstein, a feeling of awe and admiration he has called "cosmic religion."

These are facts that no one conversant with physical science would deny. Whoever keeps to the scientific method and refrains from reasoning by analogical thinking will stop here and shift the question of whether a personal creator is responsible for this seeming intelligence in nature to sacred theology and "revealed knowledge." Hume was well aware that frequently scientists and other thinkers who do not believe that scientific method can lead to knowledge of the cause of rationality in the universe may appeal to divine revelation as a source of knowledge. Hume wrote:

> A person, seasoned with a just sense of the imperfection of natural reason, will fly to revealed truth with the greatest avidity: While the haughty Dogmatist, persuaded that he can erect a complete system of Theology by the mere help of philosophy, disdains any farther aid and rejects this adventitious instruction. To be a philosophical Sceptic is, in a man of letters, the first and most essential step towards being a sound, believing Christian.[27]

This doctrine of "double truth" has greatly shaped the worldview of scientists and scholars. If we investigate the "philosophy" of scientists in the eighteenth, nineteenth, and twentieth centuries, we frequently find the type who attempts to keep scientific research and teaching free from the interference of reasoning and in agreement with revelation. This means, of course, that the scientist abandons in this field his acquired methods of careful scrutiny and accepts the results of revelation as a whole. Thomas Hobbes characterized this way of thinking by a sentence that many would call "flippant": "For it is with the mysteries of our Religion, as with wholesome pills for the sick, which swallowed whole, have the virtue to cure; but chewed, are for the most part cast up again without effect."[28]

6. Realism and Nominalism

The doctrines of Ockham and other medieval advocates of scientism have usually been presented and transmitted to later centuries as "nominalism." This expression comes from the fact that Ockham and his followers regarded only individual horses, for example, as "things" but did not agree to call "horse in general" without specification of an individual a "thing" They maintained that "horse in general" is a *name* for a class of things but not itself a "thing," as believed by "realists." Modern positivism has often been traced back to medieval nominalism, and one of the first books of the twentieth century, titled *Ockham's Razor*,[29] referred to Ockham's effort to cut all superfluous words from our language. Through such a purge, words that do not designate individual things, like a spot on a screen, but instead designate abstract entities (e.g., "inclination or force") should be declared superfluous and introduced only as auxiliary concepts that can be defined by words that designate "things."

If we remember what we said previously about philosophical and metaphysical interpretations of science, it is obvious that the medieval dichotomy between Thomism and scientism, or between realism and nominalism, appears again in the nineteenth and twentieth centuries as the dichotomy between metaphysics and positivism. We use these two terms because they occur in the history of science in Max Planck's attack on Ernst Mach. All metaphysical doctrines have important features in common with Thomism, and all positivistic lines of approach have something in common with "Machism." August Comte, the first "avowed" positivist, wrote: "The spirit of all theological and metaphysical philosophy consists in conceiving of all phenomena as analogous to the only one which is known by immediate consciousness—Life."[30] This means that analogical thinking is an essential element in all metaphysical doctrines. This holds for such divergent metaphysical systems as those of Plato, Aristotle, St. Thomas, Descartes, Leibniz, Kant, Hegel and—for that matter—Jeans and Whitehead.

7. The Situation in the Nineteenth and Twentieth Centuries

Science began to separate from philosophy, as we have learned, in the seventeenth century. This established "science proper" as a system of thought based upon experiment and logical conclusions but independent not only of "philosophy" but also of the search for intelligible principles and of the ability of our intellect to "see" the essence of things. The remaining parts of the old unified system of science and philosophy became a somewhat unorganized body of knowledge. In the seventeenth and eighteenth centuries, there was no attempt to sever the links with science and to respond to the rise of "science proper" by building up a "philosophy proper" that would be, in turn, independent of science. There is ample evidence of this attitude in the works of eighteenth-century philosophers—men like Leibniz, Voltaire,[31] d'Alembert, or, for that matter, Newton himself. Eventually, however, it became evident that after the separation from science, "philosophy" had become a rather incoherent and doubtful doctrinal body. At the end of the eighteenth century, attempts were first made to investigate how philosophy, isolated from science, was possible and what the logical and empirical structure of this "new" doctrine should be. More and more, the term "metaphysics" came to mean a system of philosophy that established its truth independently of the actual state of science. In contrast to the changing doctrines of science, "metaphysics" would become independent of these changes.

Bishop Fulton J. Sheen, a contemporary authority on metaphysics in the United States, has written the following:

> In the Middle Ages there was accepted a *distinction between science and philosophy, but not a separation.* Physics formed an integral part of philosophy as did astronomy. The basic rational exploration was to be formed in a science from which all other sciences borrowed its first principles, namely metaphysics. . . . The divorce of science from philosophy never actually took place until the time of Kant.[32]

Around 1800, Immanuel Kant, the great German philosopher, published his *Critique of Pure Reason*, which has become one of the landmarks in the history of philosophy and almost a Bible for German philosophers and scientists. He definitely attempted to build up a "philosophy proper," which would be safe against any attack from future progress in science. He wrote, for example, that

> reason is impelled by a tendency of its nature to go beyond the field of experience and to venture . . . by means of mere ideas to the uttermost limits of all knowledge. Moreover, it finds no rest until it as fulfilled its course and established a systematic whole of knowledge which exists by itself.[33]

Kant emphasized that this "systematic whole of knowledge" found beyond experience was a philosophy separated from science, a metaphysics. "Whatever discoveries might be made" he wrote with reference to metaphysics, "we should not be able to make use of them in any helpful manner *in concreto*, that is, in the study of nature."[34] The purpose of metaphysics, therefore, was not to give us theoretical knowledge about the world of observable facts; it was, rather, to save what Kant called the "practical interests of reason." After the rupture of its link with science, "philosophy" became of "practical" importance because it provided rules of human conduct. In a similar way, Kant ascribed to metaphysics a "practical" role in human life.

During the nineteenth century a separate doctrine developed, partly under Kantian influence, that was in addition to (and superior to) science. It was called "philosophy," occasionally "pure philosophy" or "metaphysics," and was taught in all the universities. Students of science soon developed the habit of disregarding these courses and frequently even looking down on this "philosophy." The students of philosophy, in turn, developed the habit of looking down upon the "special sciences" as being merely technological information that provided no "real knowledge" of the world.

This situation has not changed much. In parallel to the tremendous growth of science, there has developed an isolated philosophy or metaphysics that takes special pride in being independent of the radical changes marking nineteenth- and twentieth-century science. Let me quote some characteristic examples of this "isolationist" attitude in philosophy. Mortimer Adler, a prominent advocate of Thomistic philosophy, has written that "philosophy is a body of knowledge, not obtained by any of the specialized methods of empirical research, and consisting of truths which are not dependent for their validity upon the findings of science."[35] Bishop Sheen writes sarcastically about philosophers who take their cue from contemporary science: "To marry the spirit of any age means to be a widow in the next."[36]

It would be a great error to believe that this isolationist attitude belongs only to advocates of medieval philosophical systems. We can also quote, for example, the British philosopher, C. E. M. Joad, usually called a very "progressive" author. He attempts to justify the static character of philosophy as going against the dynamic advance of the sciences:

> If, as I believe, philosophy is largely concerned with an objective and changeless world of value and with the manner of its intrusion into the familiar world which we know by means of our senses, then the objects of philosophy are, as Plato maintained, eternal. . . . The world of "eternality" is always present to men's mind, and there is no reason why the twentieth century A.D. should enjoy a deeper insight into its nature than the fourth century B.C. Hence it is unreasonable

to expect progress in philosophy, if progress is taken to mean the accumulation of a developing body of ascertained and verifiable knowledge.[37]

Thus, as we have seen, the physical universe and its conception play different roles in human societies, including guiding moral, legal, and religious thought and behavior. These roles depend in many ways on how we conceive the relation between science and philosophy, which is the topic of the next chapter.[38]

Chapter 9

Union, Divorce, and Reunion between Science and Philosophy

1. Science and Philosophy in the British and Soviet Encyclopedias

Opinions held during a certain period in a certain area of the world are often represented in the large encyclopedias. We may take our information about the "Western world" from the *Encyclopedia Britannica*,[2] and about the "Eastern world" (or, more precisely, the Western rim of the Eastern world) from the *Large Soviet Encyclopedia*.[2]

The author of the article on "science" in the *Encyclopedia Britannica*[3] states that the "results of science" must serve as a starting point for "more penetrating inquiries." He thinks that when we have pursued science as far as it can go, we must face an

> inquiry into the ultimate nature of reality. But this inquiry lies in the province of metaphysics, and is not necessarily involved in the pursuit of natural science. Metaphysics uses the results of natural science, as of all other branches of learning, as evidence bearing on her own deeper and more difficult questions.[4]

According to this *Encyclopedia*, there is a gap between science and philosophy "because the connection between the special sciences (physics, biology, psychology, etc.) cannot at the present time be answered by science. Now, and perhaps always, it is a problem of 'metaphysical speculation.'"[5] The "isolationist" attitude of the *Encyclopedia* appears in this assertion that the connection between the special sciences cannot be explained by purely scientific methods and that we need "metaphysics" to fill the gap.

Turning now to the article on "philosophy"[6] in the same *Encyclopedia*, we read that philosophy is the "science of the whole" and that "the synthesis of the parts is something more than the detailed knowledge of the parts . . . that is gained by the man of science." Here again the affirmation of a "philosophy proper," distinct from the "special sciences," consists in the assertion that "the synthesis of the parts" cannot be achieved by the methods of the "special sciences." This is, however, by no means self-evident. Physics and chemistry, as we have seen, were once two "special sciences," which were clearly distinct from one another. Chemistry concerned "qualitative changes," while physical changes were called "merely quantitative," as not accounting for qualitative changes. However, the gap between physics and chemistry was bridged by purely scientific methods—by building up two new "special sciences" of physical chemistry and chemical physics. Therefore, to assert that the gap between the special sciences can only be bridged by a "metaphysics" that is not based on the method of science proclaims a belief in the eternal separation between science and philosophy.

If we turn to the *Large Soviet Encyclopedia*, which presents opinions approved by the Philosophical Institute of the Communist Academy in the USSR, we find that it strongly advocates a "philosophy proper" separate from the special sciences. As early as 1908, Lenin[7] himself warned against leaving "philosophical" problems to the decision of the scientists. He wrote: "There is not a single professor who has done valuable work in the special fields of physics or chemistry whose word can be believed when a problem of philosophy arises."[8] The Soviet encyclopedia, like the British one, does not believe that connections between the special sciences can be established by applying scientific methods.

The article on "philosophy" attacks the opinion that a "philosophy proper" is superfluous and that the task of ancient and medieval philosophy can be performed today by a systematization of the special sciences. The article denounces the "positivists and mechanists" who claim that "philosophy can be reduced to a synopsis of the results of contemporary science."[9] We may recall that in the *Encyclopedia Britannica*, philosophy is defined as the "science of the whole" and that the word "whole" can be interpreted in two ways. It can mean the integration of the special sciences using the methods of "science proper," or integration by other, "superscientific" means. The first conception is often called "positivistic."

The article in the *Large Soviet Encyclopedia* denounces "positivistic" groups within Russia in this way: "They declare that philosophy is a leftover from medieval scholasticism and advocate the thesis of bourgeois positivism that science is its own philosophy."[10]

The *Large Soviet Encyclopedia* asserts that formulations of "science proper" claim to result from experience and logical thinking but often contain unconscious metaphysical interpretations that have slipped into traditional scientific terminology. Similar opinions about traditional formulations of science have been expressed by philosophers and scientists who have otherwise differed greatly with one another and with the Soviet writer, such as Whitehead, Mach, and Bergson.[11] Because of the implicit philosophical content of traditional science, positivism is denounced by the Soviet encyclopedia as a "fetishization of bourgeois science."[12] In institutions of higher learning in the Soviet Union, students of science are obliged to take courses in "philosophy proper" designed to neutralize the philosophy hidden within "science proper."

According to this encyclopedia, the "results of empirical knowledge can be systematized only on the basis of dialectical materialism."[13] *This* is the "philosophy proper" taught in Soviet schools. Those scientists who refuse any philosophical integration beyond the special sciences are referred to in the *Encyclopedia* as "mechanists and positivists" who "capitulate to bourgeois science."

These encyclopedias present a certain average of the views held in their respective areas and particularly the doctrines taught in schools of higher learning. It is obvious that in a democratic country like Great Britain, printed books and lectures show a wide dispersion around this average, while in a strictly supervised country like the Soviet Union this dispersion is small. It would be a mistake, however, to believe that the opinions of the most prominent and advanced scientists are close to this average; neither in Great Britain nor in Russia is this the case.

2. "Truce" through a Naturalization of Science

In a number of cases, a philosophy considered to be independent of "science proper" has attempted to control general scientific principles. Everyone knows that the Copernican system was condemned in the seventeenth century, that the Darwinian theory of evolution was banned in the nineteenth, and that the Mendelian theory of heredity has been ostracized in the twentieth. In each case, the action was based on some "philosophical" system that seemed to be true beyond any doubt. However, in all of these cases the ban would not have done much harm were the philosophical system not supported by a political power. This type of support is not an exception but a recurrent phenomenon in intellectual history.

Until the nineteenth century, the most powerful "philosophical" systems were the philosophies backed by the churches. In the nineteenth and twentieth centuries, the influence of "science proper" increased so much that every conflict between it and "philosophy proper" weakened philosophy and did not modify science.

Philosophers themselves attempted to work out a plan for the "containment" of science that would also satisfy "science proper." One of these plans, which seems to show a high degree of logical perfection and psychological insight, is the twentieth-century version of medieval Thomism, known as Neo-Thomism.[14] Its most brilliant representative in the nineteenth century was Pierre Duhem; today, it is Jacques Maritain.

Duhem's starting point was Simplicius,[15] a Greek commentator on Aristotle who defined the task of astronomy. Since the assumption that celestial bodies moved in concentric circles around the earth did not agree with the observed facts—that is, did not "save the appearances"—astronomy presented to Plato and his contemporaries a more general problem: what circular and regular motions must we assume as hypotheses in order to save the appearances presented by the motion of the planets? The Ptolemaic theory of epicycles and the Copernican theory of heliocentric circles both saved the appearances of the planetary motions. But science cannot decide which of two hypotheses is "true" if both save the appearances. The decision must be made, according to Duhem, by nonscientific, superscientific criteria.

The organismic philosophy of Aristotle and St. Thomas decided, for example, that Copernicus must be wrong and Ptolemy right. Duhem claimed the complete independence of science from any philosophy if science restricts itself to "saving the appearances." The formulation of the "true" hypothesis, the selection of principles from which the observable facts can be derived, belongs to "philosophy proper," with the string attached that no contradiction of observed facts may follow from these principles. This established a truce between science and philosophy along the following dividing line: science remained independent in collecting facts and in describing these facts by general statements. Philosophy, in turn, independently selected the "true hypothesis" among those that "saved the appearances." This selection had to be made on the basis of "philosophy proper," or even "revealed theology."

According to Maritain, modern theoretical physics abandoned the search for real causes in order to devote itself to the translation of measurements of things into a coherent system of equations.[16] This search for real causes then became the task of "philosophy proper." Scientists are allowed to collect and describe facts to their hearts' content, but the formulation of the principles of science must conform with the principles of a philosophy independent of the advance in science. In other words, principles of science must agree with the truths of metaphysics. Maritain has said, "We shall take philosophy to mean philosophy *par excellence*, the first philosophy or metaphysics."[17] He admits that philosophy is in some ways dependent on the sciences but in the same way as a gentleman depends upon his manservant. Science collects facts for philosophy as a waiter lays the table and serves the meal to his master.

Along these lines a truce between science and philosophy can be established, but the effect of scientific research upon the general principles of science is checked. Science is, therefore, "neutralized."

3. Attempts at a Reunion by a Positive Philosophy

For different reasons, many scientists, philosophers, and educators of the nineteenth and twentieth centuries approved (and even enjoyed) a separation of philosophy and science. Scientists and philosophers alike have embraced their independence from each other as a condition for unchecked growth. Educators approved the separation because it conspicuously reconciled the free development of scientific research with the preservation of traditional religion and ethics.

We shall later discuss this situation from the humanistic and sociological point of view. But at this point we must mention that in contrast to these approving attitudes of the average scientist and the average university, a protest against this separation emerged at the beginning of the nineteenth century. The mind of the scientist strives toward unity, and educators became concerned lest the strict separation between science and philosophy lead to a permanent separation between the actual conduct and the ethical conduct of man. Such a permanent separation could justify a "hard-boiled" type of man who listens to the sermon on

Sunday and pursues his profit on weekdays without establishing any connection between these two aspects of his life.

After the Napoleonic wars, the "Holy Alliance" of Russia, Prussia, and Austria dominated Europe.[18] The trend favored by these powers of the restoration was directed toward wiping out the frame of mind that produced the French Revolution. This was done by guiding the minds of the people through a return to metaphysics, to a revival of traditional religion. But as early as the 1830s, the scientific spirit of the eighteenth century, the Age of Enlightenment, made its comeback as new bridges between science and philosophy were built, not on the basis of metaphysics but rather that of science.

The most conspicuous and elaborate attempt to restore the unity of science and philosophy was made by the French instructor in mathematics at the École Polytechnique in Paris, Auguste Comte.[19] In his book on French philosophy (1925), the American philosopher George Boas writes:

> Comte must not be imagined to be a lonely figure thinking out great ideas which were ahead of his time. His philosophy on the contrary was a much more eloquent expression of the total civilization of early nineteenth century France than that of any one man.[20]

Comte spoke for a large section of nineteenth-century scientists. In 1830 Comte published his book *Positive Philosophy* from which the term "positivism" entered the history of human thought. For Comte, philosophy was the integration of all the special sciences into one coherent system. Again, the British and the Soviet encyclopedias claim emphatically in the twentieth century that this systematization cannot be achieved by the sciences themselves. They claim that this synthesis, the "science of the whole," requires the intervention of a "philosophy proper" that stands above the special sciences. Comte's philosophy denies this claim. According to him, the systematization can be achieved by exactly the same methods by which the special sciences were built up. Comte was deeply convinced that by advancing his "positive philosophy" he restored the ancient dignity of philosophy as the system embracing all sciences:

> I employ the word *philosophy* in the sense which was given to it by the ancients, in particular by Aristotle, to denote

the general system of human tenets. . . . By adding the word *positive* I mean that I consider the object of the theories to be the coördination of observed facts, whatever system of ideas may be employed for this purpose.[21]

Comte clearly intended to use "philosophy" in the sense that it was used before the separation of philosophy from science. His expression "coördination of observed facts" means approximately what ancient and medieval philosophers called "saving the appearances," or "saving the phenomena." By adding "whatever system of ideas may be employed," he means clearly that philosophy is not interested in the search for "intelligible principles" but only in "saving the phenomena."

The "philosophy" Comte described as a system of knowledge corresponds to what St. Thomas regarded as the inferior type of knowledge that was checked by the "scientific criterion of truth," as we called it in Part I, chap. 2, §7. What became "science proper" after the divorce between science and philosophy did not take over the whole task of philosophy in the ancient sense. "Philosophy proper" or "metaphysics" should go "out of business" and eventually disappear. Comte explicitly rejected the use that was made of the term "philosophy" in the period of separation, writing, "I regret that I have been forced to adopt, for lack of another, a term like *philosophy*, which has been employed and misused in a multitude of ways."[22] For he emphatically denied the possibility of a philosophy separated from science. A later edition (1868) of the *Positive Philosophy* contained a "preface of a disciple," Emile Littré, who confirmed:[23]

> Like the ancient master (Plato) who admitted into his school only pupils who were conversant with geometry, our author (Comte) refused admission to all those who desired to go into philosophy without having passed through science.[24]

Comte intended not a mere "summary" or "synopsis" of the special sciences but a systematization, a derivation from a few general principles. We shall not discuss at this point his hypothesis that human knowledge developed from theological, through metaphysical, to its "positive" state. But it helps us to see that for Comte "positive philosophy" was no summary of observational facts recorded by science but rather their reduction to a coherent system of concepts and laws. He wrote:

The theological system reached the peak of perfection of which it was capable by substituting one unique being for numerous independent divine beings. In the same way, the final state of metaphysics consists in envisaging one great general entity, *nature*, as the unique source of all phenomena. Likewise, the perfection of the positive system will consist in the ability to regard all the various observable facts as particular instances of one general fact, for example, the fact of gravitation.[25]

Littré characterized Comte's goal as "to give to philosophy the positive method of science and to science the unifying idea of philosophy."[26]

4. The Role of "Sociology" in Positive Philosophy

As we know today,[27] general principles of science are not unambiguously determined by observed facts. The coordination of phenomena can be made in different ways or, to speak in medieval parlance, appearances can be saved in different ways. Believers in a separate philosophy ascribed to it the task and the ability to choose among different theories or principles by which observed phenomena could be coordinated. Since Comte claimed that his positive philosophy would take over all the functions of traditional "philosophy proper," it also had to account for this choice. If we consider the traditional so-called special sciences—mathematics, astronomy, physics, chemistry, and biology, in the order suggested by Comte—it is clear that none can account for a choice between two theories in physics or astronomy if both agree with observed facts. Let us consider our now-familiar example: the choice between geocentric and heliocentric hypotheses in the period of Galileo. If one refused the decision of metaphysics, as Comte did, one had to account for the actual choice by other types of argument.

If we consider again the Copernican conflict, we can easily discover by what factor the choice was determined in each individual case. One factor was certainly agreement or disagreement with the predominant metaphysical creed of the period: Aristotle's organismic philosophy. Another was agreement or disagreement with revealed religion, with the text of the Bible as interpreted by the established churches.[28] A third was agreement or disagreement with "common sense." The dominance of a metaphysical creed is closely connected with its being regarded at the time as indispensable for moral life and, in particular, as the basis of good citizenship. What is regarded as "common sense" depends on the stage of intellectual and social evolution that has been reached.

Hence, what determines the choice of a hypothesis in these doubtful cases falls under the common denomination "psychological and sociological factors." Comte did not assert that a decision between two astronomical hypotheses could be found by astronomical methods, nor even that it could be found by the united methods of all the sciences, astronomy, physics, chemistry, and biology; on the other hand, Comte refused to admit extrascientific, metaphysical reasons for decision. Since the actual decision is determined by "psychological and sociological" factors, Comte introduced a new, special science to investigate these factors that he called "sociology."[29] His scheme of the special sciences then embraced mathematics, astronomy, physics, chemistry, biology, and sociology. Originally, this arrangement of the sciences was considered open and linear. That is, each science needed all the preceding sciences as presuppositions but was independent of the subsequent ones. Biology depended upon chemistry, for example, but chemistry was not based upon any result of biology. We have learned, however, that the ultimate decisions between hypotheses in astronomy or physics are determined by sociological arguments. Therefore, although sociology is certainly dependent on physics and biology, these sciences are, in turn, dependent upon sociology. As Littré later described it:

> There is a necessary reaction between social science and the other sciences. . . . If, in the hierarchy the subsequent science depends upon the previous one, it is also true that the antecedent science experiences a useful reaction from the subsequent one. . . . Similar to an electric circuit, the philosophical circuit is closed.[30]

By this addition of sociology, Comte aimed to make positive philosophy, the system of the six special sciences, an equivalent of ancient philosophy before the separation from science. The ambiguities that required the interference of metaphysics would disappear if Comte and his school were to succeed. Littré said:

> Then all the distinction between science and philosophy disappear . . . As all science winds up in social science, there

is only one grand science, the science of mankind, which enters into everything. It contains the whole of philosophy and nothing remains outside.[31]

As Comte himself wrote in 1848, "The science of Society . . . supplies the only logical and scientific link by which all our varied observations of phenomena can be brought into one consistent whole."[32]

Whenever and wherever positive philosophy becomes the scheme of human knowledge, science again becomes identical with philosophy. Science is regarded no longer as a recording of observational facts but as a search for unifying ideas. The word "positivism" was later misused as the name for a doctrine that urged restriction to "fact-finding" and discouraged the use of "imagination."[33] But this is in flagrant contrast to the intentions of Comte, who said explicitly, "I shall show, in spite of prejudices which exist very naturally on this point, that Positivism is eminently calculated to call the Imaginative facilities into exercise."[34]

5. The "Truth" of General Principles in Positive Philosophy

In positive philosophy a general statement is "true" if all statements derived from it and that assert observable facts agree with actual observations. We learned that in the sixteenth century both the geocentric and the heliocentric theories agreed with observable facts. Is it possible they were both "true"? In the sixteenth century and before, the general opinion among scientists was that the decision could be reached unambiguously by "philosophy," which, then united with science, was a coherent system. It was regarded as possible to determine which theory agreed with "philosophy," and there was only one system of intelligible principles, one true philosophy, which practically speaking was the Aristotelian-Thomistic philosophy. If some doubted the validity of this system, they still believed that it should be replaced by another that would enjoy the same power to decide any alternatives left undetermined by experiment and observation. If the "truth" of the Aristotelian philosophy were assumed, it followed unambiguously that the geocentric theory was right, and the heliocentric theory was wrong.

This situation looks completely different from the point of view of positive philosophy. The only criterion of truth is agreement with observed facts. On this basis, it would not have been possible to decide between

Ptolemy and Copernicus. Thus one can say that the actual decision was made under the influence of "social factors," such as agreement with "common sense," agreement with the philosophy that was favored by the school, church, government, or just by public opinion. This may have been so, but this investigation yields only the psychology or sociological causes that led to the acceptance of a particular theory. It does not tell us anything about which is "true."

If we disregard social science and appeal only to astronomy and mathematics, we have to say that from the point of view of positive philosophy, both theories, geocentric and heliocentric, are "equally true." But if we add social science, we can make additional statements that are also "true" in the "purely scientific" sense and that allow a choice between these alternatives. We shall see that we can ascribe to the geocentric or heliocentric system only "relative truth." But before explaining precisely what this means, we emphasize that "relative truth" has nothing to do with any skeptical or "agnostic" attitude.

If in accordance with positive philosophy we refuse to admit metaphysical arguments, our problem is to describe exactly how the role of metaphysics in examining the ultimate truth of a theory (such as the geocentric hypothesis) can be replaced by social science and the establishment of "relative truth." Since this point is extremely important for understanding the relationship between science and philosophy, we shall explain it in a rather elementary way. We will start with this familiar example, the Copernican conflict, and we will somewhat oversimplify the issue.

The geocentric and the heliocentric theories both agreed with the observed facts—the positions of the planets on the sphere—but if we investigate the psychological and sociological effects of these two theories, we soon note that these effects can be quite different. To simplify matters, we may single out three types of interaction between a scientific (say, physical) theory and the social habits or cultural pattern of the environment. First, consider the "simplicity" of a theory. Generally, a simple theory is preferred to a complex one, but what is regarded as "simple" depends upon the habits of the period concerned. Second, consider agreement between a theory and "common sense"; we want the principles themselves to appear analogous to experiences that are familiar to us from daily life. Finally, we may consider a theory's agreement with the "predominant" philosophy of the period: the philosophy that supports ways of life encouraged by public opinion, government, or by the church.

6. The Relative Truth of Theories

If we base investigations not only in the physical but also in the social sciences, we find that in Copernicus's time, these four statements were as well confirmed as any statement from empirical science can be:

1. The geocentric and the heliocentric theories are both in agreement with the observed phenomena.
2. The heliocentric theory uses a simpler mathematical scheme.
3. The geocentric theory better agrees with commonsense experience.
4. The geocentric theory agrees with the predominant philosophy, while the heliocentric theory flagrantly disagrees.

If these four statements are "true," can we then determine whether the geocentric or the heliocentric theory is "true"? If we admit only agreement with observations as a criterion of "truth," we can conclude from statement 1 that both theories are "true." If we admit agreement with observations and mathematical simplicity, it follows from statements 1 and 2 that the heliocentric theory is "true." This has generally been the view of "scientists." If we assume that agreement with observed facts and agreement with common sense are required, it follows from statements 1 and 3 that the geocentric theory is "true"—the stand taken by empiricist philosophers like Francis Bacon. If we require agreement with observed facts and with the predominant philosophy, it follows from 1 and 4 that the geocentric theory is "true," while the Copernican system is only a "fiction."

If we express ourselves in this way, no "skepticism"[35] or "agnosticism"[36] enters the picture. But we can present this same idea differently without changing the meaning—we can say the heliocentric theory is only "relatively true." It is "true relative to" the requirement of simplicity, while the geocentric theory is "true relative to" the requirement of agreement with common sense or agreement with Aristotelian and Thomistic philosophy. In a practical sense this means that if we ask whether the heliocentric theory is "true," our question is incomplete and ambiguous unless we specify the purpose for which we want to use the theory. Is this purpose only to describe observable facts or to support the predominant philosophy? If we asked whether a certain theory of motor-driven vehicles was "true," we would be in a similar situation. If it is to be used on the ground, the theory of the automobile is "true," but if the vehicle is to fly, the theory of the airplane is "true" (and if in water, the theory of the ship).

We are often tempted to ask the blunt question: is the heliocentric theory "true" or "false"? This would make sense if there were a criterion according to which we could answer. This would be so if the mind could "see" truth as the eye can see color and shape. As a matter of fact, no human being can answer such "blunt" questions if the purposes behind them are not specified. This certainly has nothing to do with skepticism. If this were not so, we could require that the blueprint of a vehicle be drawn up without specifying whether it was to be used on the ground, in the air, or in water.

The impression that there must be an answer to such "blunt" questions, even though no one may be able to find it, comes again from our inclination to understand by means of analogy with familiar experiences. We like to look at every question as a puzzle. Someone made up the puzzle but didn't reveal the solution: in order to find it, we guess what the author of the puzzle had in mind. In Milton's *Paradise Lost*,[37] Adam asks the archangel Raphael whether the Ptolemaic or the Copernican system is true. Raphael replies, "From Man or Angel the great Architect did wisely to conceal" the answer.

This belief that scientific and philosophical problems are a kind of "puzzle" has perhaps never been described as well as by the great philosopher Henri Bergson in his introduction to the French translation of William James's *Lectures on Pragmatism*. Since this doctrine was a step in the development of positive philosophy, Bergson elucidated the new view by sharply defining the previous view:

> For the ancient philosophies there existed a world, raised above space and time, in which all possible truths had dwelt since eternity. According to these philosophers, the truth of human judgments is measured by the degree to which they are faithful copies of those eternal truths. . . . The whole work of science consists in breaking through the obstructing husk of facts in the interior of which the truth is housed like a nut in its shell.[38]

7. Positive Philosophy and Marginal Metaphysics

At the peak of their separation, after 1800, there seemed to be a choice between a "science," which accounted for restricted domains of phenomena without any attempt at integration, and a "philosophy," which offered a broad synthesis without accounting precisely for observed facts. The conflicting slogans were "departmentalized science" versus "philosophical integration." This vicious circle was broken in 1830 by Comte, whose program was "integration on the basis of science."

Among those who followed the path opened by Comte, we shall single out two for discussion. In England in 1860, Herbert Spencer began the publication of his *Synthetic Philosophy*, which for more than a decade appeared in installments.[39] In the United States, John Fiske published in 1874 his *Outline of Cosmic Philosophy*.[40] Both authors attempted to build a philosophical system upon the systematization of the special sciences, but their results were, in one characteristic respect, at variance with Comte's system. While the latter avoided statements about the "reality behind the phenomena," Spencer was guided by a familiar analogy with everyday life experience in which the word "real" has a very clear meaning—it distinguishes a "real dollar" with which we can buy lunch from an imaginary dollar that appears in our memory or in our dreams and which buys nothing. Spencer stressed that science can derive the phenomena of our experience from general statements about the distribution of masses and energies in the universe, but it cannot tell us what the reality is behind these forms and energies.

The old philosophy believed in "seeing" the underlying reality through the phenomena, but Spencer said bluntly that the reality behind the observable phenomena is "Unknowable."[41] He emphasized that because of the parlance of our daily life, we are almost compelled to believe such a reality exists, although science cannot describe it. Spencer even began his *System of Synthetic Philosophy* with a chapter about the "Unknowable," while the positive philosophy of Comte described the "Knowable." About the Unknowable, Spencer said that we know only that it exists and not what it is like; we can, according to him, make some statements about the "Unknowable," but they will be vague. The Unknowable must be ubiquitous, for example, since phenomena occur everywhere in space. We can easily see that this "unknowable behind the phenomena" is an analogy from everyday life; it is analogous to a puzzle, the solution of which has been lost. Spencer himself seems to have felt this way, writing, "Common sense asserts the existence of a reality; Objective Science proves that this reality cannot be what we think it; . . . and yet we are compelled to think of it as existing."[42] After Spencer had paid this compliment to the Unknowable, he went on to investigate the Knowable in the rest of his book and no more was said about that "reality."

We might say that Spencer's system is positive philosophy with a "marginal metaphysics." If we replace his metaphysical statement about the Unknowable with scientific statements about sociological facts, we need only follow Spencer's own interpretations. He identified his Unknowable with the ultimate reality that is worshipped by traditional religion under the name of God. The Unknowable was introduced in order to justify the statement: Spencer's system of synthetic philosophy agrees with traditional religion.[43]

If we now consider John Fiske,[44] we notice clearly how a marginal metaphysics is superimposed on a system that is otherwise very near to the those of Comte and Spencer. Fiske was invited to lecture at Harvard University on Comte's positive philosophy, which at that time enjoyed a great following among intellectuals in all countries. Comte himself did not care to formulate his principles so as to agree with the Christian religion in its traditional, largely metaphysical, interpretation. Fiske, like Comte and Spencer, defined philosophy as a science about phenomena that integrates results from the special sciences, but he attempted to demonstrate to Harvard students that this general view could be reconciled with as much of metaphysics as was necessary to allow for a liberal interpretation of Christian religion. Much as Spencer introduced the "Unknowable," Fiske declared that phenomena studied by science and philosophy (which is integrated science) are the "phenomenal manifestation of an Absolute Power." This, too, of course is an interpretation by analogy with daily life experience. It does not change anything in "science proper" and paves the way for agreement with liberal religion.

This marginal metaphysics of Spencer and Fiske is similar to Neo-Thomism's proposed truce between science and philosophy. Pierre Duhem, for example, followed the line of positive philosophy in the field of "science proper" in an even more logical and radical way than Comte himself. But he reserved the right to build up, behind, and above "science proper," a realm of reality that fits traditional metaphysical doctrine.

There is, however, one great difference between the metaphysics of Duhem or Maritain and that of Spencer or Fiske. According to Duhem and the French Neo-Thomists, science based on metaphysics should be a full-fledged system; it should accommodate the entire magnificent structure of St. Thomas Aquinas's philosophy. But the metaphysics of Spencer or Fiske was meant to be "marginal" and to contain very little elaborate argument, since what mattered to them was the "existence" of this metaphysics rather than its content.

8. Science and Philosophy after the Reunion

When Comte, Spencer, or Fiske introduced this new conception of philosophy and science, they were in fact restoring the chain between science and philosophy that had broken down in the seventeenth and eighteenth centuries and had almost disappeared after 1800. As we have learned, between 1850 and 1950 science and philosophy developed as separate human enterprises, each driven by a strong desire for independence. In the same period, the seeds of Comte's positive philosophy grew into a strong drive to integrate science and philosophy, which means, in particular, to integrate the special sciences, including the social sciences. For convenience, we shall refer to this drive as "positivistic," or as "positivism." On the other hand, when the emphasis is on philosophy as independent of science, we shall speak of "school philosophy."[45] This conception has been transmitted since 1800 through the teaching of "philosophy" as a "special science" in schools of higher learning. It has been greatly influenced by the "schoolmen" who follow the "great tradition" of medieval thought. In a recent book, *Positivism, an Essay on Human Understanding*, Richard von Mises[46] calls those who deny that a scientific approach to philosophical questions is possible (in a somewhat malicious way) "negativists." School philosophers, in this sense, are "negativists."[47] However, we shall not use this term unless a specific undertone is intended.

Now the question arises whether, after this reunion, philosophy remains something different from science. In other words, what is the distinction between science and philosophy for the positivist? There are, of course, authors who wish to remove altogether the word "philosophy" from the scientific vocabulary. Otto Neurath,[48] one of the leaders of twentieth-century positivism, recommended an "index of prohibited words" in order to make our scientific language clear and unambiguous.[49]

Union, Divorce, and Reunion between Science and Philosophy 243

Generally, however, the word "philosophy" is used by those, positivist and not, who write about the foundations and interpretations of the special sciences. But how is it used?

Comte himself, in his *Positive Philosophy*, explained simply and clearly that his book did not deal with science but with philosophy:

> Our object is to go through a course not of Positive Science but of Positive Philosophy. We have only to consider each fundamental science in its relation to the whole positive system and to the spirit which characterizes it; that is, with regard to its methods and its chief results.[50]

In other words, investigations into relations between special sciences and the system of all the sciences are philosophical investigations. Philosophy advances generalizations beyond the principles of physics or biology and that have implications in a great many special fields and even in all fields of science.

In this way, the chain linking science with philosophy reappears in a modified form. There are statements of intermediate generality from which we can actually derive observational facts. But in the attempt to link the special sciences to the "whole positive system and the spirit which characterizes it," often a type of principle is advanced from which a derivation of observable facts can be envisaged only vaguely and not actually performed. Such generalizations are often included in "philosophy" but, in contrast to "school philosophy," these generalizations in "positive philosophy" are not claimed to be "intelligible," "self-evident," or even "plausible by themselves." No principle, general as it may be, claims any truth "in its own right." Generalizations have no justification other than the expectation that they will grow and become parts of the "positive system," but neither is there reason to assume they will later be replaced by some "special science." Rather, we must expect that the "special sciences" will be replaced by universal science in which they will be special chapters. Physical chemistry, for example, is not a "special science" that split off from "philosophy" after physics and chemistry had already established themselves much earlier as "special sciences." "Physical chemistry" was in fact created by the merging of "special sciences," physics and chemistry, into a more universal science.

We see then that the borderline between science and philosophy is somewhat hazy. The "unified field theory" most people would call a "scientific hypothesis." If we consider a still more general hypothesis, however, the distinction is by no means obvious. If we consider, for example, the hypothesis that an "ether" acts to carry electromagnetic and gravitational fields, we are not so certain whether this is "scientific" or "philosophical." We may hope that observable facts may derive from the hypothesis, but then we must consider that one would have to make many additional hypotheses about what the ether is like before one could obtain observable facts.

9. The Name "Philosophy" as a Challenge

If we set up such general "philosophical" principles, we hope eventually to be able to derive observable facts, but we can never be certain that we will succeed. It may also happen that our principles will never connect through logical conclusions to observable facts. In some cases, a wide net of logically possible conclusions can be connected, but we still will not know whether this net will ever connect to the net of "scientific" statements that can be tested by experience.[51]

For example, we can draw a great many mathematical conclusions from Einstein's unified field theory, many conclusions from the hypothesis that man is a descendant of the ape, or that animals and plants have developed from inanimate matter. The question is, however, whether such long chains of conclusions will ever connect with observable scientific statements. There is little doubt that many of these "philosophical" arguments fulfill a very useful purpose. A simple example would be when a connection with the observed facts is achieved, those parts of the chain that have been worked out can be attached as a whole to the "positive system." Many conclusions have been drawn from the fundamental principles of Einstein's unified field theory that form, as has been mentioned, a system by themselves. If we derive an observable fact, we can attach the whole unified field theory to the system of science. As for the Darwinian hypothesis, the chain connecting it with observable facts is certainly not complete—the process of natural selection[52] is far from being cleared up in detail—but a great many parts of this chain are plausible. If we can find, for instance, some biochemical or biophysical basis of natural selection, this argument as a whole can be incorporated into an account of the descent of man.[53] In many respects these "philosophical" hypotheses play a role similar to that of mathematics. Mathematicians build up logical structures, often in hopes that

Union, Divorce, and Reunion between Science and Philosophy 245

these will become important building blocks in the system of science. We know, for instance, that non-Euclidean geometry was built originally as a purely mathematical structure, but later the whole structure became part of Einstein's general theory of relativity.

To sum up our argument, within positive philosophy we cannot distinguish precisely philosophical from scientific criteria of "truth." The ultimate criterion is, in every case, the scientific one: agreement with observable facts. However, for the sake of continuity of communication, the traditional name of "philosophy" is preserved in order to retain some valuable connotations of this term. From the old period of unity, the word has transmitted to us the idea of striving for a knowledge of the whole and for a coherent system of thought. In the period of separation, school philosophy upheld the idea of unified knowledge as a challenge to the "special sciences." It is, therefore, reasonable to preserve the name "philosophy" for research and education that is directed toward a logically coherent world picture that can stand the test of experience. John Dewey, in the introduction to the new edition of his *Reconstruction in Philosophy* (1948), describes the distinction between "science" and "philosophy" as follows:

> It is a case of "science" if and when its field of application is so specific, so limited, that passage into it is comparatively direct—in spite of the emotional uproar attending its appearance—as, for example, in the case of Darwin's theory. It is designated 'philosophy' when its area of application is so comprehensive that it is not possible for it to pass directly into formulations of such form and content as to be serviceable in immediate conduct of specific inquiry. This fact does not signify its futility; on the contrary, the contemporary state of cultural conditions was such as to stand effectually in the way of the development of hypotheses that would give immediate direction to specific observations and experiments so definitely factual as to constitute "science."[54]

A familiar example is the ancient atomic theory advanced by Democritus and Epicurus,[55] obviously a "philosophical" theory. Yet two thousand years later, modern chemistry and physics took up atomic theory, now obviously a "scientific" theory. It would be very wrong to deny, as many scientists do, a link between ancient philosophy and modern science.

The idea of a positive philosophy that would restore the ancient role of philosophy as an integration of knowledge has not disappeared during the hundred years that have passed since the time of Comte. We have, however, become more aware that the expressions "observable fact" and "general law," used by Comte in a commonsense way, become ambiguous and hazy when used in scientific language. The evolution of positive philosophy during the last century has been based, on the one hand, upon careful analysis of what it means to "set up general laws and to derive observable facts from them." On the other, the role of science as an instrument for influencing human conduct has been investigated in a more subtle and thorough way.[56]

Chapter 10

Science, Democracy, and the New Wave of Positivism

1. Science after the French Revolution

If we wish to understand the close connection in the public mind between the rise of mechanistic science and the rise of democracy in the eighteenth century, we can learn from what Henry Thomas Buckle writes about the origin of the French Revolution.[1] He complains that many historians are not conversant with the history of science and have neglected the tremendous impact of science on changes in society. "In Paris," he writes

> the scientific assemblies were crowded to overflowing. The halls and amphitheatres in which the great truths of nature were expounded were no longer able to hold their audiences, and in several instances it was found necessary to enlarge them.[2]

He continues: "The largest and the most difficult inquires found favor in the eyes of those, whose fathers had hardly heard the names of the sciences to which they belonged."[3]

Buckle wishes to prove

> that the intellect of France was, during the latter half of the eighteenth century, concentrated upon the external world with unprecedented zeal, and thus added to the vast movement, of which the Revolution itself was merely a single consequence. The intimate connexion between scientific progress and social rebellion, is evident from the fact, that both are suggested by the same yearning after improvement.[4]

Summarizing his argument, Buckle writes, "The hall of science is the temple of democracy."[5]

However, in the second half of the nineteenth century as science developed beyond Newtonian physics, "idealistic" interpretations of science grew in popularity. Since, as we have seen, these interpretations have a scientific meaning only if we understand them as a return to the organismic science of earlier days, there was widespread talk that modern physics was leading back from modern Newtonian science to medieval organismic science. For those forces that advocated a return to a medieval social system—in other words, to a hierarchic society instead of modern democracy—this rumor was hailed as supporting a return to the political and moral systems they cherished. Many scientists and philosophers were aware of the declining role of Newtonian physics, but they remained firm advocates of the new democratic orientation and were concerned that democracy faced the danger of losing the support of science that it had received since the heyday of Newtonian mechanistic physics.

2. Positivism in the Second Half of the Nineteenth Century (Stallo)

The task now at hand for democratic scientists was to formulate the new phase in science in a way that would recognize its nonmechanistic character but would prevent a relapse into medieval organismic science. This relapse seemed to be unavoidable (and even desirable) to a great many scientists and philosophers because the defeat of organismic science in the seventeenth, eighteenth, and nineteenth centuries has frequently been connected with the victory of mechanistic science and mechanistic philosophy. For this reason, the decline of mechanism seemed to bring about a victory of organicism. The first step in preventing this victory, therefore, was critical scrutiny of the role played by the mechanistic conception.

Although it seems slightly paradoxical, the preliminary skirmish in the fight against a renewal of organismic science consisted in "debunking" mechanistic science. We note this particularly in the work of two authors who, in the last decades of the nineteenth century, were responsible for a new wave of positivism: the American scientist and philosopher John Bernard Stallo and the Viennese Ernst Mach. In 1869 the great German physicist, physiologist, and philosopher Hermann Helmholtz said in a speech, "The object of the natural sciences is to find the motions upon which all other changes are based. I.e., to reduce all science to mechanics."[6] This would mean that unless the program of mechanistic materialism is viable, the aims of natural science in its modern sense, since Galileo and Newton, cannot be achieved. Perhaps more explicitly, the prominent physicist and physiologist Emil Du Bois-Raymond stated the following:

> Natural Science . . . is a resolution of the phenomena of nature into atomic mechanics. It is a fact of psychological experience that, whenever such a reduction is successfully effected, our craving for causality is, for the time being, wholly satisfied.[7]

Until about 1600, however, it was a "fact of psychological experience" that the craving for causality was only satisfied if a phenomenon was reduced to the organismic scheme of St. Thomas and Aristotle. In both cases, "psychological experience" was a behavior conditioned by indoctrination, according to which either only the behavior of an organism or only the behavior of a familiar mechanical device seems "intelligible" in the Aristotelian sense. According to Stallo, belief in the exclusive intelligibility of the mechanistic philosophy is, "in fact, a survival of mediaeval realism."[8] Mach and Stallo emphasized that the principles of Newtonian mechanics have no logical status that is different from other generalizations of our sense observations.

Given these formulations from Helmholtz and Du Bois-Raymond, we note that according to them the real world consists of inert masses and forces acting upon these masses—a distinction that is, after all, similar to the Thomistic distinction of prime matter and form. This distinction has given rise to a conflict between two metaphysical schools and their doctrines: "atomism," according to which only masses are "real," and the doctrine of "dynamism," according to which no masses exist—only "forces." This latter doctrine was reformulated toward the end of the nineteenth century by means of the terms "field" and "energetic." What "really" exists in the world is only a "field of forces," or "energy distributed in space," where "field" and "energy" are asserted to be something "immaterial." Prominent advocates of energetics, on the other hand, were Wilhelm Ostwald[9] in Germany, Pierre Duhem in France, and William J. M. Rankine[10] in England. Ostwald was so inclined to regard "energy" as a substance totally different from matter that he occasionally suggested that the energy of the brain has "consciousness" as a property, while "space" is a property of potential energy in a mechanical system. The *Catholic Encyclopedia* quotes in its article on materialism Ostwald's energetic refutation of materialist doctrine and writes: "Is not this Materialism pure and simple?"[11]

As a matter of fact, there is no great difference between materialism and idealism. The overthrow of modern physics in favor of medieval organismic science cannot be achieved through replacing mechanistic materialism with idealistic energetics but only by splitting matter in its commonsense meaning (a piece of wood or iron) into prime matter and form. The decline of mechanistic science would, however, cease to argue for a return to organismic science were an attack made against the Thomistic split into matter and form or, in terms of physics, the split into mass and force.

With great lucidity and vigor the American philosopher John Bernhard Stallo made this very attack. In 1881, a few years after Du Bois-Raymond's statement on mechanistic science, Stallo denounced four metaphysical fallacies that follow the line of medieval realism and analogical thinking but that are not compatible with the method of modern science. If one refrains from employing these fallacies, the conflict between "atomism" and "dynamism" soon loses its meaning. Materialism and idealism also cease to become statements about realities but only analogies with daily experience. As Stallo writes,

> Whatever diversity may exist between metaphysical systems, they are all founded upon the supposition that there is a fixed correspondence between concepts and their filiations on the one hand and things and their modes of interdependence on the other.[12]

Stallo calls the first metaphysical error the assumption "that every concept is the counterpart of a distinct objective reality, and that hence there are as many things, or natural classes of things, as there are concepts or notions."[13] He showed how this metaphysical error is committed by the mechanistic theory of our physical world, according to which the "absolutely real and indestructible elements of all forms of physical existence" are "*matter*" and "*force*."[14] If we describe observable motions according to Newton's theory of motion, in every individual description the terms "matter" and "force," or, more precisely, "inert mass" and "force" occur in the context "mass times acceleration is equal to force." But if we wish to avoid this metaphysical error, we must keep in mind that "matter" and "force" are words or concepts that occur simultaneously in every description of motion, but there is no such individual "thing" of "physical reality" that we could call "force" or "motion." This is also true, of course, for the term "energy," which occurs as a concept in descriptions of moving bodies but does not denote a "real physical object."

It makes no scientific sense, according to Stallo, to say that the real physical world consists only of matter, as metaphysical materialism holds; but it makes no more sense to replace matter by force or energy or any "immaterial substance" and to claim that "the real world is not material." Stallo's positivism rejects metaphysical idealism as well as metaphysical materialism. It maintains that science consists of principles from which observable facts can be deduced but does not contain statements about the "substance of which the real world consists." If we keep this in mind, the problem of "atomism" or "dynamism" becomes a pseudoproblem. Both assertions are as metaphysical as the Thomistic dichotomy of matter and form. None of these three can be confirmed, or even checked, by scientific methods.

We can sum up Stallo's attitude as follows: the essential point in the transition from medieval to modern science, around 1600, was not the introduction of mechanistic philosophy but rather the doctrine that science consists in a system of principles from which the observable facts can be derived. The principles themselves need not be intelligible (as organismic and mechanistic science had claimed), but they must yield observable facts by a simple system of concepts and conclusions. If we take this attitude, the decline of mechanistic science is immaterial, provided that it is replaced by a system of principles that are simple and empirically confirmable. This attitude has been called "positivism" because it is in many respects similar to Comte's "positive philosophy."

3. Positivism in the Second Half of the Nineteenth Century (Mach)

A positivism very similar to Stallo's was advocated on a broader basis and with a wider scope by his contemporary Ernst Mach, the prominent physicist, psychologist, physiologist, and philosopher. Mach was born in Moravia and worked in Prague and Vienna on the fringes of the German community. His doctrine was the cornerstone of Central European positivism, which became in the twentieth century a prominent movement conspicuously opposed to the idealism that had been prevalent in Germany. It quickly spread to Eastern Europe, to Poland, and to Russia, where the term "Machism" later became almost a household word (with a derogatory connotation in the official Soviet philosophy).

Mach's views are presented concisely in his address to the Academy of Science in Vienna in 1882. He pays homage to the builders and advocates of mechanistic science and to "the great success of the Enlightenment to which we owe our intellectual freedom."[15] He thought, however, that the essential point of modern science is not its mechanistic character but another that could still be preserved when Newtonian mechanistic science was on the decline. "It becomes physical science," writes Mach, "to secure itself against self-deception by a careful study of its character, so that it can pursue with greater sureness its true objects."[16] He thought that to regard mechanistic science as an eternal truth is just as illusory as the belief in the eternal truth of the organismic world conception.

According to Mach, the characteristic of modern science is not the acceptance of mechanism instead of organicism; the function of science "is simply the saving of experience."[17] The Copernican system was preferable to the Ptolemean because it described the same observed facts by a more convenient mathematical pattern and saved, therefore, more experience. Physical laws, according to Mach, are abridged descriptions of observed facts:

> To save labor . . . abridged description is sought. This is really all that natural laws are. . . . No human mind could comprehend all the individual cases of refraction. . . . More than this comprehensiveness and condensed report about facts is not contained in a natural law.[18]

This is far from the Aristotelian and Thomistic conception that the laws of nature describe "reality" and are somehow "intelligible." Mach speaks of the "Economic Nature of Physical Inquiry" and calls this "a clue which strips science of all its mystery, and shows us what its power really is."[19] While diametrically opposed to the Thomistic view that there are "genuine laws," this follows closely the line of David Hume and of Auguste Comte.

It would be wrong to simply say that according to Mach, science is a description of sense observations. To be exact, it is the presentation of a multitude of sense observations by a simple mathematical formula. It would also be wrong to say that Mach's "descriptions" are direct enumerations of sense observations. If this were the case, the presentation would not be economic and would not save experience. The wave theory of light describes, for example, optical phenomena by its analogy with wave motion. Mach calls this an "indirect description."[20] But it certainly does not enumerate the sense impressions we have upon observing optical phenomena.

4. The Reception of Mach and Stallo?

Mach's statement that science consists in a "description" of physical phenomena has often led to the interpretation that, according to positivism, science "describes" the physical world as a traveler describes adventures and sights. However, we must not forget that Mach speaks of an "economic" description and that this "economy" can only be achieved by "indirect" description. Mach says, "We see, thus, without difficulty, that what is called a *theory* or a *theoretical idea*, falls under the category of what is here termed indirect description."[21] In a practical sense, physical descriptions of the world have become "more and more indirect," and basic hypotheses are worded in terms increasingly remote from those with which we describe the familiar physical world of daily life.

In his book, *Knowledge and Error*,[22] Mach elaborated his views on theory formation. He particularly accentuated the dynamical role of theories. Scientists prefer theories that not only present the known facts economically but also suggest new experiments and observations by which new facts are found. He also explicitly points out that the terms of description need not be expressions that can be directly interpreted as denoting sense observations. In his paper "Space and Geometry from the Viewpoint of Natural Science," he writes that

> the operation with symbolic presentations (e.g. multidimensional manifolds) as history of science teaches us, must by no means be regarded as fruitless . . . one has to remember only the use of exponents which are negative, fractions, or even irrational.[23]

Above all, Mach and Stallo attempted to save the essential results of the scientific revolution after 1600 from perishing along with the decline of mechanistic science. What they attempted was a refinement of Auguste Comte's positive philosophy adapted to the state of science at the end of the nineteenth century. But Stallo came up against a stone wall in English-speaking countries, as did Mach in the German-speaking countries. Neither scientists, philosophers, nor laymen were prepared to accept the new positivist doctrines. One group of scientists held firmly to mechanistic philosophy of science because they were afraid that its abandonment would lead to a return of the medieval world picture. Another group, in turn, longed for the return of medieval ideas in religion and politics and hailed the decline of mechanistic philosophy. Neither group had any reason to approve the rise of a positivistic philosophy of science that would make philosophical and political interpretations of twentieth-century physics independent of whether or not one accepted the mechanistic worldview.

About twenty years after Stallo's book *The Concepts and Theories of Modern Physics* was published, Karl Pearson's *Grammar of Science*[24] followed roughly the same line as Stallo and Mach. But the climate of opinion had changed. Pearson's book was highly praised by Josiah Willard Gibbs,[25] the most original and creative theoretical physicist in America. The continued crumbling of mechanistic physics and the continued misuse of this situation on behalf of a return to medievalism produced a longing for a secure basis in modern science that would not become a victim of "antimechanistic" currents. One characteristic testimony from that period is the *Education of Henry Adams* (1918) from which one can see the strong impact of positivism on intellectual life. Speaking about Pearson's *Grammar of Science*, Adams writes:

> The progress of science was measured by the success of the "Grammar," when, for twenty years past, Stallo has been deliberately ignored under the usual conspiracy of silence inevitable to all thought which demands a new thought-machinery. Science needs time to reconstruct its instruments, to follow a revolution in space; a certain lag is inevitable; . . . but such revolutions are portentous, and the fall or rise of half-a-dozen empires interested a student of history less than the rise of the "Grammar of science."[26]

5. Conventionalism (Poincaré, Le Roy)

Henri Poincaré, the great French mathematician and theoretical astronomer, directed attention to the fact that scientific theories are "inventions" of the human mind and products of free imagination. They are "arbitrary" in the sense that they cannot be derived by logical conclusions or direct experiments. If we omit operational definitions, the principles of a theory, such as the axioms of geometry or mechanics, contain no assertions about observable facts. They are actually definitions of the terms of which they consist: in geometry, the terms "point, straight line, coincidence"; in mechanics, "mass, force velocity." In other words, the principles are conventions about how to use the terms "point," "force," and so on. For this reason, Poincaré's approach has been called "conventionalism." In order to interpret this approach in ordinary language, it is frequently said, the laws of geometry or mechanics are arbitrary conventions in the same sense as the names we give to a thing—perhaps to a dog or a baby. Poincaré wrote, for example, "The axioms of geometry are nothing but disguised definitions."[27] In a similar way it was argued that the laws of motion, the conservation of energy, and even the law of causality are arbitrary conventions.

This way of speaking was occasionally used to minimize the role of science within human thought. Édouard Le Roy wrote frankly, "Rational science is nothing but a purely formal play with written symbols without intrinsic significance."[28] From this Le Roy inferred that science cannot find "truth" and that the search for truth must use other sources—metaphysics or revelation. Poincaré himself, a positivist in the sense of Mach and Stallo, certainly did not approve. In 1905 he published the *Value of Science*,[29] a book in which he flatly rejects the inferences drawn by Le Roy. If we do not take operational definitions into account, it is certainly true that axiomatic systems are "arbitrary conventions"; but a theory consists of principles *and* operational definitions.

Such a theory is certainly arbitrary in the sense that it cannot be uniquely inferred from our observational data. If we consider, however, the purpose the theory must serve, it is certainly not arbitrary. If it does not serve the desired purpose, it is rejected. As we learned previously, the purpose may be purely technological, but it may also be moral, religious, or political. The purpose greatly limits the arbitrariness of the theory as the construction of an airplane is determined by its purpose.

By "arbitrary" Poincaré meant "not determined by logical conclusions and direct sense observations." Even the name that we give to a baby is not "arbitrary" but determined by social and religious factors.

According to Poincaré, scientific laws are disguised definitions, but by interpreting straight lines as light rays or as knotted cords, they enable us to derive as logical conclusions statements about observable facts. Hence, laws are economic descriptions of observable facts, but the descriptions contain in part the axiomatic system that would be a system of disguised definitions if it were separated from its operational definitions. Only in this sense is Le Roy's statement true and science is "nothing but" an empty system of relations between symbols. This new conception has often been called "new positivism." The new feature is the great role of formal systems. Obviously these systems cannot be derived from experience but are "free" creations of the human imagination. The word "free" means that these systems cannot be logically derived from experience. They are determined by psychological and sociological factors and by efforts to construct science as an efficient instrument in the human struggle for a good life.

6. Abel Rey and the Bankruptcy of Science

The feeling was strong at the turn of the century that the breakdown of mechanistic science and philosophy was a turning point in intellectual history. One of the best representations of this sense of crisis is *The Theory of Physics According to Contemporary Physicists*, by the prominent French historian and philosopher of physics, Abel Rey.[30] He pointed out that in the mid-nineteenth century, physicists regarded the mechanistic hypothesis as the true description[31] of physical reality—a metaphysical truth obtained by generalizing our experience. But at the end of the nineteenth century, science became "nothing but" a system of symbolic formulae, "a work of art for the amateur and a work of artisanship for the utilitarian," Rey wrote.[32] The failure of this hypothesis as a general theory of physical phenomena showed that man has no metaphysical cognition of the physical universe. "The failure of traditional mechanism," wrote Rey, "entailed the proposition: 'Science itself has failed, it is bankrupt.'"[33] According to him, this "failure of science" would have far-reaching effects on the history of ideas:

Science, Democracy, and the New Wave of Positivism   257

> Physics would lose all educational value; the positive spirit represented by physics is a false and dangerous spirit. Reason, rational method, experimental method have to be regarded as having no cognitive value. . . . The cognition of physical reality has to be investigated and presented by other methods. . . . Physics is ignorance and not knowledge of true nature. . . . The emancipation of the human mind as conceived by Descartes by means of physics is a mistake and a very dangerous one. One has to return to subjective intuition, to a mystical sense of reality.[34]

This means that with the failure of mechanistic science one has to abandon the modern conception of science and return to the Aristotelian and Thomistic conception of truth, according to which observable facts should be derived from "intelligible principles," the truth of which is seen by the eye of the intellect.[35] This can only be avoided if the mechanistic conception is replaced by the positivistic conception of Stallo, Mach, and Poincaré. Rey wrote, "To give the mind a scientific attitude, in the sense it is understood by positivism, remains the necessary and sufficient condition of intellectual sanity. Physics is the school in which one learns the cognition of things."[36] According to Rey, the mechanistic conception of science was to be replaced by the positivistic, and there is no "metaphysical certainty" beyond the methods of positivistic science. There can be opinions, anticipations, or "hunches" but no criteria of truth beyond those of positivism. This was most likely the opinion of Mach, Poincaré, and other leaders of positivism at the end of the nineteenth century.

There have been, however, positivists who envisaged the impact of their views on metaphysical insight in a completely different way. Le Roy, again, started from Poincaré's conventionalism and argued that

> the results of science are, strictly speaking, unverifiable, but altogether certain and general. We have to conclude, therefore, that the principles of science have not anything to do with true reality. . . . They are rather rules of language.[37]

Since science does not tell us anything about true reality, it cannot contradict any statement about true reality. If metaphysics claims to make such statements, they cannot be refuted by any statement arrived at by scientific method. Hence we can make metaphysical claims about the "real world," or "true reality" without being contradicted by any scientific results. This interpretation of positivism has great significance for human behavior if we apply it to the main tenets of traditional religion: the existence of God, of an immaterial soul, and free will. Although these propositions cannot be derived from positivistic science, they certainly cannot be refuted by it.

7. Duhem's Accommodation of Positivism and Metaphysics

Our discussion shows that an advocate of positivism in science in the nineteenth and twentieth centuries can defend a view similar to those of scholastics late in the fourteenth century. One can even hold a radical positivism in science but claim that beliefs in God and an immaterial soul have their source in revelation and do not disagree with positivistic science. This interpretation has been held by prominent physicists. The most instructive and brilliant example is the French physicist Pierre Duhem, one of the most lucid and dynamic advocates of positivism in physics. He presents an account not much different from that of Mach and Poincaré but in some respects even more radical and consistent. The German edition of Duhem's book was introduced and highly recommended by Mach himself.[38]

In his doctoral dissertation at Columbia University, Armand Lowinger contrasted the positivism of Comte and Duhem and their attitudes toward metaphysics and traditional theology. He wrote in 1941:

> Positivism of the Comtean type is an antimetaphysical and antitheological movement with its center of gravity not really in scientific problems on their metaphysical side, but in ambitious cultural and intellectual orientation. . . . In contradistinction, Duhem's positivism is a methodological positivism which avoids all entanglements with problems which do not lie strictly within the province of scientific methodology.[39]

Mach and Poincaré advanced a type of positivism that, like Duhem's, centered on methodological problems of science. But their general philosophy was similar to Comte's, even though they did not elaborate "ambitious" conclusions about sociology and theology.

For Duhem, however, positivism was a method used by the physicist within the boundaries of his specialty. Duhem asked *why* the physicist constructs theories as he does: using criteria of economy, consistency, and so on. Yet in a review of Rey's book, Duhem refused to admit that positivism is the ultimate truth about the physical universe. He resorts to analogical thinking of the medieval scholastics and writes, "Through an analogy whose nature escapes the confines of physics but whose existence is imposed as certain on the mind of the physicist, we surmise that it corresponds to a certain supremely eminent order." For Duhem, a physical theory is the "reflection of a metaphysics; the belief in an order transcending physics is the sole justification of a physical theory."[40] His conception is certainly metaphysical in the sense in which we use the term. We learned that such interpretations usually have practical goals and support desirable rules for human conduct. In his paper "Physics of a Believer,"[41] Duhem shows clearly and emphatically that he is a strict believer in the doctrines of the Roman Catholic Church and a radical positivist in physics. Anyone who believes in a doctrine that will improve individual as well as social life but also believes in the great value of positivistic science will not fail to see in Duhem's conception a thoroughly satisfactory theory.

Above all, Duhem aimed to separate science and religion. As the great French physicist Louis de Broglie explained,

> Pierre Duhem held firmly to separating physics from metaphysics. It was not that he, a convinced Catholic, rejected the value of metaphysics; he wished to separate it completely from physics and to give it a very different basis, the religious basis of revelation.[42]

He saw the boundaries separating science, metaphysics, and religion as did the scholastics of the late fourteenth century, as did Roger Bacon, William Ockham, and the school of nominalism in general. Duhem believed strongly that a positivistic conception of science would remove any possibility of conflict between science, philosophy, and religion.

As we saw previously (e.g., in Part I, chap. 4, §6), metaphysics enters physical science usually by introducing the expression "physical reality," which always carries a metaphysical interpretation. But according to Duhem, metaphysics and religion are essentially different from science because they speak of "reality," while physics does not. In "Physics of a Believer," Duhem wrote:

> What is a metaphysical proposition, a religious dogma? It is a judgment bearing on an objective reality, affirming or denying that a certain real being does or does not possess a certain attribute. Judgments like "Man is free," "The soul is immortal," "The pope is infallible in matters of faith"—are metaphysical propositions or religious dogmas; they all affirm that certain objective realities possess certain attributes.[43]

If one accepts Duhem's positivism, no principles of theoretical physics involve statements about objective reality. "In itself and by its essence," he declared, "any principle of theoretical physics has no part to play in metaphysical and theological discussions."[44]

By scientific methods, one cannot decide which type of positivism one should adopt. Mach and Poincaré adopted a type that was different from Duhem's because they did not believe as firmly in the church as a guide for human behavior, but they did not differ from him in their scientific conclusions. We have here a good example of what we learned previously about metaphysical interpretations: they originate in controversies about what kind of human conduct is desirable and not in controversies about how to apply the method of science.

Chapter 11

The Vienna Circle

Moritz Schlick, Rudolf Carnap, and Otto Neurath

1. The Turning Point in Positivism

The positivists of the nineteenth century, Ernst Mach, Henri Poincaré, and others, worked to eliminate metaphysics from science. They did not deny, however, that philosophy in the metaphysical sense might have some legitimate task outside of science. Mach, for example, repeated again and again that he did not claim to be a "philosopher" in the technical sense of the word. Pierre Duhem even maintained that his positivism paved the way for a true metaphysics.

In the twentieth century, however, positivists became more confident and demanding. They were no longer satisfied to exclude metaphysics from science and to cultivate it in a "reservation" as a "special field of knowledge." Twentieth-century positivists denied metaphysics any value for cognition in a scientific (or nearly scientific) sense. They regarded metaphysics as belonging to theology or poetry and maintained that there is no method for cognition of the objective world besides that used in successful research in fields like physics or biology—briefly, the "method of the special sciences." Their goal was to undo the nineteenth-century creation of a "philosophy proper" that was independent of "science proper." They wanted to join science and philosophy by throwing a bridge across from the scientific side of the abyss. The time had arrived, they believed, for the dream of Auguste Comte to come true.

This feeling of a decisive turn in the long history of philosophy was particularly strong in the Vienna Circle, a group that came into being around 1930, approximately one century after the appearance of Comte's

Positive Philosophy. Moritz Schlick, often regarded as the leader of the Vienna Circle, expressed this confidence in the introductory article of the journal that inaugurated the new movement.[1] In "The Turning-Point in Philosophy," Schlick wrote:

> I am convinced that we are in the middle of a final turn in philosophy. I am convinced that the sterile conduct of philosophical systems is settled. The present period, so I claim, possessed already the methods by which the idleness of all these conflicts can be shown; what matters is only to apply them to these conflicts resolutely.[2]

What are the "good grounds" and "new methods" that provided this Viennese group with so much confidence? Undoubtedly this new confidence rested on developments in mathematical and physical science that had taken place around 1900.

In twentieth-century science, a great many fundamental principles were abandoned that were believed to have been proved by metaphysical insight and, therefore, eternally valid—whatever changes in mathematics and physics might occur. This includes the abandonment of the axioms of Euclidean geometry, the pattern of classical (Aristotelian) logic, and Newtonian mechanics with its axioms about space, time, and force. Even the traditional form of the law of causality was fundamentally changed. Along with these changes came a breakdown of commonsense analogies as principles of knowledge and a breakdown in the belief in metaphysics as the foundation of science.

Twentieth-century positivists even had misgivings about nineteenth-century positivism. In Mach's emphasis on sense perception as the foundation of science and in Poincaré's doctrine that science consists of conventions, they saw attempts at explanation by commonsense analogies and, therefore, metaphysical interpretation. They understood that Mach's views are often interpreted to mean "the real world consists of sense-impressions"—a statement supporting the metaphysics of "idealism." On the other hand, to say that "behind" sense impressions is an objective world, a "real" world, would support the metaphysics of "realism" according to which terms in physics like "mass," "force," "electric charge," and so on denote "real" objects. With "positivism" seemingly facing the danger of encouraging "idealistic" or "realistic" metaphysics, the twentieth-century positivists asked for a thorough reconstruction of philosophy of science. In order to avoid even the appearance of a relation to a philosophical school, they named the resulting philosophy the "scientific world conception."

2. Logical Positivism and the Theory of Correspondence

Originally, Schlick attempted to avoid idealism, realism, and even positivism. As this "scientific world conception" was actually worked out, however, it became clear that it was essentially a refinement of Mach's and Poincaré's positivism, a presentation more consistent and logically satisfactory. For this reason, A. E. Blumberg and H. Feigl proposed the name "logical positivism" for this twentieth-century positivism. They wrote:

> The new logical positivism retains the fundamental principle of empiricism, but profiting by the brilliant work of Poincaré and Einstein on the foundations of physics and Frege and Russell in the foundations of mathematics, it has attained a . . . unified theory of knowledge in which neither logical nor empirical factors are neglected.[3]

The origin of this doctrine in twentieth-century science can be seen from the fact that it was first taught in Schlick's small book on Einstein's theory of relativity. He flatly rejects statements according to which

> only the intuitional elements, colours, tones etc., exist in the world. We might just as well assume that elements or qualities which cannot be directly experienced also exist. These can likewise be termed 'real,' whether they be comparable with intuitional ones or not.[4]

Schlick recommends applying the term "real" as it is applied in science. He writes:

> For example, electric forces can just as well signify elements of reality as colors and tones. They are *measurable* and there is no reason why epistemology should reject a criterion for reality which is used in physics.[5]

For Schlick the world picture presented by physics is a system of symbols that gives us our knowledge of reality. He flatly rejects the "strictly positivistic" conception according to which only intuitional elements are real, while others like electric force are merely "auxiliary concepts." According to the "scientific world conception" that Schlick advocates, "this antithesis . . . between conceptions that denote something real and those which are only working-hypotheses finally becomes unbearable."[6] The system of symbols including the operational definitions denotes as a whole the physical reality, but we cannot say of an individual symbol whether it denotes something real or an auxiliary hypothesis. It would be awkward (as Schlick says) to say that "the pencil in my hand is to be regarded as real whereas the molecules which compose it are to be pure fictions."[7]

Schlick suggests that "every conception which is actually of use for a description of physical nature can likewise be regarded as a sign of something real."[8] In formulating this criterion for the validity of a theory, Schlick takes his cue from the way science tests the validity of theories. The typical method in theoretical physics consists in finding the value of a physical quantity (such as Planck's constant, h) by different chains of conclusions and measurements. For example, one computes the value of h from the hydrogen spectrum and then from the photoelectric effect. If the values obtained by these two methods were different, the system of symbols and propositions describing physical phenomena would contain logical contradictions. Hence, we can say: a theory is valid for a certain set of physical phenomena if (and only if) all the systems of concepts and propositions that can be derived from the theory and the phenomena are logically compatible. Schlick calls this condition "a unique correspondence . . . between conceptions and reality."[9]

This definition replaces the Aristotelian-Thomistic doctrine, according to which "correspondence of theory and reality" is the correspondence of an image to its original. "If such a unique correspondence exists," writes Schlick, "it is possible, with the assistance of the network of propositions in the theory to derive the successive steps in the phenomena of nature, e.g., to predict the occurrences in the future."[10] This conception of reality is obviously adapted to the technological tasks of science. According to this view, every special science consists of a system of propositions, which are relations among basic symbols and contains operational definitions by which symbols are connected to symbols of direct sense observations. Theoretically, to every set of observable phenomena belongs a special science. However, we ordinarily use "special science" to denote one of the traditional sciences, such as physics, chemistry, or biology. Each corresponds to a set of phenomena for which man has successfully set up a theory. But there is no objection to regarding as a special science any theory that coordinates some range of observable phenomena.

3. Philosophy as Activity and the Unified Picture

According to Schlick, "cognition" of a certain domain is setting up a scientific theory about this domain. "The whole of the sciences," he writes in "The Turning-Point in Philosophy," "is the system of cognition." Therefore, there cannot be a "philosophical cognition" that is different. "There is no realm of 'philosophical truth' beyond the special sciences; philosophy is not a system of propositions; it is no science." According to Schlick, it is a "system of acts, namely the activity by which the meaning of propositions is recognized and revealed. By philosophy statements are clarified, by science verified."[11] The meaning of a proposition is recognized when we know what observable facts can be derived from this proposition. To "clarify" a proposition of science through philosophical activity means to find out what these observable facts are.

With these considerations in mind, we can easily understand how Hans Hahn introduced the new "scientific world conception" to scientists and philosophers of the period, around 1930. He wrote, "The name 'scientific world conception' is to be a conception and a delimitation. What can be said meaningfully is a statement of a special science.' And, he continued, "To work in philosophy means merely to examine the statements of the special sciences as to whether they have the clearness and meaning which is ascribed to them by the representatives of the special sciences, or whether they are pseudo-propositions." In addition, work in philosophy aims "to debunk all statements which pretend to have a meaning of a different kind, superior to the meaning in the special sciences, and to show that they are 'pseudo-statements.'"[12]

One could easily believe that by this restriction from a "system of knowledge" to an "activity of clarification," the role of philosophy is downgraded but this was not the intention. The founders of the "scientific world conception" knew very well that the starting point for the work of twentieth-century physics was Mach's, Einstein's, Bohr's, and Heisenberg's clarification of the propositions by which eighteenth-

and nineteenth-century physics had described the motions of bodies. Schlick writes:

> The decisive advances of science are always of the type that they are clarifications of the fundamental propositions. Only people who are gifted in philosophical activity succeed in this kind of world. All great scientific investigators are great philosophers.[13]

There seems a sharp contrast between this "task of clarification" and the "task of integrating knowledge into a coherent picture" often assigned to philosophy as its main work. This contrast has often been described as a distinction between "analytic" and "speculative philosophy." Obviously clarifying propositions is a less important than creating an integrated picture of the universe, and it looked as if the advocates of a "scientific world conception" as well as other supporters of "analytic philosophy" had abandoned the old, proud dream of philosophy to create a coherent worldview and restricted themselves to a modest goal: the analysis of propositions. But this contrast arises only if one looks at the "scientific world conception" in a perfunctory way. Within the Vienna Circle, Otto Neurath stated unequivocally that the new philosophy had not abandoned the search for a unified world picture but would work on it with renewed vigor using the methods of twentieth-century science. He writes:

> Some persons proposed to use the term 'philosophize' for an activity which makes concepts and statements clear. . . . If one takes the thesis seriously that in the field of knowledge one only has to deal with scientific statements, the most comprehensive field of statements must be that of unified science.[14]

Neurath continues, "If one does not care to avoid the term 'philosopher,' one may use it for persons engaged in unified science." It seems not to matter much whether we define "philosophy" as the activity of clarification or the building of a unified science. "It is common to all these persons," writes Neurath, "that they do not join scientific statements with a second type of specific 'philosophical' formulation."[15]

The Vienna Circle eventually defined "philosophy" as it was defined by Comte or, if we believe Comte, as it was envisaged by Aristotle. However, we must now ask how the activity of clarifying the propositions of the special sciences leads eventually to the construction of a unified science. If we investigate the meaning of mathematical concepts (like point, straight line, etc.), our results are propositions that in fact do not belong to mathematics. For we are led into interpretations of "straight lines" by light rays or other physical objects and eventually into problems of the special sciences, as we examine the roles actually played by mathematical propositions in the life of men. This has been carefully and lucidly pointed out by Neurath since the beginning of the Vienna movement. "One cannot separate the clarification of concepts from the scientific world to which it belongs. It is inseparably intertwined," he wrote in 1931.[16]

4. Cross-connections among the Sciences

If we attack any problem in the philosophy of science, we soon note that the "special sciences" are purely conventional entities; if we investigate the relation between science and philosophy, we must always take "science" to mean "unified science." The traditional special sciences must not be taken as the basis for solving fundamental problems. For Neurath proved convincingly that scientific problems posed by nature lead necessarily to the concept of unified science and that not even the simplest question can be solved, or even formulated, by individual sciences:

> One can separate from each other different kinds of special laws, e.g., chemical, biological, sociological, *but one cannot say that the prediction of a concrete individual fact depends upon a special type of law.*

This is obvious from very elementary examples.

> Whether a forest situated at a particular place will burn at a certain time does not depend only upon the weather, but also upon whether or not men by their actions will interfere. The actions resulting from man's interference can only be predicted if one knows laws of human behavior. This means that *there must be a possibility of connecting occasionally all kinds of laws.* All laws, whether they are chemical, climatological, or sociological laws, must be regarded as *parts of a system of unified science.*[17]

From the point of view of unified science, all propositions can be regarded as propositions of the same type: statements about facts, expressed in the language of daily life. Neurath vigorously and directly rejected the ancient and medieval conception that valid statements are true pictures of reality, asserting that

> *statements are compared with statements*, not with "experiences," not with a "world" nor with anything else. All these meaningless duplications belong to a more or less refined metaphysics and are therefore to be rejected. Each new statement is confronted with the totality of existing statements that have already been harmonised with each other. *A statement is called correct if it can be incorporated* in this totality. What cannot be incorporated is rejected as incorrect.[18]

One always tries to fit new propositions into the system, but if we find more and more propositions that are hard to fit, we modify the system to such a degree that the new propositions can be admitted. "*Within* unified science," Neurath writes, "there are important tasks of transformation."[19]

This includes mathematical or logical transformations. In mathematical geometry, arbitrary rules of transformation are laid down in the axioms. We understand what we mean by "two equivalent propositions" or "a tautological statement." But when we apply geometry to empirical science, we cannot prove that two "logically equivalent" propositions lead to two physically equivalent propositions. In unified science, we must introduce an "operational" definition of what we mean by the "equivalence of two propositions." Neurath describes how one can check this equivalence by investigating the impact of systems of imperatives upon men. The imperatives contain different kinds of propositions, A, B, ... For example, "If A is true, do this and this." One can then check the change of reaction that occurs when A is replaced by a different proposition, B. If this replacement does not change the reaction, we may say that "A and B are equivalent." Such statements are an essential part of a theory and must be checked by experiment like any other hypothesis or axiom.

5. Changes in the Science of Meaning

In the doctrine of the Vienna Circle, no point has caused so much excitement as its criterion for what is meaningful and, perhaps even more so, for what is meaningless. If we apply Neurath's criterion, a "proposition is meaningful" if it changes the reaction to an imperative to which this proposition is added. If there is no change, the proposition is meaningless. We easily see that the reaction to a proposition depends on how the recipient is conditioned. With Neurath the "science of meaning" would be an empirical science and, at that, a very complex and difficult one. It would be a part of physiology and sociology or, more generally, of "behavioral science." Like every "empirical science," this "science of meaning" can also be considered as a record of sense observations in which, just as in physics or biology, the attempt has been made to coordinate the sense observations by a pattern that consists of symbols (i.e., by a theory). To give an example: the theory of weather that leads to the predictions published in newspapers yields statements about directly observable facts, but these predictions are not absolutely reliable. If we replace this theory with one that is "more theoretical" and consists in differential equations for "temperature at a certain point," "velocity of air at a certain point," and so on, a prediction consists in the integration of these differential equations. It is therefore reliable but does not tell us much about directly observable facts.

Nonetheless, in a great many sciences, the introduction of such an abstract pattern or theory has been extremely important since correlations have been found among observable facts that are very close to the symbolic pattern. The most impressive examples are in geometry, mechanics, and other parts of theoretical physics. Now the attempt has been made to construct a "theory of meaning" along similar lines, to introduce patterns by means of which we can find out using clear-cut, logical operations which propositions have meaning and which do not. In the Vienna Circle, the most successful efforts toward a theory of meaning have been made by Rudolf Carnap. His first attempt closely followed Mach's positivism as he defined the meaning of a proposition by translating it into statements about direct sense observations. Neurath called propositions of this type (e.g., "Otto sees a red quadrangle") "protocol statements."[20]

In his principal book, *The Logical Syntax of Language*, Carnap points out that Maxwell's equations, which contain the basic theory of electromagnetism, cannot be translated into protocol statements, "although, of course, sentences of protocol form can be deduced from the Maxwell equations . . . ; in this way, the Maxwell theory is empirically tested."[21] Carnap suggests, therefore, defining as meaningful all systems of propositions from which protocol statements can be deduced. A concept like "electromagnetic charge" is meaningful if there are meaningful propositions in which it occurs. Carnap offers as counterexamples concepts like "entelechy" or "vital force," which occur in the writings of vitalistic biologists. According to Carnap, there are no laws that can be empirically tested in which these concepts occur. No protocol statements can be derived from them; hence, they are meaningless. However, we should point out, the mere fact that a concept cannot be defined in terms of observational concepts does not justify declaring it meaningless. What is necessary is only the fitness of a concept to serve as a building stone in a system from which protocol statements can be logically derived. (As a matter of fact, as we mentioned previously [Part II, chap. 10, §6], even Mach mentioned occasionally that there are scientific propositions, like the wave theory of light, that are not directly translatable into sets of protocol statements. Science, however, makes use of such propositions because protocol statements can be logically deduced from them. They are, in Mach's language, "indirect descriptions.")

Carnap takes aid and comfort from characteristic features of twentieth-century physics. He emphasizes that physicists are now inclined to choose a method that does not begin to build a theory up from protocol statements. Their method "begins at the top of the system. . . . It consists in taking a few abstract terms as primitive signs and a few laws of great generality as axioms."[22] The general theory of relativity and quantum mechanics are obvious examples. Taking his cue from this method, Carnap calls a theory "meaningless" if no protocol statement can be logically derived from it. In this category belong, according to Carnap, all systems of traditional metaphysics: realism and its opposite, idealism; the same judgment holds, according to Carnap, for solipsism and even for positivism, if we take it to mean that "nothing real exists except sense-impressions."

6. The Vienna Circle and the Pragmatics of Metaphysics

Like Schlick, Carnap sees the work of the Vienna Circle, the new "scientific world conception," as a decisive turn in intellectual history:

> The difference between our thesis and that of *earlier antimetaphysicians* should now be clear. We do not regard metaphysics

as "mere speculation" or "fairy tales." The statements of a fairy tale do not conflict with logic, but only with experience; they are perfectly meaningful, although false.

Metaphysics is not "*superstition*"; it is possible to believe true and false propositions, but not to believe meaningless sequences of words.[23]

A great many authors—not only philosophers but those from various walks of life—strongly resent this statement that metaphysical propositions are "meaningless." Everyone knows the significant role that metaphysical creeds like idealism or materialism have played in shaping people's behavior. It seemed absurd that these powerful weapons in political and religious battles should be called "meaningless." But Carnap recognized the meaning of metaphysical propositions in this pragmatic sense, stating that

> the pseudo-propositions of metaphysics do not describe states of affairs (in the physical world) . . . they express attitudes towards life. . . . What is objectionable in metaphysics is the fact that it makes use of a form of expression that is misleading. The form of metaphysical discourse is a system of propositions that looks like a scientific theory (like geometry or mechanics). The system pretends to have a theoretical content although it has none.[24]

Since an "attitude towards life" certainly influences human actions, metaphysical propositions have meaning if we understand "meaning" in Neurath's sense. We also remember (Part I, chap. 6, §5) that John Dewey emphasized again and again that metaphysical propositions are not about the objective reality of the physical universe but about human aspirations. Essentially, what Carnap calls "attitudes toward life" is not very different from what Dewey called "human aspirations." Practically, the Vienna Circle shared the opinion of the pragmatists about the "meaning of metaphysics." The salient point is that metaphysical propositions about the physical universe are actually meaningful propositions about human behavior: in other words, they are propositions of sociology. It would be wrong, however, to say that metaphysical propositions are "not cognitive, but emotional," for propositions about human behavior are as cognitive as those of geometry or mechanics.

The point is, again, that according to the Vienna Circle metaphysical propositions about the physical world are meaningless within the system of physical concepts. But they have meaning within the wider "universe of discourse" that embraces physical and sociological concepts. In this way, the circle's "scientific world conception" agrees basically with Comte's positive philosophy. As a matter of fact, it agrees with the great Immanuel Kant, who declared in his more "positivistic" mood that metaphysics has no speculative meaning, only a practical one.

Many great efforts have been made to develop a logically coherent theory of meaning. Carnap and his collaborators have attempted to define "meaningful concepts," "meaningful sentences," and to build on this basis criteria according to which a scientific theory can be proved to be meaningful or not. In other words, they have looked for criteria of "cognitive significance" apart from the merely emotional significance ascribed to metaphysical systems. Although these efforts have been successful and important in many cases, it has been difficult to build a highly general theory of meaning. The main difficulty has been that every logical theory of meaning must confront this larger, pragmatic conception of meaning that was introduced by men like Neurath.

7. Cognitive Significance and Scientific Value

These difficulties are aptly described by Carl G. Hempel, one of the most prominent authors to work on a logical theory of meaning. He concludes "that a satisfactory criterion of cognitive significance cannot be reached . . . by means of specific requirements for the terms which make up significant sentences."[25] Hempel stresses that cognitive significance can only be attributed to a theory as a whole. From our presentation of geometry, mechanics, and twentieth-century theories, we have learned that the meaning of a theory cannot be judged unless the operational definitions of the symbols are added to the axioms. In full agreement with these realities of science, Hempel writes that "the decisive mark of cognitive significance in such a system appears to be the existence of an interpretation . . . in terms of observables."[26] A scientific theory is all the more valuable to the scientist the more observable facts can be deduced from the axioms and operational definitions, from interpretations in terms of observables. Whether these facts agree or disagree with actual observations, in any case, the theory reveals facts about the physical world that had previously been unknown. The more suggestions for actual observations a theory provides, the greater its cognitive significance. As Hempel writes, "Cognitive significance . . . is a matter of degree."[27] He identifies "cognitive significance" with what is called here "scientific significance" and offers several characteristics to help us estimate the degree of cognitive significance in a theory. Perhaps the most relevant characteristic is "the systematic, i.e. explanatory and predictive power of the systems in regard to observable phenomena."[28]

To summarize, we could say—with a grain of salt—the cognitive significance of a theory is proportionate to its scientific value. Then, Carnap's statement that traditional metaphysics lacks cognitive significance would mean that metaphysical doctrines are theories without scientific value. However, the more refinement we achieve in a logical theory of meaning, the clearer it becomes that general problems cannot be solved without going back to the study of actual science, to the logical and empirical aspect in philosophy of science, and particularly to the "science of science."

This discussion of cognitive significance has recently been treated by W. V. O. Quine, who began his work taking his cue from Carnap. He has certainly been one of the most thorough and broadminded researchers in this domain of logic. His result agreed with Hempel and recognized that the criteria to determine whether a theory is meaningful or not do not represent "yes" or "no" answers but rather serve as a statement of degree. Quine recognized that criteria of cognitive significance are closely connected to criteria of cognitive synonymy, or equal significance. They can only be formulated as criteria for the value of a scientific theory or formulated by using the concept of empirical confirmation. Quine writes:

> As an empiricist I consider that the cognitive synonymy of statements consists in sameness of the empirical conditions of their confirmation. A statement is analytic when its operational condition of verification is, so to speak, the null condition.[29]

This comes very close to Neurath's formulation, according to which an analytical (tautological) proposition has, as a part of a given order or imperative, no effect upon a human reaction to the order. Quine stressed, as did Neurath, that from the pragmatic point of view, conceptions of "cognitive significance" and "synonymy" are inseparable.

Chapter 12

Pragmatism

1. Pragmatism (William James, Charles S. Peirce, and John Dewey)

We noted that the end of the nineteenth century, the period around 1900, was marked by the breakdown of Newtonian science and the rise of new theories like relativity and quantum theory. These no longer followed Newtonian physics and were, for this reason, not "intelligible" according to nineteenth-century criteria. We learned, moreover, that at the same time new philosophical presentations of science were developed by Ernst Mach, Henri Poincaré, Pierre Duhem, and others that changed completely what it is to be regarded as intelligible. While these "positivistic" movements took shape in Europe, a movement similar to European positivism in many respects took place in the United States. The school of pragmatism adopted criteria of meaning and truth actually very similar to those of positivism.

Pragmatism's founder, Charles S. Peirce, as early as 1878 had described what he understood as the "conception of an object." While Thomistic philosophy considered the conception to be the portrayal of an object, Peirce emphasized an object's observable effects. "Our conception of the effects," he wrote, "is the whole of our conception of the object."[1] Similarly, Peirce did not take "belief" to mean a portrayal of reality. According to him, 'The essence of a belief is the establishment of a habit; and different beliefs are distinguished by the different modes of action to which they give rise."[2] These definitions of "meaning," "conception of an object," and "belief," lead us to the result that the principles of science are a system of linguistic forms that serve as a program of action in man's relation to nature. Peirce writes:

> Imaginary distinctions are often drawn between beliefs which differ only in their mode of expression. . . . From all these sophisms we shall be perfectly safe so long as we reflect that the whole function of thought is to produce habits of action.[3]

As an example, he discusses the conception of "force." There is a difference between presentations of mechanics in which "force" is defined by an observable quantity, acceleration, and one in which force is a "real entity," quasi-mental in character. He says further:

> Whether we ought to say that a force *is* an acceleration or that it *causes* an acceleration, is a mere question of propriety of language, which has no more to do with our real meaning than the difference between the French idiom "*Il fait froid*" and the English equivalent "*It is cold.*" Yet it is surprising to see how far this simple affair has muddled men's minds. In how many profound treatises is not force spoken of as a 'mysterious entity,' which seems to be only a way of confessing that the author despairs of ever getting a clear notion of what the word means![4]

Peirce insists that different metaphysical interpretations of science are not actually differences in opinion but differences in words. "Imaginary distinctions," he writes, "are often drawn between beliefs which differ only in their mode of expression." But, he adds wisely, "the wrangling which ensues is real enough."[5] The "reality" in the fight over verbal expressions comes from the role these expressions play in the formulation of political and religious events.

2. Peirce's Pragmatism and Positivism

There is no doubt that Peirce's views on meaning, conception, belief, and truth are similar to those of Comte, Mach, Pearson, Poincaré, and other "positivists." Pierce's emphasis on the linguistic nature of metaphysical differences was even ahead of nineteenth-century positivism and a predecessor of twentieth-century "logical" positivism. We must therefore attend to Peirce's efforts to draw a sharp dividing line between his pragmatism and the positivism of men like Comte, Mach, and Pearson. He raised two objections to positivistic criteria for the validity of a hypothesis. First, he attacked Comte's requirement according to which, to quote Peirce's formulation, "no hypothesis ought to be admitted, even as a hypothesis, any further than its truth or falsity is *capable* of being directly perceived."[6] If this requirement were accepted, no hypothesis about the structure of the atom or waves of light and, for that matter, no hypotheses about past events, could be accepted. According to the method of present-day science, a hypothesis is accepted if the conclusions drawn from it are confirmed by direct sense perception, even though the truth of the hypothesis itself cannot be "directly perceived."

The second objection is based upon an interpretation of Mach's and Pearson's doctrines, according to which they claim that a hypothesis or theory is an economic or convenient description of facts already observed. This would not be in line with the actual procedure of science, which aims not to describe facts already known but to yield new facts that lead to new experiments and observations. Peirce wrote, for instance, in his review of Pearson's *Grammar of Science*,[7] "Professor Pearson tries to persuade us that prediction is no part of science, which must only describe sense-impressions."

In his review of Mach's *Science of Mechanics*,[8] Peirce attacked the doctrine that scientific theories are nothing but economic descriptions of observations, writing "Mach pushes his idea so far as to see no value in science except as an economy." Peirce rejected this and replaced it with the "pragmatistic" requirement of finding new facts that were previously unknown. Peirce occasionally formulated this property of a theory by stressing that a theory reveals a part of objective reality; if it were only a description of sense impressions, as the "positivists' claim, it could not predict the outcome of experiments that have never been tried before. Peirce saw in this "strenuous insistence upon the truth of scholastic realism (or a close approximation to that)" an important characteristic distinguishing his pragmatism from positivism.[9]

It is perhaps of sociological interest to recall that for Duhem, a radical positivist in physics, any statement about objective reality is no longer science but belongs to metaphysics or, rather, theology. In this sense, Peirce's pragmatism is essentially positivism but contains a modicum of theology. If we carefully study the relation between Peirce's pragmatism and positivism, we easily see that the positivists did not regard a scientific hypothesis or theory as a mere economic description of known observations but strongly stressed the role of scientific theories as instruments for the discovery of new facts. This becomes especially clear in Ernst Mach's book *Knowledge and Error*. In the chapter "Hypothesis," he wrote, "The essential function of a hypothesis consists in its fitness to lead to new observations and experiments by which our assumptions will be confirmed, refuted, or modified, briefly speaking, by which our experience will be enlarged."[10] Mach explains this without introducing metaphysical expressions like "objective reality." He proceeds in a strictly pragmatic way.[11]

We have learned the sociological role of the term "reality." When scientists like Duhem and Peirce introduce this term, they intend to give the laws of physics a moral value and to make them an instrument for the guidance of human behavior. Duhem makes it completely clear that within theoretical physics the concept of "reality" is irrelevant.

3. James's Pragmatism and Metaphysics

Among the pragmatists, Peirce certainly made the greatest contribution to the philosophy of science and was in many respects ahead of his time. Yet the philosopher who had the greatest effect on contemporary thought among educated men and women was William James, whose thinking was oriented more toward psychology and human behavior than toward physical science. He lucidly analyzed the ways in which science is interpreted philosophically, finding the roots of metaphysical interpretation in the desire to understand principles of human knowledge, not only in a conceptual and abstract way but also by analogy to simple aspects of our familiar experience. James wrote:

> All philosophers . . . have conceived of the whole world after the analogy of some particular feature of it which has particularly captivated their attention. Thus, the theists take their cue from manufacture, the pantheists from growth. For one man, the world is like a thought or a grammatical sentence in which a thought is expressed. For such a philosopher the whole must logically be prior to the parts. . . . All follow one analogy or another.[12]

Positivists like Mach had stressed that science is an economic order established among our sense impressions. But the psychologist and pragmatist James raised the question of why and to what purpose man wants to establish such a rational and economic system. As early as 1879, James wrote that

> the sentiment of rationality, our pleasure at finding that a chaos of facts is the expression of a single underlying fact is like the relief of the musician at resolving a confused mass of sound into melodic and harmonic orders.[13]

James can be seen as responding to positivists like Comte, Mach, and Poincaré, who made great efforts to "purge" science of metaphysics by defining as precisely as possible scientific criteria of validity. They excluded from science all propositions or theories from which one cannot derive propositions that can be directly checked by scientific means of observation.

Men like Duhem allowed for a realm of metaphysics beyond the realm of science. While more demanding in the request for observational checks than any "pure positivist," they allowed that propositions about objective reality could connect with science only by chains of analogies. Although these metaphysical statements were certainly made to serve a practical purpose, such as supporting ethical and religious rules, this metaphysical system was kept apart from science as much as possible.

Since important activities of man are guided by this kind of metaphysics, these "pure positivists" have often been accused of being blind to matters of ethics and religion and restricting themselves to teaching of mere technological rules of behavior. Again, the "positivists with a metaphysical fringe" divide life into two compartments, science and morality, separated by a watertight curtain. In this way, science would play a part only in the less important area of human activity. James and the pragmatists always objected to this kind of separation and attempted to make the realm of moral and religious behavior as much a topic of science as the realm of physical and biological phenomena.

4. Dewey and Political Interpretations of Science

One of the most prominent pragmatists, John Dewey, strongly emphasized the sociological aspect and came close to the original line of "positivism." If we add his doctrine to those of Mach and Poincaré, positivism becomes an integral worldview that fulfills the program first envisaged by Comte. If we ask how philosophical interpretations of science originate, Dewey answers bluntly that they do not come "out of intellectual material, but out of social and emotional material."[14] The social values and tradition that we inherit are supported by the interpretation of science and obtain, in this way, science's blessing. The philosophical interpretation of science has been presented as the search for "ultimate reality" behind the science that is built upon sense experience. But while "under disguise of dealing with ultimate reality," writes Dewey, "philosophy has been occupied with the precious values embedded in social traditions."[15]

As an example, Dewey treats two theories about the origin of the state that were presented as generalizations of our experience with actual states. According to Aristotle, the state is a product of nature, but according to eighteenth-century theories (such as those of Jean-Jacques Rousseau) the state is the product of a contract among the citizens. It is difficult to prove the contract theory by "scientific methods," but there is no doubt that this theory supports liberal and democratic views in politics, while the other serves easily to support conservatism and efforts to preserve the "natural structure" of society.

Dewey did not restrict this doctrine to the social sciences. He pointed out that philosophical interpretations of physical science—that is, the most general principles of physics, astronomy, or biology—also originate in the desire to support useful views in ethics, religion, and politics. His most obvious and instructive example is the transition from medieval organismic science to modern mechanistic science. He characterizes the medieval conception as follows:

> The earth, though at the centre, is the coarsest, grossest, most material, least significant and good (or perfect) of the parts of this closed world. It is the scene of maximum fluctuation and vicissitude. It is the least rational and therefore the least notable, or knowable . . . ; it offers the least to reward contemplation, provoke admiration and govern conduct.[16]

The celestial bodies are very different and of a much nobler nature. Dewey also directs our attention to the great role this division into distinct classes has played in the philosophy of the biological sciences:

> Just as we naturally arrange plants and animals into series, ranks and grades, from the lower to the highest, so with all

things in the universe. The distinct classes to which things belong by their very nature form a hierarchical order. There are castes in nature.[17]

Dewey tells us which order of human society corresponds to the medieval philosophy. "The universe is constituted," he writes, "on an aristocratic, one can truly say a feudal, plan. Species, classes do not mix or overlap—except in case of accident, and to the result of chaos."[18] Dewey describes the physical universe of the Middle Ages in such a way as to serve directly as an example for human society to imitate:

The universe is indeed a tidy spot whose purity is interfered with only by those irregular changes in individuals which are due to the presence of an obdurate matter that refuses to yield itself wholly to rule and form. Otherwise, it is a universe with a fixed place for everything and where everything knows its place, its station and class, and keeps it.[19]

It is easy to see how this theory of the physical universe could be used practically to justify racial segregation.

In the physical science of Newton and his successors, the general picture of the universe is completely different. Newton's laws of motion are valid without modification for celestial and terrestrial bodies. "The earth," writes Dewey, "is not superior in rank to sun, moon, and stars, but it is equal in dignity, and its occurrences give the key to the understanding of celestial existences."[20] And this new philosophy would no longer support a feudal system. "The net result," he writes, "may be termed, I think, without any great forcing, the substitution of a democracy of individual facts equal in rank for the feudal system of an ordered gradation of general classes of unequal rank."[21]

5. A New Development: Scientific Empiricism

In the first half of the twentieth century, other groups besides the Vienna Circle developed nineteenth-century positivism along similar lines. We may name as representatives Hans Reichenbach in Germany, Joergen Joergensen in Denmark, Arne Næss in Norway, Louis Rougier in France, Tadeusz Kotarbiński in Poland, and many others. A group that advocates closely related views is the great British School of Analytical Philosophy, including such outstanding men as Bertrand Russell, George Moore, Max Black, Gilbert Ryle, and others. Ludwig Wittgenstein, Karl Popper, and Friedrich Waismann have formed a bridge between the Viennese and the British groups. We cannot survey all forms of positivism, so we will restrict ourselves to examples that have had a significant impact on the philosophy of science.

The "scientific world-conception" of the Vienna Circle was introduced on the American scene, as it was mentioned in Part II, chap. 11, §2, by Feigl and Blumberg under the name of "logical positivism." Many people resented the political and religious connotations historically connected with the term "positivism" and suggested instead the name "logical empiricism." This term is, in some ways, the most "neutral" one. It describes the generally accepted method of modern science that consists in logical deduction from principles and empirical confirmation of the results.

Logical empiricism is in many respects very close to the pragmatism taught by men like Peirce and Dewey. So naturally the problem arose of how to define the mutual positions of logical empiricism and pragmatism. Charles Morris[22] and Ernest Nagel[23] have been greatly interested in the original doctrine of the Vienna Circle and possibilities of using it as a building block to consolidate and enlarge the structure of original pragmatism. According to logical empiricism, science walks on the two legs of theory and observation. But according to pragmatism, Morris writes, "science walks on three legs of theory, observation, and practice."[24]

There is, of course, practice in theory and observation. Setting up axioms and handling them by logical conclusions and transformation is, of course, a practice. Operating measuring instruments to confirm theories is a practice, too. But the "practice" in pragmatistic interpretations originates in the undeniable fact that, according to Morris, "science is part of the practice of the community in which it is an institution, ministering—however indirectly—to the needs of the community and being affected—and very directly—in its development by the community of . . . which it is a part."[25]

One need of the community, according to these groups, was to replace traditional metaphysical creeds with scientific discourse. This was done in two steps. First, rationalistic and realistic metaphysics were removed from the traditional presentations of science; second, the intellectual significance of practice had to be accounted for. "The first expansion," Morris writes, "was made by logical empiricism, the second by pragmatism."[26] As a result of interaction and cross-fertilization between logical empiricism and pragmatism, a philosophy of science has emerged for which Morris suggests the name "scientific empiricism." Its evolution and growth have been conspicuous in the United States, but it has exerted a strong influence in all countries where the tradition of idealistic or realistic metaphysics has not been strong enough to frustrate the new unity between science and philosophy.

6. The Meaning and Significance of Bridgman's Operationalism

The best way to characterize "scientific empiricism" is perhaps to discuss the work of two prominent representatives: the physicist Percy William Bridgman and, in the next section, the philosopher Ernest Nagel.

Bridgman actually worked in the physics laboratory, where he mainly researched the behavior of material under extremely high pressure. He did not devote much time to reading philosophical books, although he likely absorbed indirectly some views of positivists and pragmatists. He was, however, eager to fully understand the meaning of the operations he performed in the laboratory. His work entailed carrying out a program of action, the question that concerned him being how this system of planned action connected to the system called a physical theory. In many respects Bridgman was like Ernst Mach, who was also primarily an experimental physicist. Their primary task was not the confirmation of "intrinsically clear" theories but rather the coordination of physical operations performed in their laboratories.

Bridgman stressed the point that the conspicuous increase of interest in the philosophy of science around 1900 was not whimsy or fashion but was connected with fundamental changes in the principles of science. In his first and impressive book, Bridgman wrote:

> The growing reaction favoring a better understanding of the interpretive fundamentals of physics is not a pendulum swing of the fashion of thought towards metaphysics, originating in the upheaval of moral values produced by the great war, or anything of the sort, but is a reaction absolutely forced on us by a rapidly increasing array of cold experimental facts.

This reaction, a rather new movement, was without doubt initiated by the restricted theory of relativity of Einstein.[27]

We remember that Schlick's "scientific world-conception" and the logical positivist movement can be ascribed to the same cause. Bridgman even assumed that to "the younger generation of physicists, born since the special theory of relativity was formulated, . . . many of our considerations . . . will appear trite and uninteresting, because in absorbing the new physical theory they absorbed also implicitly the new philosophy of science."[28]

As in logical empiricism, the central point of Bridgman's doctrine is the *theory of meaning*. "We do not know the meaning of a concept," he later formulated his doctrine, "unless we can specify the operations which were used by us or our neighbor in applying the concept in any concrete situation."[29] This "concept of meaning," by which Bridgman defines the "meaning of a concept," is certainly close to the theory of meaning introduced by pragmatists and logical empiricists, particularly to the behavioristic flavor given to the concept by Peirce and Neurath. Yet there is an element in Bridgman's concept that makes it somewhat different. We may say that the flavor of the physics laboratory tinged his concept of meaning, that he meant by "operation" the actual behavior of the physicist during research, and had in mind the many factors of an operation that must be planned to achieve a desired result.

Bridgman found that "operational definitions" of highly general physical concepts can be extremely complicated. The operation, for example, to measure the velocity of a moving body can only be performed if the motion is "smooth." If one formulates the first law of thermodynamics, one has to introduce concepts of "supply of mechanical work" and "heat supply." The operations by which these concepts are defined are distinct from one another only if the phenomena take place in a simple and smooth way. If, for example, we examine what happens around Joule's wheel in a liquid,[30] we are at a loss to measure in each volume element the heat supply and the mechanical work separately from each other. Bridgman found in a similar way that under general and complicated conditions there is no operation by which to clearly distinguish pure heat radiation.[31] In one of his latest papers, he discusses the impossibility of distinguishing by any operation whether in an electric field the force acting upon a charge is due to action at a distance or to an electric field.[32]

Not all operations necessary for the definition of a concept are manipulations of physical instruments. If we define, for example, the stress inside an elastic body of the ψ-function (de Broglie's wave function) in wave mechanics, we must perform what Bridgman calls "paper-and-pencil operations" in order to fit them into the behavioristic pattern. "Among the paper-and-pencil operations are to be included all manipulations with symbols, whether or not the symbols are the conventional symbols of mathematics." Besides paper-and-pencil operations, Bridgman also introduces "verbal operations." He writes:

> Civilised man lives . . . in a verbal world of his own making; in this verbal world he exhibits patterns of behavior which he finds not less compelling than the patterns forced upon him by . . . the physical world.[33]

Man's strong urge to formulate the general principles of science as analogies to commonsense concepts is, for Bridgman, a characteristic of our verbal behavior: "The sort of physical concept which [man] finds it profitable or at least congenial to use is usually determined to a certain extent by his verbal demands, as disclosed by verbal experiments."[34] According to Bridgman, we can perform verbal experiments by asking, for example, "Would I say thus and thus in such and such a situation?"[35] Given Bridgman's belief that the distinction between "action at a distance" and "field action" cannot be demonstrated by an "instrumental operation," "we must . . . recognize that the distinction which physicists actually do make between these two concepts is verbal, and the corresponding operations are verbal operations."[36] Similarly, Bridgman discusses the concept of light as a thing that travels through space. "Instrumentally," he says, "light only departs and arrives. . . . Until new sorts of experimental fact are discovered it seems to me that the concept of light as a thing traveling remains a predominantly paper-and-pencil concept, mostly verbal in character."[37]

Although physicists make great use of paper-and-pencil operations, they have agreed, as Bridgman puts it, "in imposing one restriction on the freedom of such operations, namely that such operations must be capable of eventually, although perhaps indirectly, making connections with instrumental operations."[38] Of course this agrees with the doctrine accepted by positivists since Comte: it is irrelevant what type of symbols occur in physical laws, provided one can derive from them statements about observable facts.

Again, Bridgman consistently stressed that his operational theory of meaning developed along with twentieth-century physics. In his lecture, "Philosophical Implications of Physics," he said this about Einstein's handling of physical concepts:

> There was in the first place a realization that the paradoxes involved primarily questions of meaning and that the common sense meanings of the physical terms such as length and time were not sharp enough to serve in the new physical situations. . . . The attitude towards meaning which eliminated the paradoxes of relativity theory has been carried over by the physicist into all the rest of physics, particularly into the new realm of quantum phenomena, where it is absolutely essential to any valid thinking at all.[39]

This usefulness of the new theory of meaning in physics generated in Bridgman's mind the idea that the same method of operational analysis may shed light on other fields of human knowledge and activity. In the disputes about social, political, or religious topics, "comparatively few of these terms find their meanings through simple, direct 'objective' operations in the 'external' world in a way that the instrumental operations of the physicist find their meaning."[40] In such discussions verbal operations often play the main role. Bridgman thinks that a good understanding of twentieth-century physics can provide us with an understanding of verbal operations as a part of verbal behavior in general; that "modern man must have to be educated to cope with his verbal environment" and should acquire "a knowledge of the results of an operational analysis of the important abstract terms used by the culture in which he lives."[41] As an example, Bridgman discusses "the state and our attitude towards it." If we make use of science's concept of meaning—the operational point of view—"it will be impossible to think of the state as some super thing or even super person with an existence of its own."[42]

However, "from the long-range point of view," Bridgman writes, "the most revolutionary of the insights to be derived from our recent experience in physics . . . is the insight that it is impossible to transcend the human reference point."[43] In quantum theory one clearly learns that we cannot describe the objective physical world in terms of "things," if we mean by that what we call "things" in commonsense language. We can only coordinate our subjective experience by a symbolic system of axioms while our experience remains the ultimate system of reference. This insight will be perceived to be "more revolutionary than the insight afforded by the discoveries of Galileo and Newton or of Darwin."[44] It will teach us the impossibility of the attempts made by the human race "to transcend its own reference point by the inventions of essences and absolutes and realities and existences."[45]

7. Nagel's Contextualistic Naturalism

Ernest Nagel is a contemporary American author who has built up a philosophy of science imbued with the spirit of positivism and pragmatism. Nagel presents[46] a lucid and impressive account of the Vienna Circle, British analytical philosophy, and the great advances that had been made in liberating the philosophy of science from traditional metaphysics. However, Nagel also perceived a serious shortcoming of these European doctrines: their failure to look at themselves in a historical perspective. In other words, European positivism and analytical philosophy had both emphasized the logical and observational aspects of science but had neglected what Charles Morris had called the "practical" or the "pragmatical aspect." Nagel worked to fill this gap by stressing the relationship between European positivism and American pragmatism. He analyzed with great care the work of Peirce and of Dewey and emphasized that the interaction of positivism and pragmatism formed the main characteristics of contemporary American philosophy of science.

Nagel enumerates the philosophical trends long established in the United States in the nineteenth and twentieth centuries: some surviving Calvinist theology, rosy-hued philosophy of progress, absolute idealism in many forms, a militant Thomism, and the recently imported doctrines of phenomenology and existentialism. Yet the movement that has emerged from positivistic and pragmatic thought "appears to express more adequately than any of these the dominant temper of American life."[47] This philosophy, again, has been called "scientific empiricism." Nagel himself suggests the name "contextualistic naturalism"[48] and claims that "it is without question America's most significant contribution to philosophic intelligence."[49]

He provides an instructive sketch of this philosophy in his paper "Philosophy and the American Temper":

Pragmatism is the matrix from which contextualistic naturalism emerged. Contextualistic naturalism is not a finished intellectual edifice, and perhaps it will never develop into one. For a cardinal thesis is the essentially incomplete character of existence, in which no overarching pattern of development can be discerned. Contextualistic naturalists exhibit a profound distrust of philosophic systems which attempt to catch, once for all, the variegated contents of the world into a web of dialectical necessity.[50]

For contextualistic naturalists, "the familiar distinction between appearance and reality does not play any role in their thinking about nature, since the term 'reality' designates no inherently basic substance in the world, but at best only a humanly valuable phase of existence."[51] This agrees entirely with modern positivism, which regards "reality" as a concept that has meaning only in the context of sociological statements about human predilections. Nagel calls this "naturalism" to emphasize the refusal to introduce "supernatural" arguments into philosophy of science. But in order to draw a line between his views and some metaphysical kinds of "naturalism" in which nature is introduced as an agent by "analogical" chains of thought, Nagel adds the qualifying term "contextualistic." This addition reminds us that according to positivistic and pragmatistic criteria, any concept or statement has significance only in a certain context. Nagel strongly stresses the point that the best work on the philosophy of science is done by those "who have interpreted the meaning of theoretical constructions in terms of their manifest functions in identifiable contexts."[52]

Having seen in the last two chapters how logical positivism and pragmatism have evolved and how some philosophers emphasize their similarities, we shall move on in the next chapters to the last general type of philosophy in our time, namely materialism and its different versions.

Chapter 13

Mechanistic and Dialectical Materialism

1. Mechanistic Materialism

There are two types of facts in our daily lives that man has always regarded as most familiar: the behavior of our fellow men or of animals and the motion of simple mechanical devices, such as wheels, pulleys, levers, or launched stones. The philosophy of Aristotle or St. Thomas Aquinas treated facts in the physical universe by their analogy with the behavior of organisms, which means, briefly, to ascribe to the motions of physical bodies purposes directing them—as purposes direct the motion of men and animals. When this philosophy failed to be helpful, after 1600, when advances of modern science were increasing, two ways of arriving at a more useful philosophy were tried. The first was to refrain from the use of analogies as much as possible and to dismiss the question of whether a system of principles is "intelligible," provided that it is practical.

There was, however, a second way. For even in ancient Greece some philosophers had rejected the organismic theory of the universe and started from principles formed as analogies to the motions of simple mechanisms. The leading group that took this approach was the school of Epicurus, whose impressive doctrine is presented in the famous poem by Lucretius.[1] This school was silenced in the Middle Ages and did not attain any strong influence until Newton showed that the laws governing simple mechanisms in our daily lives are also fit to account for the motions of celestial bodies, of planets, and even comets. On the basis of these achievements of seventeenth- and eighteenth-century science, Epicurean philosophy revived and gained new strength. Its mechanistic world conception contrasted with the organismic view and developed in the eighteenth and nineteenth centuries into a philosophy according to which observable facts are to be derived from the principles of Newtonian mechanics and physics. The theory that all observable facts can be so derived has been called "mechanistic materialism." If we recall Comte's view that to find a way out of the vicious circle between observations and theories man took refuge in a theological theory of observed facts, "mechanistic materialism" can be seen as a second way out of Comte's "vicious circle."

Like positivism, this view rejects the belief that the observable phenomena of nature can be derived from statements about purposes. However, positivism subjected the principles of science to only two criteria: that all conclusions drawn from them should agree with observations and that the principles and derivations should be "simple" and "convenient," putting aside any requirement that these principles should be "intelligible" or "plausible." Materialism, on the other hand, required that these principles not only yield the observable facts but that they also be derivable from Newtonian mechanics or, for that matter, some set of physical laws. Within mechanistic materialism, however, there have always been two trends. Some scientists advocated the materialistic theory only because it was best for deriving observable phenomena; this group was, of course, close to positivism. However, other philosophers and scientists believed that the laws of Newtonian mechanics are somehow more intelligible and more plausible than the observable facts of physics. For them, "mechanistic materialism" becomes "metaphysical materialism"—something closer to Thomism than positivism. These materialists proceed, as Aristotle recommended, from "unclear" sense observations to intelligible principles, in this case Newtonian laws of motion. And the intelligible basis of all observable phenomena is the motion of one substance: matter. In this respect, of course, this materialism is the very opposite of Thomism, in which the intelligible basis of phenomena is the agglomerate of prime matter and substantial form.

2. La Mettrie's Materialism

The main point of mechanistic materialism is to ban concepts like form, soul, or inclination from science and philosophy, that is, to ban all analogies taken from the behavior of organisms. A vivid illustration of this is the eighteenth-century bible of materialism, Julien Offray de La Mettrie's

Man a Machine. Given the poor reputation of Epicureanism and other forms of mechanistic materialism among the representatives of religion, La Mettrie pointed out that there cannot be any contradiction between knowledge based upon the observation of nature and knowledge based upon revelation, provided we accept only revelation actually confirmed by witnesses. He wrote:

> If there is a God. He is the author of nature as well as of revelation. . . . If there is a revelation, it can not then contradict nature. By nature only can we understand the meaning of the words of the Gospel, of which experience is the only interpreter.²

His main point is that man's nature cannot be grasped by philosophy but only by the methods of the natural sciences. "Experience and observation," he writes, "should therefore be our only guides here. Both are to be found throughout the records of the physicians who were philosophers and not in the works of the philosophers who were not physicians."³ Since the human body is more accessible to scientific methods than the soul, our author writes that only "by trying to disentangle the soul from the organs of the body, so to speak, can one reach the highest probability concerning man's own nature, even though one can not discover with certainty what his nature is."⁴ All the evidence of our senses points toward the theory that man follows laws established by physical science. "Let us conclude boldly," La Mettrie writes, "that man is a machine and that in the whole universe there is but a single substance differently modified."⁵ This substance is, of course, matter. Physical laws govern the behavior of matter; hence all phenomena of inanimate bodies, of planets, animals, and man, are governed by the same laws.

If we examine the theory of knowledge underlying this materialistic theory, we note that, according to La Mettrie, sense observations teach us that we cannot trust reason. This doctrine

> is not the work of prejudice, nor even of my reason alone; I should have disdained a guide which I think to be so untrustworthy, had not my sense, bearing a torch . . . induced me to follow reason. . . . Experience has thus spoken to me in behalf of reason and in this way I have combined the two.⁶

La Mettrie kept himself more to the positivistic (scientific) aspect of materialism than the metaphysical. "The brain," he writes, "has its muscles for thinking as the legs have for walking. . . . The nature of motion is as unknown as that of matter."[7] La Mettrie does not believe that we have metaphysical insight into the nature of matter and motion, but he is elated that his materialism provides mankind with a definite (although sober) truth about man and his position in the universe. "Such is my system," he writes, "or rather the truth unless I am much deceived. It is short and simple. Dispute it now who will."[8]

Perhaps the most shocking point in this materialistic system was its absolute refusal to admit purpose and inclination in nature, except as an effect of the human brain. The indignant rejection of the "man a machine" theory is quite understandable if we consider that the belief in the purposiveness of the universe is a foundation of traditional religion and ethics.

3. Purposiveness in Nature

Even the famous French writer Voltaire, a bitter enemy of the church and political groups sympathizing with the church, strongly opposed La Mettrie's system. In his *Philosophical Dictionary*[9] under the heading "God—Gods," he devotes section IV to d'Holbach's *System of Nature*,[10] which, along with *Man a Machine*, represented eighteenth-century materialism. Voltaire attacks the materialistic view that the behavior of living organisms can be understood by assuming that they consist of the same matter as inanimate bodies and are moved by the same forces. Voltaire writes the following:

> Matter has extent, solidity, gravity, divisibility. I have all this as well as this stone: but was a stone ever known to feel and think? If I am extended, solid, divisible, I owe it to matter. But I have sensations and thought—to what do I owe them? Not to water, not to mire—most likely to something more powerful than myself. Solely to the combination of the elements, you will say. Then prove it to me. Show me plainly that my intelligence cannot have been given to me by an intelligent cause. To this are you reduced.[11]

Though described by many as "irreligious" and even blasphemous, Voltaire was actually eager to prove the existence of an "intelligent cause" because he was convinced that the belief in such a cause is essential as a rational basis of moral behavior. Therefore he regarded its existence as proven until its nonexistence could be proven. He even seems to have approved the Thomistic method of proving the existence of objects by analogies. "If you had never seen a gunner," he writes, "and you saw the effects of a battery of canon, you would not say that it acts entirely by itself."[12] This is exactly the Thomistic inference from the world as a machine to the existence of a machine maker. It is instructive to note that by using analogical thinking, the picture of the world as a machine leads to the existence of an intelligent cause, while without admitting this type of thinking "man a machine" became the slogan of mechanistic materialism.

We can see from Voltaire's *Philosophical Dictionary* that even for a man who despised the church and strongly opposed medieval philosophy, one feature of materialism seemed to him not only wrong but repellent: the denial of purposiveness in nature. In the eighteenth century, and much later too, this denial implied that the universe is governed by two types of uniformities: causal laws and statistical laws. The most difficult problem materialism faced was the origin of organic life in the universe. Since no known physical laws could account for the emergence of organisms among inanimate matter, the only way left for materialists was to assume that organisms originated by chance. The idea had already existed in the Epicurean school that by the irregular, zig-zag movement of atoms, occasionally chance combinations of atoms would arise, stick together by hooks, and eventually become configurations.

However, there have always been arguments claiming to prove that the probability of such a configuration forming is so small that it can be regarded as practically impossible. Hence we need "purposiveness" or a "superior intelligence" to account for the existence of organisms among dead matter. As a matter of fact, a similar argument was used even by Newton in his theory of the planetary system. Observation shows that all planets circulate in the same plane and in the same direction. It would be extremely improbable that this coincidence could take place by chance. Therefore, Newton argued, it must be the effect of an intelligent cause.[13]

Arguments and computations to prove that the rise of organisms by chance is extremely improbable were very common in the eighteenth century and used in the struggle against materialism. La Mettrie, in his plea for materialism, argues against the existence of a supreme being on the basis of an extremely small chance for organisms to emerge from inanimate matter. In his *Man a Machine*, we read: "To destroy chance is not to prove the existence of a supreme being, since there may be some other thing which is neither chance nor God—I mean, nature."[14] He wants to say that from the lack of any chance for the emergence of organisms, one cannot conclude the intervention of an intelligent being—only the existence of laws of nature unknown before. It is instructive to compare this view with that of a prominent contemporary biologist who studies organic evolution. George Gaylord Simpson writes, "Current studies suggest that it would be no miracle, not even a great statistical improbability, if living molecules appeared spontaneously under special conditions."[15] Not unlike La Mettrie, he continues:

> That is not to say that the origin of life was by chance or by supernatural intervention, but that it was in accordance with the grand, eternal physical laws of the universe. It need not to have been miraculous, except as the existence of the physical universe may be considered as a miracle.[16]

The probability of a certain series of events cannot be calculated unless we know the pertinent laws of nature. If we disregard all laws of nature, the falling of a stone or the circulation of planets around the sun are very improbable.

4. Materialism Refuted?

Materialists like La Mettrie believed that there are physical laws that make the emergence of organisms probable and plausible; but we cannot prove that they are "organic laws" that act for a purpose. The main tenet of materialism is that biological phenomena, including human behavior and human history, are directed by natural laws—a type of physical law. In the eighteenth century, the concept of physical law was determined by Newton's mechanics. In the nineteenth century, when laws of thermodynamics and of electromagnetic fields emerged, this raised questions as to whether including these types of laws abandoned—or, at least, weakened—materialism. The situation became even more critical in the twentieth century with laws of relativity and quantum theory—laws that differed greatly from the Newtonian type. In each of these cases, one could say with some justification that "materialism is refuted" because the meaning of "physical law" had to be broadened in order to justify the assertion that, for example, human actions are governed by physical laws.

Whether an individual scientist or philosopher preferred to broaden the concept of physical law or to drop materialism altogether depended solely upon his belief in the usefulness or harmfulness of materialism as a guide for human action. There is no doubt that materialism weakened philosophical arguments (such as those of Thomism) on behalf of traditional religion and, indirectly, of traditional ethics. But leaders of eighteenth-century materialism eagerly pointed out that the main doctrines of Christian ethics were not harmed but emphasized by *Man a Machine*. La Mettrie even used biblical language, such as "man comes from dust and returns to dust." He said believing that men, animals, plants, and even stones are machines makes these beings our equals and supports the idea of universal love. The characteristic of materialism is thus not a certain system of ethical values but the argument by which the acceptance of those values is supported.

Similarly, for many authors, the essential point of mechanistic materialism is not the acceptance of a certain type of law but the rejection of "supernatural" influence. Such influences can only be defined by enumerating examples because a general definition cannot be given unless we have a general definition of "law of nature." John Dewey, Sidney Hook, and Ernest Nagel wrote in 1945 that "apparently . . . only those can call themselves non-materialists who maintain that causal efficiency resides in . . . unexpressed wishes, silent prayers, angelic or magical powers, and the like."[17] In this sense the most convincing "refutation of materialism" would be experiments like Joseph Banks Rhine's on extrasensory perception.

On the other hand, there is "materialism" in the sense of "reductive materialism," a doctrine that would "reduce" all statements about nature to statements about physical phenomena. As Dewey, Hook, and Nagel write in the paper previously quoted,

> Reductive materialism may be taken to maintain that every psychological term red, fear, pain, etc. is *synonymous with*, or *has the same meaning* as, some expression or combination of expressions belonging to the class of physical terms, e.g. the

word 'red' has the same *meaning* as the phrase 'electromagnetic vibration having a wave-length of approximately 7100 Angstroms.'[18]

5. Materialism versus Positivism

Before we turn to dialectical materialism, we will recall some of the sociopolitical uses of philosophical interpretations. We saw on several occasions how philosophical interpretations have been used to support desirable ways of human life. Governments or other institutions that believe they have a mission to improve human behavior will, quite naturally, favor philosophical interpretations of science that support their goal. What "favoring" means, in practice, depends upon the degree to which a government is willing and able to pressure its citizens. The situation in totalitarian countries is particularly instructive. In such countries every human activity, science as well as religion, is regarded as an instrument to direct human behavior and interaction between the philosophical interpretation of science in question and the goals of political groups becomes obvious. These countries can serve therefore as a magnifying glass for the study of this interaction under different circumstances where it is rather mild and often hidden from the scientist whose attention is directed toward only the logical and empirical aspects of science.

For example, in the Soviet Union the philosophical interpretation of science has been strictly regimented by the government and the ruling party. We shall therefore direct our attention to the official philosophical interpretation of science that is taught everywhere, from the secondary schools to the universities and technological institutes.

This official philosophy is known as "dialectical materialism" or "diamat," in abbreviated form, and we shall attempt a short presentation to understand the favored interpretations of science. If we look into a textbook of dialectical materialism from Russian schools of higher learning, the first thing that we notice is a sharp distinction between "materialistic" and "idealistic" interpretations of science. We learn that, in practice, materialism supports the political goals of the Soviet government and its ruling party, while idealism opposes these goals and supports the goals of bourgeois governments and parties. He who denies a deep cleavage between these two basic philosophies is, politically speaking, a "neutralist." He refuses to take sides in the struggle between "good" and "evil."

As we have learned, the positivists vigorously deny the cognitive significance of the difference between the materialistic and the idealistic interpretations. According to Carnap,[19] neither has a scientific meaning, and therefore no decision between them can be reached by scientific methods. But in the Russian textbook by Mitin, we read the following:

> Behind the verbal trumpery of the numerous philosophical systems, behind the variety of colorful labels attached to their teaching by the philosophers, is hidden a long-lasting and embittered fight between the two principal lines in philosophy: materialism and idealism . . . The pretension to stay outside both tendencies, to stay "above" them, to build up some new philosophy that is neither materialistic nor idealistic is a maneuver, used by some contemporary bourgeois philosophers, either with the purpose of covering up their belonging to idealism. or, in some case, in order to veil the shamefaced fear of others. who do not dare to admit publicly that they are materialists.[20]

The basis of "idealism" in Soviet philosophy is the doctrine that natural objects are "animistic," that in every plant or rock is a "spirit" and the behavior of all objects—rock, plant, or man—can be understood from the analogy with human behavior. Mitin's textbook says: "Idealism originated as a product of the limited and ignorant ideas of the aboriginal savages. The evolution of scientific knowledge in connection with the evolution of the means of production was bound to lead to the complete triumph of materialism and the elimination of all idealistic conceptions."[21] Although this crude, animistic idealism has been destroyed by the advance of science, disguised forms of idealism remain in our time. The reason it survives, according to our textbook, is "the division of society into classes [and] the rule of the bourgeois class in capitalistic society which encouraged idealistic theories and doctrines because they seemed to serve some of the interests of that class."[22] This service was rendered, according to the Soviet doctrine, by the support that idealistic philosophy provided for traditional religion, whose leaders, in turn, supported the Czarist regime in Russia.

The philosophical interpretations of science most frequently denounced as "disguised idealism" are mechanism (mechanistic materialism) and positivism. In order to understand the background of these claims, it is helpful to situate dialectical materialism among the other attempts to interpret science philosophically. We previously described (e.g., in Part II, chap. 10) the situation at the end of the nineteenth century, when mechanistic philosophy seemed no longer able to satisfactorily interpret actual science. The failure of the mechanistic interpretation provided an opportunity for a comeback of the medieval organismic philosophy connected with political, social, and religious views that representatives of nineteenth-century science considered obsolete. Nineteenth- and twentieth-century positivism, we saw, offered a way to acknowledge the failure of mechanistic philosophy without approving a return to medieval views.

According to positivistic views, mechanistic interpretation is only one of many possible pictures of our physical observations. Science is a coordination of experience by means of a system of axioms that consist in relations between symbols. It is equally scientific to start from propositions about electric charges as it is from mechanical masses, chemical affinities, or gravitational attractions. Statements about energy are no less scientific than statements about matter. For according to positivism, electric charges, mechanical masses, chemical affinities, gravitational attractions, energy, and matter are all symbols that we use to coordinate our sense observations. It is fairly correct to say that dialectical materialism was originally introduced as a method of solving the same problem that positivism originally aimed to solve. Diamat abandons mechanistic philosophy (mechanicism), but it sticks to materialism. This is done by generalizing the concept of matter.

While originally "matter" was identified with mechanical mass in Newtonian physics, diamat distinguishes between "matter in the physical sense," identical with mechanical mass, and "matter in the philosophical sense." This second sense embraces all physical quantities belonging to "objective physical reality," that is, which can be measured as mechanical mass can be measured. Electric charge in this sense is "matter"; electromagnetic energy is "matter in motion," but this motion is different from local, mechanical motion: when it became obvious, toward the end of the nineteenth century, that moving electric charges behave like moving mechanical masses, diamat interpreted electromagnetic energy as "matter in motion." This interpretation was materialistic but not mechanistic, since "matter in motion" is not a mechanical mass.

6. Soviet Attacks against Positivism

Positivism replaces mechanistic science with a system of relations between symbols from which statements about sense observations can be derived. According to diamat, this means to interpret physical phenomena by mental phenomena (sensations). In this diamat sees idealism replacing mechanistic materialism, since science becomes a system of statements about mental states. Diamat itself would instead interpret all phenomena as statements about "matter in motion." Since positivism and diamat are two roads issuing from the same crossroad, the failure of mechanicism, it seems understandable that dialectical materialism from the start has chiefly attacked positivism. Today the authoritative presentation of dialectical materialism is Lenin's book *Materialism and Positivism*,[23] which is essentially an attack upon the "reactionary" philosophy of Mach. In every diamat textbook, much of it is directed against positivism, for which Lenin and others have used the names "empiriocriticism" (introduced by Avenarius), and "Machism." In his book, Lenin writes:

> Despite the pretentions of Mach and Avenarius to overthrow materialism *and* idealism as metaphysical systems, they remain idealists in the purest sense. The assertion that only the sensations (sense observations) can be thought of as existing reflects the subjective idealism subsisting in their heads.[24]

From the writings of Lenin's enthusiastic followers, it is clear that this book was intended to be a deadly blow against all lines of positivism. Ralph Fox writes,

> With intense polemical fire it destroys not only the Machian faith, but also those varieties of it which were then becoming fashionable—Henri Poincaré in France, the pragmatists in the United States, the various schools of mathematical physicists produced by the crisis in modern physics.[25]

While positivists described discoveries by introducing new symbols and new relations between them, diamat saw "electric charge" as a new form of matter and electromagnetic energy as a new motion of matter. Lenin writes, "Science has succeeded in discovering new forms of matter, new forms of material motion." He describes electromagnetic theory "reducing the old and familiar forms of matter to the new ones."[26] Mitin, in his previously quoted textbook, follows Lenin's distinction between the philosophical and the physical conception of matter: under the philosophical conception, one understands what acts upon our sense organs and produces sensations. According to Lenin, "Matter is the objective reality that is given to us in our sensations."[27] Matter is what exists outside and independent of our consciousness. The physical conception of matter, on the other hand, considers what the objective world is like from the angle of contemporary science. While the philosophical concept of matter involves relations between being and knowing, between subject and object, the physical concept has to do with the structure of matter.

By introducing this *philosophical* conception of matter, diamat hopes to keep the materialistic interpretation of science alive, even though mechanistic materialism has been abandoned. If this failure of mechanistic science is not to push philosophical interpretation back into its medieval, pre-Newtonian form, the principal choice for twentieth-century scientists has been between one of the positivistic doctrines and the generalized materialism of diamat.

In order to avoid the "idealistic and reactionary" consequences of positivism, Soviet doctrine has consistently denounced positivism, in the broadest sense of the word, as politically and morally undesirable. In the article "Logical Positivism" in the *Short Philosophical Dictionary*, we read:

> Logical positivism, or logical empiricism, is one of the dominant movements in the contemporary philosophy of the Anglo-American bourgeoisie. It is a form of subjective idealism proper to the period of rotting capitalism. This reactionary idealist school was born in Austria in the twenties of the twentieth century: the so-called "Vienna Circle" was a direct prolongation of Machism. The survivors of this Circle have presently found refuge in the United States. . . . The basic functions of positivism consist in 1) falsification of the conclusions of science and the perversion of its meaning and content in favor of idealism, 2) empiricist limitations of scientific cognitions in order to favor religion and its pretension that science is not competent to judge mystical, ethical, and

aesthetical cognition. . . . Without refuting the principles of materialism, logical positivism eliminates the most important problems under the pretext that they are pseudo-problems.[28]

7. The Conversion of Mass and "Star-Spangled" Operationalism

It is instructive to illustrate diamat's attacks on positivism by an example from physical science, namely the conversion of mass into energy according to Einstein's law, $E = mc^2$. This law has been frequently cited in support of idealistic metaphysics, since according to it matter can be converted into a nonmaterial substance. According to the positivist interpretation presented in this book, mass (m) and Energy (E) are symbols by means of which statements about sense observations can be derived, and these statements can be checked by experiments. According to diamat, this interpretation supports idealism and, indirectly, traditional religion; it should be replaced by a materialistic interpretation.

Consider the article "Mass and Energy," published in 1954 in a French journal of dialectical materialism. Its author, Jean Druan, attacks the assertion that matter can be "converted" into energy, writing "to admit such a transformation would mean to admit that besides matter 'something else' exists but is different from matter; it would exist besides matter, namely energy."[29] Positivism has barred the infiltration of spiritualistic metaphysics into the interpretation of science by insisting upon the operational definition of "matter" and "energy." The author quotes a passage from the *Encyclopedia of Unified Science*, which presents the views of logical positivism:

> The annihilation of mass has been interpreted occasionally as a refutation of materialism and a support of spiritualism. This can be done only if one uses this language without having in mind the operational meaning of the word and sentences. (The statement) 'Matter can be annihilated and converted into energy,' sounds strange to the man of average school training, and smacks at least of spiritualism. . . . If we introduce the operational meaning of 'matter' and 'energy,' the annihilation of matter and its conversion into energy is not

at all obscure, and has nothing to do with the experiments of the spiritualists which demonstrate "dematerialization."[30]

The words "spiritualism" and "dematerialization" are used in their commonsense meaning, as they occur in the description of tricks performed by "mediums." But the French author presenting diamat continues:

> The choice is very clear. If we consider a certain number of observed phenomena that have been studied by nuclear physicists, we are left with the choice of being either spiritualists or of aligning ourselves under the "star spangled banner" of operationalism.[31]

The author further explains that "operational or logical positivism recognizes as real only the results of measurements. . . . The concepts of 'truth' and 'reality' are eliminated from their language."[32]

By tagging operationalism as an American philosophy, the author reflects the political antagonism between the United States and Soviet Russia on a philosophical level as an antagonism between positivism and materialism. Since diamat rejects operational definitions as a means of avoiding spiritualism in physics, a materialistic solution of the "crisis" must be found. This solution is again to introduce a generalized "philosophical" conception of matter. As "electric charge" was declared by Lenin to be a "form of matter," Druan describes "energy" as a "form of matter." Then, we need not state that mass "disappears," but only that what is called "conversion of mass into energy" is actually the conversion of one form of matter (mass) into another (energy). A lump of matter in the physical world of diamat has mass m and energy E, between which there exists the relation $E = mc^2$.

The Russian physicist Kuznetsov[33] writes that this relation expresses the "connection between mass and energy in a piece of matter" and not the possibility of a mutual transformation. According to dialectical materialism, the indivisible tie between matter and movement admits no "transformation of matter into energy." By the explosion of the uranium bomb, no "matter disappeared" but matter changed its form from mass into energy. This "materialist" interpretation differs from the "operational" only by a change in terminology. But advocates of diamat see an advantage in the fitness of its terminology to better formulate a certain political creed. When the "conversion of mass into energy" is presented in the terminology of diamat, matter's prestige is enhanced, and "star-spangled" operationalism is replaced by Soviet materialism. The language of science is more compatible with the language in which the political and religious goals of this group are formulated.

Chapter 14

The Laws and Politics of Dialectical Materialism

1. Dialectical versus Mechanistic Materialism

According to mechanistic materialism, the world consists of masses in the Newtonian sense that are acted on by forces and obey the laws of mechanics as formulated in the science of mechanics. There exists also the more general conception of materialism, according to which the universe consists of matter whose changes obey laws of physical and chemical science. As we have seen, dialectical materialism (or diamat) includes a "philosophical" concept of matter besides the physical one and does not assume that this matter always obeys physical and chemical laws. The matter of which human bodies consist obeys biological and sociological laws that, according to diamat, cannot be reduced to physical and chemical laws. The reducibility of the behavior of organisms and human societies to mechanisms is strongly rejected by Soviet philosophy and is denoted, one could even say "branded," as "mechanicism." However, the assumption that biological and sociological laws are not reducible to physics but are "autonomous" is in some respects similar to the teachings of idealistic philosophers. Many Soviet philosophers have uttered opinions that were branded by advocates of diamat as "idealistic."

These "idealistic deviations" have been described as "Menshevik Idealism," which ties these deviations to the Menshevik Party, originally formed within the moderate wing of the Socialist Party before it split into Bolsheviks and Mensheviks. The first decade of the Bolshevik regime saw fairly lively discussion between mechanistic and idealistic interpretations of diamat. Since according to the Marxist doctrines every philosophical conflict of opinion is connected with political conflict, a government that does not tolerate opinions different from those of the ruling party does not tolerate conflicts in the philosophical interpretation of science. In December 1930, Stalin spoke to the Communist cell in the Institute of Red Professorship (Division: Science and Philosophy).[1] In his paper, "Stalin and Science," Ernst Kolman described how "Comrade Stalin and the Central Committee of the party gave to the workers on the philosophical front concrete directions that helped them to correct mistakes and to guide their work into a path that fits the requirements of life."[2]

The point was, of course, to keep science on the narrow path between deviations toward either mechanism or Menshevik idealism. If we look into the *Large Soviet Encyclopedia* under the heading "Philosophy," we see that it is important for the government to keep scientists and philosophers on the right track:

> The victory of socialism in the USSR manifested itself even in the domain of ideology. Attempts to revise Dialectical Materialism were made under the hostile influence of the bourgeoisie and petty bourgeoisie. They found their expression in "Mechanicism" and "Menshevik Idealism," which were exposed and liquidated under the direct leadership of Stalin.[3]

In Mitin's textbook on diamat, prescribed in Russian universities in the 1930s, mechanistic materialists are accused of facilitating idealistic and spiritualistic philosophies: this is because by restricting science to mechanics, the new phenomena of atomic and nuclear physics seem to defy the application of scientific methods and invite spiritualistic explanations. To an even greater degree, the same accusations have been launched against mechanicism for its treatment of individual and social behavior. As early as 1845, Marx and Engels pointed out[4] that changes in human society cannot be treated by methods of mechanics. If a restriction to mechanicism is imposed on the scientist, social events are excluded from the application of the scientific methods and must therefore surrender to idealistic and theological arguments. Mitin writes the following in his textbook about contemporary mechanicists:

> They reduce the quality of autonomous laws (social, biological, etc.) to purely quantitative laws derived from mechanical laws, to the principle of equilibrium of forces acting in mutually opposite directions. . . . On the present level of evolution mechanistic materialism disorganizes the fight of materialism

against idealism and facilitates the fight of idealism against us, by producing a cleavage between the discoveries of contemporary science and materialistic philosophy.[5]

Mechanicism is often accused of leading ultimately to idealistic metaphysics by training the objective laws of social evolution to the opinions of men, their views, and the ideas prevailing in a certain society.

2. Diamat and Philosophy

On these grounds, Soviet philosophy since 1930 has regarded mechanistic materialism as a major enemy. The official doctrine has been that in connection with the new stage of social relations and the new level of evolution of science, a new form of materialism has developed—the philosophy of dialectical materialism—that rejects the reduction of biological and social laws to mechanical laws. It is well known that scientists actually working in experimental science or mathematical analysis are not generally interested in philosophy and may even look askance at it. This attitude has diminished the willingness of Russian scientists to accept the new philosophy and has made them susceptible to positivism and mechanicism. These "mechanists" are often blamed for cultivating among scientists the belief in the self-sufficiency of science and the uselessness of philosophy. The prominent Russian physicist Sergei Vavilov noted, "Dialectical materialism in physics has to contend with three adversaries: with mechanical, metaphysical materialism, with idealism of all possible shades, and with philosophical indifferentism."[6]

Most Western scientists believe that science is independent of philosophy and will eventually correct its own mistakes. But the official Soviet doctrine calls this "bourgeois science" and holds that science needs guidance from philosophy. "Among the professors," writes Lenin, "who are capable of valuable contributions in the special field of physics or chemistry, there is not a single one from whom we could trust one word when he speaks about philosophy."[7] (A fervent advocate of Thomistic philosophy would probably agree with this.) About the professors of philosophy and the professors of science who write about philosophy, Lenin wrote:

> There is hardly one contemporary professor of philosophy who is not, directly or indirectly, employed in overthrowing

> materialism . . . They pretend now to refute materialism from the standpoint of "recent" or "modern" positivism, natural science, etc.[8]

Prominent Russian physicist Abraham Fyodorovich Ioffe published a paper on the occasion of the twenty-fifth anniversary of Lenin's book, saying in part:

> Occasionally physicists like Bohr, Heisenberg, Schroedinger, etc. tell us in their popular writings about the philosophical generalizations of their scientific work. Then their philosophy is frequently the effect of the social conditions under which they live, and of the social orders which they carry out, consciously or unconsciously.[9]

Ioffe believes that Heisenberg's theory is properly materialistic and that it is the best approximation to reality that is possible today. It is important to note, however, that while a "materialistic theory" is in the best possible "agreement with reality," in strictly scientific language a theory must agree with experiments and observations. This use of the "reality" makes materialism a doctrine similar in substance to Aristotelian-Thomistic doctrine, according to which a true theory is a picture of reality.

3. Diamat and Realism

Dialectical materialism describes the relation between physical reality and human thought by the "theory of images." Mitin's textbook says:

> According to this doctrine, our intuitions and conceptions are not only *produced* by objective things, but they *portray* them. Intuitions and conceptions are not produced by self-evolution of the subject (as idealists maintain); they are not hieroglyphs (as agnostics think); but they are pictures, images, copies of things. . . . Our cognition becomes more and more precisely, more and more profoundly a picture of the material world. There are no limits set to our ability to know the world, but there have been in every case, historically determined limitations to our approximations to absolute truth.[10]

A well-rounded picture of the place of materialism in the philosophy of science that is officially taught to Russian students of science comes from a recent physics textbook that has been introduced in Russian universities:

> The common-sense materialism of the scientists is not sufficient; the future scientists have to get training in genuine philosophy. . . . The great majority of physicists have always been "common-sense materialists." However, the weakness of common-sense materialism consists in its unconsciousness and in its inability to exert philosophical judgment about the experiental data of science. This lack has led to the result that a part of the bourgeois scientists, under the influence of the reactionary ideology of the ruling class, has attempted repeatedly to exploit physical discoveries for the confirmation of idealistic views. We meet such attempts particularly in periods of great new discoveries when the principles have to be revised and new ones are not yet sufficiently clarified, e.g., at the end of the nineteenth and the beginning of the twentieth century.[11]

Students are taught that Lenin exposed and refuted these "reactionary" attempts. In particular, they are warned against current interpretations of the statement "matter disappears," used to describe electromagnetic theory at the end of the nineteenth century and the conversion of matter into energy at the beginning of the twentieth. The textbook says:

> Only those properties of matter disappeared which had been believed to be absolute, unchangeable, elementary (like impenetrability, mass, inertia, etc.) and which now, it appears, are relative and lacking only to some states of matter. But the only property of matter which is relevant to the philosophical concept of matter is that of being an objective reality, to exist outside our consciousness.[12]

Here again we see that philosophical "realism" is the fundamental doctrine of Soviet philosophy, which is, in this respect, close to Aristotelianism and Thomism. Nominalism and positivism are their targets in common.

4. The Dialectical Laws

We stated that diamat dropped mechanistic materialism and retained a more general type of materialism, but that presents dialectical materialism as the sum of its negative characteristics and its abandonment of mechanistic philosophy. We now ask, in a positive way, what kind of laws replaced the abandoned mechanical laws? The answer is simply that the most general laws in nature are no longer mechanical but "dialectical." What does this mean? In the realm of mechanistic materialism, physical, biological, and social phenomena should be described, in principle, by one and the same system of general laws: the laws of mechanics. If, however, mechanistic philosophy is abandoned, there is no longer a tie between the laws that govern these three domains. The "dialectical laws" introduced by diamat are valid in all three domains and provide, as mechanistic materialism did previously, general principles valid in the physical, biological, and social sciences. They became the new tie between the natural sciences, the social sciences, and the humanities.

The three dialectical laws are: 1) the identity of opposites, 2) the transition from quantity to quality, and 3) the negation of negation. For the scientist, the laws may seem obscure because they use language very different from the language of modern science. The best way to make them understandable is perhaps the historical way. Their origin is in the idealistic philosophy of the great German philosopher Georg Wilhelm Friedrich Hegel, who started from the assumption that there is an exact correspondence between the system of concepts in the human consciousness and the system of concepts that can be used to describe the physical universe. Empirical philosophy regards concepts in the human mind as constructed for the deliberate purpose of systematizing the facts sense-experience. Hegel's idealistic philosophy, however, assumed that the conceptual system in the human mind can be constructed by logical methods from one fundamental concept—without consulting experimental and observational investigations. This means that the conceptual system underlying physical science can be constructed from one fundamental concept, which Hegel called the "absolute idea" or the "absolute spirit." Since the universe consists of mechanical, physical, and organic phenomena, the fundamental concepts by means of which these phenomena are systematized—the concepts of mechanical, physical, and biological science—can and must be logically derivable from the fundamental concept of philosophy, the absolute spirit.

Walter Terence Stace offers the most lucid presentation of this philosophy, writing that[13]

> the universe is (according to Hegel) nothing but the content of consciousness. . . If we admit [this], we are bound to admit also the objectivity of concepts or universals. For then the object *is* wholly and solely the object as we know it. And we know it only as a congeries of universals. And if we accept this view, we are committed to an objective idealism. The identity of knowing and being is, in fact, the basic principle of . . . idealism.[14]

By "content of consciousness," the idealistic philosopher means the fundamental concepts that systematize our commonsense experience, including being, quality, quantity, measure, and so on. According to Hegel, from these concepts the concepts of science, like physics or biology, can be logically derived directly without requiring actual scientific research in these fields. In other words, the concepts of physics can be derived from the concepts of commonsense experience without making use of physical theorems in the modern sense. Perhaps the most famous example is Hegel's derivation of the concept of measurement, one of the fundamental concepts in physical science.[15]

5. Quantitative and Qualitative Changes

Being contains the three grades of quality, quantity, and measurement. Quality is the character identical with "being"—so identical that a thing ceases to be what it is if it loses its quality. Quantity, on the contrary, is the character external to being and does not affect the thing at all. Thus a horse remains what it is whether it be greater or smaller; and red remains red, whether it be brighter or darker. Measure, the third grade of being, is the union of the first two and is a qualitative quantity. All things have their measure, that is, the quantitative terms of their existence; their being so or so great does not matter within certain limits, but when these limits are exceeded, things may cease to be what they are.

Hegel strongly emphasized that this independence of quality from quantity exists only within certain limits. For example, the temperature of water, within certain limits, has no influence upon its liquidity; but with the increase or diminution of temperature of the liquid water there comes a point when this state of cohesion suffers a qualitative change, and water is converted into steam or ice. A quantitative change takes place, apparently without any further significance: but there is something lurking behind, and a seemingly innocent change of quantity acts as a kind of snare to catch hold of the quality.

According to Hegel, it can be deduced from the concept of being that from increasing quantitative changes eventually qualitative changes must result. Since this rule has its roots in the general nature of our concepts, it must be just as true in politics as in physics. As Hegel writes, "The same principle may be applied in political science, when the constitution of a state is regarded as independent of . . . the extent of its territory, the number of its inhabitants, and other quantitative points of the same kind."[16] He means, of course, that a quantitative increase in territory must eventually result also in qualitative changes, like a change in the constitution. A large state like Germany cannot be qualitatively of the same type as a small state like Switzerland.

All three dialectical laws are equally valid in physics, biology, and sociology because observable phenomena are systematized by our fundamental concepts of knowledge. In one sense, these laws played, in Hegel's idealism, the same role as the laws of motion in materialistic philosophy. If we assume, as mechanical materialism does, that for all phenomena motion of material bodies is primarily responsible, the laws of motion must be valid, not only in what we call mechanics but also in all physics, chemistry, biology, and even sociology. Hegel's dialectical laws then replaced Newton's laws of motion as the theoretical basis that is common to all human knowledge. For when it became clear that Newton's laws of motion could not account for all observable facts in physics, biology, and sociology, Marx and Engels gave Hegel's idealistic philosophy a "materialistic turn."

They regarded objective reality as "matter" and interpreted Hegel's dialectical laws as laws about material changes. Their "dialectical materialism" did not attempt to reduce all qualities into quantities, as mechanistic materialism had done. On the contrary, diamat insists that matter undergoes "qualitative changes." Hegel's law of the transition from quantity into quality was now interpreted as a law about the behavior of matter: when matter experiences quantitative changes without changes in quality, this will not go on forever. When the quantitative changes are sufficiently large, a sudden qualitative change—a "jump" into a different quality—takes place. When water evaporates the quality of liquidity "jumps" at a certain point into the quality of gaseousness. In contrast to the mechanistic view, the behavior in the liquid state cannot be determined from the gaseous behavior, and vice versa. Different qualities are governed by different laws; but for every transition to another quality the dialectical law is applicable.[17]

6. Social Change and Natural Science

The universal validity of dialectical laws was particularly important for Marx and Engels because they applied to changes in social and economic structure. In a society based upon private ownership, there can be great quantitative change when single proprietors own more and more property, but the quality of the society (sometimes called capitalistic) remains unchanged. However, according to dialectical law there must come a point when the society undergoes qualitative change and becomes a collectivistic (Communistic) society in which property is no longer owned by individuals. In this way, Marx and Engels predicted the Communist revolution as a qualitative change in society that had been based on private ownership.

The idea of social revolution is linked with the idea of qualitative change in natural science. While there is no great value for physics in formulating a law of evaporation as a "jump" from quantity into quality, it is of great value for shaping human behavior to advance a theory according to which the Communist revolution and the evaporation of water appear as two instances of the same general law: the dialectical transition from quantity into quality. The great value of such a theory for teaching a certain political behavior becomes clear if one considers that, according to diamat, Communist society is qualitatively different from capitalistic society, and no law advanced by economists and sociologists in capitalistic society can be applied to a Communistic society.

In some ways, the value of a philosophy like dialectical materialism is similar to that of older metaphysical systems like Thomism. Thomistic principles like "no body can move unless it is moved by another body" or "all uniformities in nature are due to genuine laws" are not very helpful for understanding mechanics, or for that matter, physics or biology. They are, however, of great value for the demonstration of theological principles, like the existence of God. Since they do not logically contradict any established principle of physical science, it is reasonable to accept them as useful for indoctrination in sound religious and moral behavior.

Mitin, in his textbook on dialectical materialism, points out that production in the Soviet Union is not only quantitatively larger than in the capitalistic countries but that it is qualitatively different. He writes: "We have a tempo in the evolution of production that has never existed before, because the USSR represents a new quality in conditions of production."[18] He emphasizes explicitly that the new order does not obey the same laws as the old. "The quality of a thing is inseparably connected with its very being. . . . The qualitative determination of a thing manifests itself by the specific laws which it obeys."[19] Mechanists, he points out, attempt to produce such a tight link between the philosophy of science and political philosophy that, practically, they deny the possibility of social revolutions. "The mechanists," he writes, "deny that qualities possess an objective character. But this denial implies, as a necessary conclusion, the denial of an evolutionary process by jumps."[20]

The practical political value of the dialectical laws is so strong in the Soviet Union that statesmen and politicians will always advance their arguments in the form of dialectical laws. In the *History of the Communist Party*, the basic text for the indoctrination of party members, Stalin writes, "Dialectics does not regard evolution as a simple process of growth in which quantitative changes do not lead to qualitative changes, but as a transition from hidden quantitative changes to conspicuous qualitative changes."[21] For example, in the evolution of the capitalistic society the changes are at first merely quantitative and hidden; the social revolution grows invisibly to the untutored eye until manifesting itself in a jump that recalls Hegel's theory of qualitative change, like the evaporation of water. In the same way, according to Marx, the coming change into collective property lurks behind the appearance of private property.

In a recent issue of a Russian popular science magazine, in an article entitled "Mechanists, and the Servants of Imperialism," we read the following:

> Mechanism strives to explain all natural and social phenomena in terms of the laws of mechanics, denying the qualitative diversity of existence and reducing all differences among objects and processes to quantitative differences. Mechanism obliterates the borders between distinct spheres of natural order, which ought to be investigated by different natural and social

sciences: it rejects the most important position of dialectics: the autonomy of matter; discontinuity of development; and the coming into being of new forms.²²

It is instructive that Soviet philosophy has this feature in common with a great many philosophical trends:²³ it also disapproves of the quantitative approach of modern science and maintains that science oppresses the most essential trait of nature: quality. These objections from Russia sound in some ways similar to the requirements of Thomistic metaphysics and several types of idealistic metaphysics, with their strict divisions between natural and social sciences.

Conclusion

Einstein's Philosophy of Science

1. The Positivistic Basis

If we contemplate the interpretations of science given by various philosophical "schools," we can see clearly that the "school" to which an individual scientist adheres is not determined to any great extent by the result of his or her research as a scientist but rather by religious or political predilections to support a certain philosophical interpretation. Perhaps it would be instructive to clarify this situation by presenting a philosophical interpretation given to physical laws not by "schools" but by an individual scientist.

We shall use as an example one of the greatest physicists of our time, Albert Einstein. Although the cleavage between positivistic and metaphysical concepts has often been called fundamental, Einstein has been claimed as an adherent on both sides of the gap. When we read his obituary on Ernst Mach,[1] we gain the impression that Einstein has been a clear-cut positivist:

> Science is nothing else but the comparing and ordering of our observations according to the methods and angles which we learn particularly by trial and error. . . . As results of this ordering abstract concepts and the rules of their connection appear. . . . Concepts have meaning only if we can point to objects to which they refer and to the rules by which they are assigned to these objects.[2]

Einstein refused to interpret Mach as an idealist or solipsist, as have both Thomists and dialectical materialists, while stressing the influence of Hume and Mach upon his thinking. In presenting the theory of relativity in 1921, Einstein wrote:

> The object of all science, whether natural science or psychology, is to co-ordinate our experiences and to bring them into a logical system. . . . The only justification of our concepts . . . is that they serve to represent the complex of our experiences; beyond this they have no legitimacy. I am convinced that the philosophers have had a harmful effect upon the progress of scientific thinking in removing certain fundamental concepts from the domain of empiricism, where they are under our control, to the intangible heights of *a priori*.[3]

These passages could have been written by positivists in the old sense, like Comte, Mach, or Stallo.[4] Occasionally Einstein called a system of propositions from which no statement about sense experience can be deduced "metaphysics" or "empty talk."[5] In this passage he uses the language of the "logical positivists" like Rudolf Carnap, who have declared metaphysics to be "meaningless."[6] There was, however, one point in the "positivism" of Comte and Mach with which Einstein did not agree. This positivism was often characterized by the claim that science is a registration of "sense observations." If we accept these words in their commonsense meaning, they imply that science describes our sense observations in the same way as a reporter describes some event (a fire, or a meeting, for instance) for the "record." Einstein emphasizes instead the active, creative work of the scientist. He does not "record" sense observations, he "produces a theory." As a matter of fact, the "new positivism" of Poincaré and the "logical positivism" of Carnap have always insisted that scientific theories are products of the human mind. According to this view, the general concepts and propositions of science certainly cannot be "deduced" from sense observations.

Einstein disagrees with "positivists" who require that "all those concepts and propositions which cannot be deduced from the sensory raw material are, on account of their 'metaphysical' character, to be removed from thinking."[7] The principles of the general theory of relativity are obviously not "deduced" from sense observations; neither are the principles of quantum theory. If we are clearly aware of these facts, it is easy to find, by careful analysis, that the same thing holds for the principles of Euclidean geometry and Newtonian mechanics. The new twentieth-century physics, particularly the theories of Einstein and Niels Bohr, have led scientists and philosophers to the general view that science is a product of human imagination: it is to be "checked" by sense observations but cannot be "deduced" from sense observations.

2. The Metaphysical Basis

In his Herbert Spencer Lecture, Einstein said clearly that philosophers of science of the eighteenth and nineteenth century

> were, most of them, possessed with the idea that the fundamental concepts and postulates of physics were not in the logical sense free inventions of the human mind, but could be deduced from experience by "abstraction"—that is to say, by logical means. A clear recognition of the erroneousness of this notion really only came with the general theory of relativity.[3]

He called the systems of axioms on which science is based "metaphysical" because the principles are neither statements about experiences nor statements logically "deduced" from experiences. If we understand "metaphysical" in this negative sense, Einstein's opinion is not really different from the logical positivists' view of the logical structure of theories. However, if we consider what we have called (Part I, chap. 2, §10) "philosophical interpretations of science," Einstein's interpretation is clearly different from the interpretations given by logical positivists, pragmatists, and operationalists. This means that Einstein's proposed commonsense interpretation of this "science of science" seems to support a more desirable behavior of the scientist than did the positivistic interpretation. Every scientific theory "invents" a set of "simple" principles from which the immense complexity of observable phenomena can be logically derived. The possibility of such "inventions" is what makes science possible.

We say that a domain of observational facts is "comprehensible" if we succeed in deriving them from a few simple principles. Now the question naturally becomes: How does it happen that the physical universe, or the world in general, is "comprehensible"? This question can scarcely be answered by a scientific theory. Einstein said, "The most incomprehensible thing about the world is that it is comprehensible."[9] For Thomistic philosophy, the answer is simple. The laws of nature are regarded as genuine laws given by a lawgiver who is a "rational being" with a mind similar (although superior) to the human mind. So, of course, natural phenomena must obey laws that are "comprehensible" as well to the human mind. Einstein does not join Thomists in assuming a "personal lawgiver." But he would like to say that the "universe is rational" or, in other words, the laws of nature can be formulated by the mathematics that the human mind has been able to construct. "I am convinced," writes Einstein, "that we can discover by means of purely mathematical constructions the concepts and the laws connecting them with each other which furnish the key to the understanding of natural phenomena."[10] He would not say that the validity of a physical theory can be proved by mathematics; he would say rather that it helps us to find a theory, which must be confirmed by experience.

"Experience," writes Einstein, "remains, of course, the sole criterion of the physical utility of a mathematical construction. But the creative principle resides in mathematics."[11] Occasionally, he even expresses himself in a way close to the language of ancient and medieval metaphysics. "In a certain sense," he writes, "I hold it to be true that pure thought is competent to comprehend the real, as the ancients dreamed."[12] This "dream of the ancients" was, of course, what we have described earlier as "seeing with the intellect," or "metaphysical insight." But his qualification "in a certain sense" brings us back to the "positivistic doctrine" that mathematical construction is an instrument for the invention of physical theory but not a criterion of its validity. This heuristic role of theoretical construction does not necessarily lead to metaphysical interpretations, according to which the physical world must agree with our mathematical constructions or, as some philosophers and scientists (for example, James Jeans)[13] have expressed it, the creator of the world must be a pure mathematician. One can certainly interpret the helpfulness of mathematical constructions in physics by commonsense analysis, which leaves us with the "positivistic" view of nature. We may quote as an example what Percy W. Bridgman writes about the role of simple mathematical laws in physical science. He distinguishes between two attitudes: "One is that there are probably simple general laws still undiscovered, the other is that nature has a predilection for simple laws."[14] Bridgman rightly emphasizes that simple means only "simple in terms of our human concepts." However, "our concepts are . . . hazy and do not fit nature exactly."[15] Hence there is little presumption in assuming that nature has a

> predisposition to simplicity as formulated in terms of our concepts. . . . The known laws of nature are "simple" if we consider only a limited range of facts. . . . It does not seem so very surprising that over a limited domain, in which the most important phenomena are of a restricted type, the conduct of nature should follow comparatively simple rules.[16]

The laws of nature seem simple because they are first approximations. The higher approximations are unknown to us and may be far from simple. However, this presentation of "simple" laws and their role in science was not satisfactory to Einstein. It does not provide sufficient mental stimulation to search general laws of nature.

3. The Analogical-Religious Basis

We may also give philosophical interpretations that make use of commonsense analogies that are different from Bridgman's "positivistic formulations" but nonetheless compatible with the results of all experiments performed and recorded by physicists. Einstein, in contrast to Bridgman, prefers the analogy to blueprints that reasonable beings have made of the physical world. He prefers it because of its higher emotional value. It gives strength to the physicist in his efforts to invent theories and to the student in his efforts to comprehend them. The analogy supports the belief that the "world is comprehensible," which is, for the physicist, identical with the belief that simple mathematical laws are to be found for describing natural phenomena. Einstein writes explicitly that "the faith in the possibility that the regulations valid for the world of existence are rational,' or, in other words, the belief that the world is "comprehensible to reason" belongs to the sphere of religion. "I cannot conceive," he writes, "of a genuine scientist without [this] profound faith."[17] He even maintains that the faith in the comprehensibility of nature is the center of genuine religiosity. "In this sense, and in this sense only, I belong in the ranks of devoutly religious men."[18]

Einstein speaks of a religious interpretation of science. If one formulates the most general principle in this "science of science," the "theory of theories," as belief in the existence of a system of simple mathematical propositions man can find and from which he can derive the immense complexity of observable facts, one formulates the chief tenet of "cosmic religion." By using this analogy, we cannot add anything to our knowledge. Religion cannot give us any insight, according to Einstein, but it can give us strength. Einstein formulated this opinion in a brief but memorable sentence: "Science without religion is lame, religion with science is blind."[19]

Notes

Chapter 1

1. Kurt F. Reinhardt *The Existential Revolt: The Main Themes and Phases of Existentialism* (Milwaukee: Burce, 1952), 2.

2. Adlai Stevenson, "A Purpose for Modern Woman," *Women's Home Companion* (September 1955): 29–31.

3. Archibald MacLeish, "Why Do We Teach Poetry?" *American Monthly* (March 1956) 197, no. 3: 43–53.

4. MacLeish, "Why Do We Teach Poetry?," 51.

5. MacLeish, 51.

6. Aldous Huxley, *Science, Liberty and Peace* (London: Chatto & Windus, 1950), 29. Originally published in 1946.

7. Frank quotes the preface of the second edition of *The Birth of Tragedy*. See, for example, Friedrich Nietzsche, *The Birth of Tragedy*, in *The Complete Works of Friedrich Nietzsche*, vol. 1, ed. Oscar Levy, London: George Allen & Unwin Ltd., New York: MacMillan (1886/1923), 3: "What I then laid hands on, something terrible and dangerous, a problem with horns, not necessarily a bull itself, but at all events a new problem: I should say today it was the *problem of science* itself—science conceived for the first time as problematic, as questionable."

8. Frank refers to the classic story of Prometheus. See Aeschylus, *Prometheus Bound*, trans. Marion Clyde Wier (New York: Century, 1916).

9. Frank refers here to the "Two Cultures" debate sparked by Charles Percy Snow who first published his "The Two Cultures" in the *New Statesman*, October 6, 1956.

10. A similar formulation can be found in Bergson's preface to the French translation of James's *Pragmatism*, which was translated into English recently as "On the Pragmatism of William James: Truth and Reality," in *Henri Bergson—Key Writings*, ed. Keith Ansell Pearson and John Mullarkey (New York and London: Continuum, 2002), 267–73.

11. Peirce introduced the idea of pragmatism in his "How to Make Our Ideas Clear?" originally published in 1878. In *Philosophical Writings of Peirce*, ed. Justus Buchler (New York: Dover, 1955), 23–41. We note that Peirce's article did not contain the word "pragmatism."

12. See, for example, Korzybski's *Science and Sanity: An Introduction to Non-Aristotelian Systems and General Semantics* (Lancaster, PA: Science Press, 1933) (second and enlarged edition 1941). Korzybski established the Institute of General Semantics, which in 1943 began publishing the movement's official journal, *ETC: A Review of General Semantics*. In that journal Frank published his paper "Science Teaching and the Humanities," *ETC* 4 no. 1 (1946): 1–24. About the Vienna Circle's relation to the Significs movement, see the essays in Gerard Alberts, Luc Bergmans and Fred Muller, eds., *Significs and the Vienna Circle: Intersections* (Dordrecht, The Netherlands: Springer, forthcoming).

13. *Christian Science Monitor*, February 4, 1956.

14. *New York Times Magazine*, February 12, 1956.

15. Huxley, *Science, Liberty and Peace*, 28–29.

16. Huxley, 28.

17. Huxley, 27.

18. The notion of 'conceptual scheme' is prominent in James B. Conant's *On Understanding Science, An Historical Approach* (New Haven, CT: Yale University Press, 1947). For more on Frank's personal and professional relationship with Harvard's president Conant, see George Reisch, "Pragmatic Engagements: Philipp Frank and James Bryant Conant on Science, Education, and Democracy," *Studies in East European Thought* 69, no. 3 (2016): 227–44.

19. Huxley, *Science, Liberty and Peace*, 27.

20. Huxley, 27.

21. Huxley, 28.

22. Neurath argued similarly that the sciences are already unified in certain ways. Like Frank, he claimed that in order to make even simple predictions, such as "The forest will burn," one must utilize different sciences like physics, chemistry, biology (about the nature of the wood and the forest); meteorology (about rain and wind); and even social sciences like sociology and economics (to know the location, preparedness, and determination of the fire department). See, for example, Neurath's article from 1938 for the first monograph of the *International Encyclopedia of Unified Science*, "Unified Science as Encyclopedic Integration," in *Foundations of the Unity of Science*, ed. Otto Neurath, Rudolf Carnap and Charles Morris, vol. 1 (Chicago: University of Chicago Press, 1971), 1–27. See also Neurath, "Unified Science and its Encyclopedia," in *Otto Neurath: Philosophical Papers, 1913–1946*, ed. Robert S. Cohen and Marie Neurath (Dordrecht, The Netherlands: Reidel, 1937/1983), 172–82. In fact, Frank cites this illustration of a forest fire in his chapter about the Vienna Circle (see Pt. 2. chap. 5, §4).

23. See Charles Morris, *International Encyclopedia of Unified Science*, vol. 1: *Foundations of the Theory of Signs* (Chicago: Chicago University Press, 1938); Rudolf Carnap, *International Encyclopedia of Unified Science*, vol. 1: *Foundations of Logic and Mathematics* (Chicago: University of Chicago Press, 1938). Both are reprinted in *Foundations of the Unity of Science*, vol. 1 (Chicago: University of Chicago Press, 1971).

24. Besides briefly mentioning Copernicus, Frank's manuscript at this point does not contain any detailed presentation of "a very old" example.

25. This paragraph originally started with the following sentences that Frank struck from the manuscript: "These studies of the relations between the person of the scientist and the symbols of science have been called 'pragmatic' in contrast to 'semantic.' We have to be aware that there is also a very important study of the relation between the symbols themselves and this study is referred to as 'syntax' or 'logic.' Altogether the triadic scheme of science implies three fields of study: observation, semantics, and pragmatics."

26. See Plato, *The Republic*, trans. Paul Shorey, vol. 2, bks. 6–10, Loeb Classical Library no. 276 (Cambridge, MA: Harvard University Press, 1943).

27. Frank's manuscript originally contained "determinism" but he struck "ism" and wrote above it "ation," which accounts for the odd use of "determination" in this context.

28. Frank defined "school philosophy" elsewhere as "a world-conception that has become rooted in the educational system through centuries-old tradition." See his "Physical Theories of the Twentieth Century and School Philosophy," in *Modern Science and its Philosophy*, Cambridge, MA: Harvard University Press, 1949), 90–121. This paper was originally published in 1929–1930.

29. J. B. S. Haldane, *The Marxist Philosophy and the Sciences* (New York: Random House, 1938).

30. Jerome Frank, *Fate and Freedom* (New York: Simon and Schuster, 1945).

Chapter 2

1. Ralph Waldo Emerson, *Conduct of Life* (Cambridge, MA: Riverside, 1860), 223.

2. Emerson, *Representative Man* (Boston: Houghton, Mifflin, 1876/1884), 62. Emerson's text includes a remark that Frank included at first but later struck from his manuscript: "So there is a science of sciences,—I call it Dialectic,— which is the Intellect discriminating the false and the true."

3. Emerson, *Representative Man*, 62.

4. Emerson, *Nature* (Boston: James Munroe, 1836), 41–42.

5. Pitirim Sorokin, *The Crisis of Our Age* (New York: E. P. Dutton, 1942), 98.

6. Sorokin, *Crisis of Our Age*, 98–99.

7. Herbert Samuel was the leader of the British Liberal Party between 1944 and 1955.

8. Herbert L. Samuel, *Essay in Physics* (New York: Harcourt, Brace, 1952), 123. Frank reviewed Samuel's book *'Newton Didn't Talk About Why and How': Essay in Physics by Herbert Samuel* in the *New York Times*, February 17, 1952.

9. Samuel, *Essay in Physics*, 20.

10. Samuel, 25.

11. Samuel, 25.

12. Samuel, 117.

13. Nicolas Berdyaev, *Solitude and Society*, trans. George Reavey (London: G. Bles, Centenary, 1938), 14.

14. Berdyaev, *Solitude and Society*, 14.

15. Mark Borisovich Mitin, *Dialehticheskii i istoricheskii materialism* (Moscow: Philosophical Institute of the Communist Academy, 1933), 57.

16. A more precise translation than Frank's reads: "In this way contemporary idealists, with all their might and main, purge science . . . of its content, its meaning, its truth." We thank Andrey Maidansky for this and subsequent translations from Russian.

17. Frank noted here, at the end of the paragraph, "At this point we are using the names of philosophical doctrines, materialism, idealism, positivism, in the vague ways in which they are used by educated men who are not specialists in philosophy. We shall discuss later on the meaning of these terms more elaborately."

18. Francis Bacon, *Advancement of Learning*, ed. Joseph Devey (New York: Collier, 1901), 150.

19. Bacon, *Advancement of Learning*.

20. Bacon, 150–51.

21. The passage is from Bacon's *A Description of the Intellectual Globe* (*Descriptio Globi Intellectualis*), in *The Philosophical Works of Francis Bacon*, ed. J. H. Robertson (London, Routledge, 1905), 677–702 (originally published in 1653). In his *Philosophy of Science: The Link between Science and Philosophy* (Englewood Cliffs, NJ: Prentice-Hall, 1957), Frank also cites a part of this passage and refers in a footnote (p. 365, n 36) to "Francis Bacon, *Descriptio Globi Intellectualis* [. . .]. See *The Philosophical Works of Francis Bacon*, edited by Ellis and Spedding (London, 1857)." The edition of Spedding and Ellis was, however, in Latin. A later edition and translation of Bacon's writings (based on the Ellis and Spedding edition) contains a different translation than the one given by Frank (see *The Philosophical Works of Francis Bacon*, ed. John M. Robertson [New York: E. P. Dutton & Co., 1905], 685). The same translation appears in William Whewell's *History of the Inductive Sciences—From the Earliest to the Present Times*, vol. 1 (Cambridge: Deighton, 1837), 387. Given that Frank referred to Whewell's *History* in his *Philosophy of Science*, he may have taken the

translation from there. See, in addition, Frank, "Philosophical Uses of Science," *Bulletin of the Atomic Scientists* (April 1957): 125–30, 127.

22. Bacon, *Description of the Intellectual Globe*, 684.

23. Bacon, *Theory of the Earth (Thema Coeli)*, in *The Philosophical Works of Francis Bacon*, ed. John M. Robertson (New York: E. P. Dutton & Co., 1905), 703–9, 708.

24. Bacon, *Theory of the Earth*, 708.

25. This claim is closely related to Plato's *Timaeus*, in which rational understanding of nature is possible, according to Timaeus, because the world itself is "a creature with life, soul, and understanding [. . .] through the providence of God." See *Plato: Timaeus and Critias*, trans. A. E. Taylor (London: Methuen), 1929), 27, 30c.

26. Gregory's memoranda are found in W. G. Hiscock, *David Gregory, Isaac Newton and Their Circle: Extracts from David Gregory's Memoranda 1677–1708* (Oxford, 1937), 29.

27. Frank added this note here: "This term of scholastic philosophy is discussed in Part II, Ch. 7, section 6, 'On Angels and Genuine Laws.'"

28. Thomas Aquinas, *Summa theological*, trans. Fathers of the English Dominican Province (New York: Benziger Bros., 1948), Question 16, Article 1, Answer.

29. Aquinas, *Summa theological*, Question 16, Article 2, Answer.

30. Aquinas, Question 16, Article 2, Answer.

31. Aquinas, Question 16, Article 2, Answer.

32. Jacques Maritain, *A Preface to Metaphysics: Seven Lectures on Being* (New York: Sheed & Ward, 1939), 84. In Maritain's book, instead of "probable" we find "possible."

33. Henri Bergson, *An Introduction to Metaphysics*, trans. T. E. Hulme (New York and London: G. P. Putnam's Sons, 1903/1912), 2.

34. Bergson, *An Introduction to Metaphysics*, 2–3.

35. Filmer S. C. Northrop, *The Logic of the Sciences and the Humanities* (New York: Macmillan, 1948), 82.

36. Northrop, *Logic of the Sciences*, 87.

37. Bertrand Russell, *A History of Western Philosophy* (London: George Allen and Unwin, 1947), 53.

38. Édouard Le Roy, "Science et Philosophie," *Revue de Métaphysique et de Morale* 7 nos. 4–5 (1899): 375–425, 503–62, 708–31, 424. Italics in the original.

Chapter 3

1. Charles S. Peirce, "Philosophy and the Sciences: A Classification," in *Philosophical Writings of Peirce*, ed. Justus Buchler (New York: Dover, 1955), 60–73, 66.

2. Peirce, "Philosophy and the Sciences."
3. Peirce, 67–68.
4. Peirce, 61.
5. Peirce, 72.
6. Peirce, 68.
7. Maritain, *Preface to Metaphysics*, 29.
8. Edmond Goblot, *Le Systeme des Sciences* (Paris: A. Colin, 1922), 87.
9. Goblot, *Le Systeme*, 87–88.
10. Goblot, 88.
11. Goblot, 88.
12. Henry Vincent Gill, *Fact and Fiction in Modern Science* (Dublin: Gill, 1943), 21.
13. Gill, *Fact and Fiction*, 21.
14. Jacques Maritain, *Distinguish to Unite or the Degrees of Knowledge* (London: G. Bles, 1937), 4.
15. Alfred N. Whitehead, *Adventures of Ideas* (Cambridge: Cambridge University Press, 1933/1961).
16. Whitehead, *Adventures*, 122–23.
17. Whitehead, 148.
18. Whitehead, 148.
19. Whitehead, 148.
20. Jacques Maritain, *The Range of Reason* (New York: Charles Scribner's, 1952), 11. Italics in the original.
21. Maritain, 8.
22. Maritain, 8. Italics in the original.
23. Maritain, 8. Italics in the original.
24. Maritain, 9.
25. Gerald B. Phelan, *Saint Thomas and Analogy* (Aquinas Lecture, 1941) (Milwaukee: Marquette University Press, 1941), 8.
26. See Whitehead, *Adventures of Ideas*, chap. 7, esp. 115.
27. Whitehead, *Adventures*, 120.
28. Whitehead, 116. This edition reads: "The order of nature expresses the characters of the real things which jointly compose the existences to be found in nature."
29. Werner Heisenberg, *The Physicist's Conception of Nature* (London: Hutchinson, Scientific and Technical, 1958), 24. Original emphasis.
30. Heisenberg, *Physicist's Conception*, 29. Original emphasis.
31. Heisenberg, 29. Original emphasis.
32. Heisenberg, 23. Original emphasis. The edition reads, "in the course of history."

Chapter 4

1. On the development of Frank's ideas about simplicity in the 1940s and 1950s, see Amy Wuest, "Simplicity and Scientific Progress in the Philosophy of Philipp Frank," *Studies in East European Thought* 69, no. 3 (2016): 245–55.

2. Discussions of the sociological criteria were not typical in philosophy of science books in and around the 1950s. Frank's somewhat misleading claim might be explained in several ways. Perhaps he was thinking of discussions held in Boston and Cambridge by his colleagues (such as James Conant, Robert Cohen, and Marx Wartofsky) who broadly shared Frank's sociological interests. Alternately, Frank perhaps did not follow closely the latest trends in American philosophy of science and had in mind pragmatist authors (like Dewey). Or, Frank had in mind something entirely different, especially given his use of "philosophy of science" in quotation marks in this chapter but not in the others; perhaps his "philosophy of science" is just "sociology of science" in disguise. Evidence for the second option includes that there is no mention in this manuscript of Hans Reichenbach's (or the late Carnap's) technical philosophies of science. Hilary Putnam lodged a similar observation (and objection) in his review of Frank's *Philosophy of Science: The Link Between Philosophy and Science*. "Anyone who still thinks that the issue in philosophy of science is between 'operational definition' and 'metaphysical interpretation' might enjoy reading [Frank's] book," Putnam wrote, and continued, "Afterward, he should learn some real philosophy of science" (see *Science* 127, no. 3301 [1958]: 750–51, 750).

3. On "Apparent" and "Real," see Frank, *The Law of Causality and Its Limits*, trans. Marie Neurath and Robert S. Cohen (Dordrecht, The Netherlands: Kluwer, 240–46). Originally published in 1932.

4. Alfred North Whitehead, *The Principle of Relativity with Applications to Physical Science* (Cambridge: Cambridge University Press, 1922), 6.

5. On chance philosophy, see Frank, "Foundations of Physics" in *International Encyclopedia of Unified Science* (Chicago: University of Chicago Press), chap. 1 and note 1, where he quotes Whitehead again.

6. Dialectical materialism is discussed in more detail in Part II.

7. Frank noted for this and the following shorter quotation only: "See the article 'Formalism' in the *Large Soviet Encyclopedia*, vol. 58."

8. Frank might have in mind Vladimir Ivanovich Nevski (or Nevskii, or Nevskiy), a well-known critic of Bogdanov and the author of an appendix to the second edition of Lenin's *Materialism and Empirio-Criticism* entitled "Dialectical Materialism and the Philosophy of Dead Reaction." See also Albert Einstein, *Geometrie und Erfahrung* (Berlin: Julius Springer, 1921) (for the English translation, see "Geometry and Experience," in *The Collected Papers of Albert Einstein*, Vol. 7: *The Berlin Years: Writings, 1918–1921*, trans. Alfred Engel (Princeton,

NJ: Princeton University Press, 2002), 208–222. Frank noted in the manuscript only the journal and the year as "*Pod Znamenem Marxizma*, 1923."

9. Mitin, *Dialehticheskii i istoricheskii materialism*, 56. A more precise translation reads: "Science has nothing to do with the *representation* of objective things and their relations, but with physico-mathematical signs, which are *free inventions*, with symbols which are assigned to the relations between subjective impressions. From combining these signs by means of mathematical equations, they derive new combinations which they denote by new signs, etc. The development of science consists, according to the Machists, in amending the construction of that system of signs."

10. Mitin, *Dialehticheskii i istoricheskii materialism*, 56. A more precise translation reads: "Here it becomes particularly clear how their *subjective idealism goes hand in hand with mechanism*."

11. Sergei I. Vavilov, "The New Physics and Dialectical Materialism," *Modern Quarterly* 2, no. 2 (1939), 146–54, 151. Frank noted here: "From 1945, he was President of the Academy of the USSR. According to the *Soviet Encyclopedia*, 2nd edition, vol. 6 (1951), 'He devoted a great deal of attention to the problems of the philosophy of science. In a series of papers he revealed the greatness of the ingenious scientific work of Lenin and Stalin. He fought against the idealism of foreign physicists and philosophers.'"

12. Abraham F. Ioffe, "Razvitie atomisticheskikh vozzrenii v XX veke," *Pod Znamenem Marksizma* 4 (1934): 52–68, 66.

13. The source of this quotation is undetermined. It is also uncertain whom Frank has in mind. One likely "Führer of the German university teachers" is Walter Schultze (1894–1979), the so-called Reichsdozentenführer, who led the *Nationalsozialistischer Deutscher Dozentenbund* between 1935 and 1944. On the other hand, the *Reichsministerium für Wissenschaft, Forschung und Volksbildung* responsible for German universities was founded in 1934, and Bernard Rust became Reichminister. Talks by Rust include "Nationalsozialismus und Wissenschaft" (*Das nationalsozialistische Deutschland und die Wissenschaft. Heidelberger Reden von Reichs-minister Rust und Prof. Ernst Krieck*, Hamburg 1936, 9–22) and "Freiheit und Ordnung. Rede bei der Kundgebung der Studenten auf dem Ehrenhof der Universität am 28. Juni 1937" (*Wissenschaft und Glaube*, Oldenburg i.O., Berlin 1938, 20–27). Since Frank quotes Rust in chapter 6, Frank may have him in mind here. Although this quotation does not match Rust's papers to the letter, it matches their spirit.

14. Frank quotes Kunstmann, Rust, and also Göring in his article on the "Philosophical Uses of Science," *Bulletin of the Atomic Scientists*, April 1957, 125–30, 128.

15. Heinrich K. Kunstmann, *Rede zur Feier der Immatrikulation, gehalten in der Aula der Neuen Universität Heidelberg, 23. Nov. 1936* (Heidelberg: Carl Winter, 1937), 6. Kunstmann was not a regular professor with a habilitation

at Heidelberg; he was a medical doctor from Pforzheim (near Heidelberg) and became an honorary professor only in 1934 at the Faculty of Medical Sciences.

16. Kunstmann, *Rede zur Feier der Immatrikulation*, 7.

17. V. I. Lenin, *Materialism and Empirio-Criticism: Critical Comments on a Reactionary Philosophy* (Moscow: Foreign Languages, 1909/1947), 355. The published translation reads, "*Not a single one* of these professors, who are capable of making valuable contributions in the special fields of chemistry, history, or physics, *can be trusted one iota* when it comes to philosophy" (emphases in original).

18. The passage could be found, for example, in Herbert Dingle's "Copernicus," *Observatory* 65 (1943): 38–57, 53: "For my part, I have always felt about hypotheses that they are not articles of faith, but bases of calculation, [so that], even if they be false, it matters not so long as they exactly represent the phenomena of the motions. [. . .] It would therefore seem an excellent thing for you to touch a little on this point in the Preface. For you would thus render more complacent the Aristotelians and theologians, whose contradiction you fear." See Frank, 'The Philosophic Meaning of the Copernican Revolution," in *Modern Science and its Philosophy* (Cambridge, MA: Harvard University Press, 1949), 219–20.

19. Andreas Osiander, "Preface to *On the Revolutions*," in *Nicholas Copernicus on the Revolutions*, trans. Edward Rosen (London and Basingstoke, UK: Macmillan, 1978), xvi. This edition reads: "For these hypotheses need not be true nor even probable. On the contrary, if they provide a calculus consistent with the observations, that alone is enough." The title of Copernicus's book in English is *On the Revolutions of the Heavenly Spheres* (see, e.g., the translation by Charles Glenn Wallis, *Great Books of the Western World* 16, [Chicago: William Benton], 497–838).

20. Cardinal Robert Bellarmine's letter from April 4, 1615, to Father Paolo Antonio Foscarini.

21. The passage is to be found in Arthur Koestler's *The Sleepwalkers: A History of Man's Changing Vision of the Universe* (New York: Macmillan, 1959), 454–55, where it reads: "It seems to me that your Reverence and Signor Galileo act prudently when you content yourselves with speaking hypothetically and not absolutely. . . . For to say that the assumption that the Earth moves . . . is a very dangerous attitude and one calculated not only to arouse all Scholastic philosophers and theologians. . . . To demonstrate that the appearances are saved by assuming the sun at the centre and the earth in the heavens is not the same thing as to demonstrate that *in fact* the sun is in the centre and the earth in the heavens." See also Pt. II, chap. 6, n. 1 (p. 336).

22. Alfred O'Rahilly, *Electromagnetic Theory: A Discussion of Fundamentals* (London and New York: Longmans, Green, 1938), 374.

23. P. J. McLaughlin, "Review of O'Rahilly's Electromagnetics," *Studies: An Irish Quarterly Review* 27, no. 108 (1938): 656–66, 665.

24. Duhem's letter is published in Réginald Garrigou-Lagrange, *God, His Existence and His Nature: A Thomistic Solution of Certain Agnostic Antinomies*, vol. 2 (St. Louis and London: Herder Book, 1949), 449–50. Italics in the original.

25. Duhem's letter in Garrigou-Lagrange, *God, His Existence and His Nature*, 451.

26. Garrigou-Lagrange, *God, His Existence and His Nature*, 448. Italics in the original.

27. Frank wrote, "To prefer the 'reality of the earth's movement' would mean to advance in astronomy to advance in moral conduct." Since the thrust of Frank's discussion is to contrast, and not to join, astronomical and moral goals, we have rewritten this sentence accordingly.

28. Frank, "Zeigt sich in der modernen Physik ein Zug zu einer spiritualistischen Auffassung?," *Erkenntnis* 5 (1935): 65–80, 71. Frank refers to Planck's pamphlet of 1913, *Positivismus und reale Aussenwelt* (Leipzig: Akademische Verlagsgesellschaft).

29. See, for example, Curt John Ducasse, *The Method of Knowledge in Philosophy* (The Howison Lecture for 1944), University of California Publications in Philosophy, vol. 16, no. 7 (Berkeley and Los Angeles: University of California Press, 1945). Reprinted in *American Philosophers at Work: The Philosophic Scene in the United States*, ed. Sidney Hook (New York: Criterion, 1956), 207–224 and *Philosophy as a Science* (New York: Oskar Piest, 1941).

30. Curt John Ducasse, "Philosophy and Natural Science," *Philosophical Review* 49, no. 2 (1940): 121–41, 139. Italics in the original.

31. Ducasse, *Method of Knowledge in Philosophy*, 218.

32. Frank noted, "The first answer was advocated by Margenau, the second one by Werkmeister." Frank likely refers to Henry Margenau's *The Nature of Physical Reality: A Philosophy of Modern Physics* (New York: McGraw-Hill, 1950); and to William H. Werkmeister's *The Basis and Structure of Knowledge* (New York: Harper and Brothers, 1948). Frank discusses this question and these sources in more detail in his *Philosophy of Science: The Link Between Science and Philosophy*, Chapter 10, "Metaphysical Interpretations of the Atomic World," especially pages 246–249.

33. Frank discusses this in more detail in his *The Law of Causality and its Limits*, chap. 4, §24.

34. Joseph Goebbels, *Der Angriff* (*The Attack*) (Munich: Zentralverlag der NSDAP, 1940). Frank noted, "The book contains essays which were written (from 1927), before Goebbels entered the government" but did not specify page numbers. He quoted the same passage in his biography of Einstein (again without page numbers): *Einstein, His Life and Times* (New York: Alfred A. Knopf, 1947), 250.

35. Frank refers to Nikolai Iakovlevich Danilevskii's major work from 1895. The book was recently translated into English by Stephen M. Woodburn and was published under the title *Russia and Europe: The Slavic World's Political*

and Cultural Relations with the Germanic-Roman West (Bloomington: Indiana University Press, 2013). Frank noted, "In the *Soviet Encyclopedia*, 2nd ed. vol. 13 (1952)," described as "reactionary and enemy of Darwinism."

36. Frank refers to a 1934 issue of the *Zeitschrift für die Gesamte Naturwissenschaft*, but we were unable to determine the source of the following two quotations.

37. *Large Soviet Encyclopedia*, 65 vols. (Moscow: OGIZ, 1926–1947).

38. *Large Soviet Encyclopedia*, vol. 65 (Moscow: OGIZ, 1931), 17. The author of the article "Aether" is Boris Hessen ("Gessen" in Russian spelling), the father of externalism in the philosophy of science. He was arrested and executed in 1936.

39. *Large Soviet Encyclopedia*, vol. 43 (Moscow: OGIZ, 1939), 616.

40. *Large Soviet Encyclopedia*, 616.

41. *Large Soviet Encyclopedia*, 617.

42. *Large Soviet Encyclopedia*, 618.

43. *Large Soviet Encyclopedia*, 618.

44. *Large Soviet Encyclopedia*, 496. Frank noted, "The authors are M. Mitin and A. Schczelow." The author of the article was in fact Jan Sten (Jānis Stens), who was arrested in 1936 and executed the next year. His article "Philosophy" was signed by Mitin and his two assistants. This story came to the light in the 1960s when Sten's widow charged Mitin with plagiarism and caused a scandal.

Chapter 5

1. Frank noted: "The doctrine of historical materialism has its roots in Marx, Karl, *Critique of Political Economy*, 1859, and Engels, F., *The Development of Socialism from Utopia to Science* (New York, 1894) [and] *Historical Materialism* (New York, 1917)." See Marx, *A Contribution to the Critique of Political Economy*, trans., N. I. Stone (Chicago: Charles H. Kerr, 1859/1904), and Engels, *Development of Socialism from Utopia to Science*, trans. D. de Leon (New York: National Executive Committee of the Socialist Labor Party, 1900). We could not identify any edition of the third text Frank noted.

2. Georg Lukács, *History and Class Consciousness: Studies in Marxist Dialectics*, trans. Rodney Livingstone (Cambridge, MA: MIT Press, 1923/1971), 68. This edition reads, "For the proletariat truth is a weapon that brings victory; and the more ruthless, the greater the victory. This makes more comprehensible the desperate fury with which bourgeois science assails historical materialism: for as soon as the bourgeoisie is forced to take up its stand on this terrain, it is lost. And, at the same time, this explains why the proletariat and only the proletariat can discern in the correct understanding of *the nature of society* a power-factor of the first, and perhaps decisive importance."

3. Engels's letter to Franz Mehring from July 14, 1893. Quoted in Robert K. Merton, "Sociology of Knowledge," in *Twentieth-Century Sociology*, ed. Georges Gurvitch and Wilbert E. Moore (New York: Philosophical Library, 1945), 366–405, 391–92. Frank's referred to a British 1947 edition of Marx's *Selected Works*, most likely: *Selected Works*, ed. V. Adoratsky (London: Lawrence and Wishart, 1945).

4. Merton, "Sociology of Knowledge," 391.

5. Max Weber, *The Protestant Ethic and the Spirit of Capitalism*, trans. Talcott Parsons (New York: Charles Scribner's Sons, 1905/1950).

6. Merton, Robert, "Karl Mannheim and the Sociology of Knowledge," in *Social Theory and Social Structure*, rev. ed. (New York: Free Press, 1941/1957), 489–508, 494.

7. Karl Mannheim, *Ideology and Utopia*, trans. Louis Wirth and Edward Shils (New York: Harcourt, Brace, 1929/1954), 68–69.

8. Merton, "Karl Mannheim and the Sociology of Knowledge," 493.

9. Mannheim, *Ideology and Utopia*, 69.

10. Merton, "Sociology of Knowledge," 370–71.

11. For further details and contextualization of Frank's sociology of knowledge, see Thomas Uebel, "Logical Empiricism and the Sociology of Knowledge: The Case of Neurath and Frank," *Philosophy of Science* 67 (2000): 138–50; and Adam Tamas Tuboly, "Philipp Frank's Decline and the Crisis of Logical Empiricism," *Studies in East European Thought* 69, no. 3 (2016): 257–76.

12. Frank made a similar comment about economic factors and the role of funds also in his essay, "The Role of Authority in the Interpretation of Science" *Synthese* 10 (1956): 335–38, 335.

13. Lukács, *History and Class Consciousness*, 5. The English translation reads: "It goes without saying that all knowledge starts from the facts. The only question is: which of the data of life are relevant to knowledge and in the context of which method? The blinkered empiricist will of course deny that facts can only become facts within the framework of a system—which will vary with the knowledge desired. . . . In so doing he forgets that however simple an enumeration of 'facts' may be, however lacking in commentary, it already implies an 'interpretation.' Already at this stage the facts have been comprehended by a theory, a method; they have been wrenched from their living context and fitted into a theory."

14. Mannheim, *Ideology and Utopia*, 246.

15. Mannheim, 246.

16. Lukács, *History and Class Consciousness*, 5. The published English translation reads: "seek refuge in the methods of natural science, in the way in which science distills 'pure' facts and places them in the relevant contexts by means of observation, abstraction and experiment. They then oppose this ideal model of knowledge to the forced constructions of the dialectical method."

17. It must be noted that in the previous section Frank celebrated Lukács for his knowledge of "new positivism" and because Lukács seemed to accept the "theory-ladenness" of facts. At this point, Frank seems to object to Lukács for not considering "new positivism" and thus remaining "able to prove that natural science furnishes objective results." In fact, however, Lukács presents the previous quotation only as the possible answer of the bourgeois scientist (or, in his words, the "opportunist"). He goes on to say that the seemingly objective character of the natural sciences is only an illusion enforced by external conditions He writes in the paragraph following this quotation (on the same page from which Frank quoted him), "Opportunists always fail to recognise that it is in the nature of capitalism to process phenomena in this way." To see that Lukács indeed accepted an early version of the theory-ladenness of facts we point out that in the German original, instead of "processing the phenomena" he spoke about "die Erscheinungen in dieser Weise zu produzieren." (That is, capitalism *produces* the phenomena in its own way, and does not simply process it.)

18. Marx, *A Contribution to the Critique of Political Economy*, 12.

19. Mitin, *Dialehticheskï i istoricheskii materialism*, 40.

20. Lukács, *History and Class Consciousness*, 10–11. The translation from the later English edition reads: "When the ideal of scientific knowledge is applied to nature it simply furthers the progress of science. But when it is applied to society it turns out to be an ideological weapon of the bourgeoisie. For the latter it is a matter of life and death to understand its own system of production in terms of eternally valid categories: it must think of capitalism as being predestined to eternal survival by the eternal laws of nature and reason "

21. Mannheim, *Ideology and Utopia*, 274.

22. Mannheim, 274. Italics in the original.

23. The "German School of Wissenssoziologie" usually refers to the classic works of Karl Mannheim, Max Scheler, and Wilhelm Jerusalem, although Frank might have Georg Lukács in mind as well. Frank never quotes Scheler and Jerusalem (although he certainly knew the latter's work from the Viennese years); on the other hand it should be mentioned that both Mannheim and Lukács were originally Hungarian scholars who emigrated to Germany in the 1910s and published their major works in German.

24. See Merton, "Karl Mannheim and the Sociology of Knowledge."

25. Frank refers to Comte's published lectures delivered between 1830 and 1842 as *Cours de philosophie positive*. These were somewhat freely translated and condensed by Harriet Martineau as *The Positive Philosophy of Auguste Comte* (London: George Bell & Sons, 1896).

26. This refined picture of facts is further discussed by Frank in his essay "Einstein, Mach, and Logical Positivism," in *Albert Einstein, Philosopher-Scientist*, ed. Paul A. Schilpp (New York: Harper & Row, 1949), 271–86.

27. Frank likely refers to the puzzle he introduced in section 7, namely that while it might be easy to identify sociological influences within social scientific knowledge, many adherents of the early sociology of knowledge, like Mannheim, thought that natural scientific knowledge poses an exception.

Chapter 6

1. This passage in Arthur Koestler's *The Sleepwalkers: A History of Man's Changing Vision of the Universe* (454–55) reads: "It seems to me that your Reverence and Signor Galileo act prudently when you content yourselves with speaking hypothetically and not absolutely, as I have always understood that Copernicus spoke. *For to say that the assumption that the Earth moves and the Sun stands still saves all the celestial appearances . . . is to speak with excellent good sense and to run no risk whatever. Such a manner of speaking suffices for a mathematician . . . [i]f there were a real proof* that . . . the Sun does not go round the Earth but the Earth round the Sun, then we should have to proceed with great circumspection in explaining passages of Scripture which appear to teach the contrary. . . . But I do not think there is any such proof *since none has been shown to me*. To demonstrate that the appearances are saved . . . is not the same thing as to demonstrate *in fact* the sun is in the centre and the earth in the heavens . . . in case of doubt one may not abandon the Holy Scriptures as expounded by the holy Fathers."

2. M. B. Mitin, "Twenty-Five Years of Philosophy in the U.S.S.R.," *Philosophy (The Journal of the British Institute of Philosophy)* 19, no. 72 (1944): 76–84. The following quotations are from page 81.

3. While German scientists under Hitler pursued the creation of an atomic bomb, they did not successfully create one, as Frank's reference to "the blessings of the atomic bomb" may seem to imply.

4. The exact source of these quotations is unknown to us. Frank might refer here to the so-called Alsos Mission for which Goudsmit served as chief scientific advisor. The mission was an American effort during World War II to gather intelligence about German research in nuclear weapons, as well as chemical and biological research.

5. Lenin, *Materialism and Empirio-Criticism*, 355. The official translation is: "*Not a single one* of these professors, who are capable of making valuable contributions in the special fields of chemistry, history, or physics, *can be trusted one iota* when it comes to philosophy." Frank cites this translation in his "Philosophical Uses of Science," 128.

6. Frank may quote Rust from Leo Nitschmann's article, "Einstein entsinnlichte den Kosmos. Die Relativitätstheorie als kulturgeschichtliches Ereignis," *Die*

Zeit 50, December 16, 1954. The German original reads, "Der Nationalsozialismus ist nicht wissenschaftsfeindlich, sondern nur theorienfeindlich."

7. We were unable to identify the source of this quotation.

8. Hans Frank, "Der Nationalsozialismus und die Wissenschaft der Wirtschaftslehre," *Schmollers Jahrbuch* 58, no. 2 (1934): 641–50, 646.

9. Frank refers to the Soviet Union and nations under Soviet influence in the 1940s and 1950s, including East Germany, Hungary, Romania, Yugoslavia, Czechoslovakia, and the Baltic States. China and North Korea were also deemed totalitarian, but Frank does not appear to have them in mind here.

10. Michael Polanyi, *Science, Faith and Society* (London: Oxford University Press, 1946), 28. In 1957 Frank participated with Polanyi and Mortimer Adler in a conference on "Science and Philosophy" (see *Bulletin of the Atomic Scientists* 13, no. 4 (April 1957): 114–130.

11. Polanyi, *Science, Faith and Society*, 27. Frank's quotation omits the words "a judicial decision by" that we include here.

12. Polanyi, 53–54.

13. Polanyi, 57.

14. Frank might have in mind Rev. Michael P. Walsh, who was chairman of the Department of Biology at Boston University in the late 1950s. We are unable, however, to identify the source of this quotation.

15. Frank refers indirectly here to the annual Conference on Science, Philosophy, Religion and Their Relation to the Democratic Way of Life, in which Frank participated regularly. See especially his book *Relativity, a Richer Truth* (Boston: Beacon, 1950).

16. Although in his notes Frank claimed that these quotes are from Jerome Frank's abovementioned *Fate and Freedom*, we are unable to identify the exact source of these two quotations.

17. Charles Malik, *War and Peace: A Statement Made before the Political Committee of the General Assembly, November 23, 1949* (Stamford, CT: Overlook, 1950), 24.

18. Malik, *War and Peace*, 24.

19. Malik, 24.

20. Northrop, *Logic of the Sciences and the Humanities*, 342. In the book one finds "deductively formulated normative social theory. It is not verified by the methods of natural science applied to social facts."

21. Northrop, 355.

22. Northrop, 343.

23. The connection Frank seems to be referring to consists in objective sense impressions that are in some sense shared by different individuals.

24. Northrop, *Logic of the Sciences and the Humanities*, 353–54.

25. Northrop, 354–55.

26. John Dewey, *Reconstruction in Philosophy* (New York: Henry Holt, 1920), 25–26. We take Frank's meaning to be "philosophical pictures of the universe" at the beginning of the paragraph.

27. See Ernst Topitsch, "Kosmos und Herrschaft. Ursprünge der 'politischen Theologie.'" *Wort und Wahrheit: Monatsschrift für Religion und Kultur* 10, no. 1 (1955): 19–30 and "Society, Technology, and Philosophical Reasoning," *Philosophy of Science* 21, no. 4 (1954): 275–96.

Chapter 7

1. Alfred North Whitehead, *Science and the Modern World* (New York: Pelican Mentor, 1948).

2. Carl L. Becker, *The Heavenly City of the Eighteenth-Century Philosophers* (New Haven, CT, and London: Yale University Press, 1932).

3. Becker, *Heavenly City*, 21.

4. Becker, 21.

5. Becker, 22 and 26.

6. Archbishop of Boston from 1907 until his death in 1944.

7. William Cardinal O'Connell, *Recollections of Seventy Years* (Boston and New York: Houghton Mifflin, 1934), 23.

8. O'Connell, *Recollections*, 324.

9. Frank's phrasing of this sentence is ambiguous. We take him to mean that Pope Leo would be unsurprised by contemporary science, which remains "speculative" in the sense that it has "little reverence" for Thomistic methods (see the next sentence). We inserted "any" to render Frank's remark more idiomatic and not an assertion that physics has in fact become "less inclined to speculation" since the turn of the century.

10. O'Connell, *Recollections*, 323.

11. O'Connell, 323.

12. James A. McWilliams, *Physics and Philosophy: A Study of Saint Thomas' Commentary on the Eight Book of Aristotle's Physics*, Washington, DC: Catholic University of America Press, 1946), 26.

13. Edward F. Caldin, "Modern Physics and Thomist Philosophy," *Thomist* 2, no. 2 (1940): 208–26, 213.

14. Caldin, "Modern Physics," 214.

15. Rev. Joseph T. Clark, "Toward an Acceptable Philosophy of Science," *Bulletin of the Association of Jesuit Scientists* 28, no. 3 (1951): 74–86, 83.

16. Clark, "Toward an Acceptable Philosophy," 85.

17. Clark. 85.

18. Richard P. Philipps, *Modern Thomistic Philosophy: An Explanation for Students*, vol. 1: *The Philosophy of Nature* (Westminster, MD: Newman Bookshop, 1934), 38.

19. Philipps, *Modern Thomistic Philosophy*, 38.

20. In a note Frank referred readers to Phelan, *St. Thomas and Analogy*, but we were unable to locate this or a similar principle in the text.

21. Frank presumably is referring to theories of transubstantiation of the Eucharist.

22. John J. Colligan, *Cosmology: A Text-Book for Colleges* (New York: Fordham University Press, 1936).

23. Clark, "Toward an Acceptable Philosophy," 85.

24. Étienne Gilson, *The Christian Philosophy of St. Thomas Aquinas*, trans. Laurence K. Shook (New York: Random House, 1956), 178.

25. Thomas Aquinas, *The Summa Contra Gentiles of Saint Thomas Aquinas. First Book*, trans. English Dominican Fathers. London: Burns Oates and Washbourne, 1924), 24.

26. Aquinas, *Summa Theologica*, Question 50.

27. Aquinas, *Summa Theologica*, Article 2, Answer. The English translation in this edition reads: "But one glance is enough to show that there cannot be one matter of spiritual and corporeal things. For it is not possible that a spiritual and a corporal form should be received by the same part of matter, otherwise one and the same thing would be corporeal and spiritual. Hence it would follow that one part of matter receives the corporeal form, and another receives the spiritual form. Matter, however, is not divisible into parts except as regarded under quantity; and without quantity substance is indivisible. . . . Therefore it would follow that the matter of spiritual things is subject to quantity; which cannot be."

28. Aquinas, *Summa Theologica*, Question 53.

29. Aquinas, Question 53, Article 3, Objection 1.

30. Aquinas, Question 53, Article 3, Reply to Objection 1. The English translation in this edition reads: "The swiftness of the angel's movement is not measured by the quantity of his power, but according to the determination of his will. . . . The time of the angel's movement . . . will have no proportion to the time which measures the movement of corporeal things."

31. Aquinas, Question 52, Article 2, Answer.

32. Aquinas, Question 53, Article 3, Reply to Objection 2.

33. Frank called this conception of laws (Part I, chap. 3, §6) "the doctrine of laws as imposed."

34. James A. McWilliams, *Cosmology: A Text for Colleges*, rev. ed. (New York: Macmillan, 1937), 148.

35. See, for example, Charles Dickens, *The Pickwick Papers* (Garden City, NY: International Collector's Library, 1944).

36. McWilliams, *Cosmology*, 149. This edition reads, "*Uniformity* of activity is not in itself a law, it is the *effect* of a law."

37. McWilliams, *Cosmology*, 149. In this edition, "imposed" and "endowed" are italicized.

38. McWilliams, 151.
39. McWilliams, 150.
40. McWilliams, 151–52.
41. McWilliams, 152. The original text begins with "Therefore."
42. Phelan, *Saint Thomas and Analogy*, 1.
43. *The Catholic Encyclopedia*, vol. 1 (New York: Encyclopedia Press, 1907), 449. In the *Encyclopedia* we read: "Two objects are related to the other not by a direct proportion, but by means of another and intermediary relation: for instance, 6 and 4 are analogous in this sense that 6 is the double of 3 as 4 is of 2, or 6 : 4 :: 3 : 2."
44. Phelan, *Saint Thomas and Analogy*, 8.
45. Frank's reference to Part IV of the *Discourse* reads: "René Descartes, *Discourse on the Method of Rightly Conducting the Reason and Seeking Truth in the Sciences*. Original published in 1637. 'I concluded that I might take as a general rule the principle that *all the things which we clearly and distinctly perceive are true*.'"
46. We presume that Frank refers here to Maritain's talk at the first Conference on Science, Philosophy, and Religion in New York City, 1940. See Maritain's "Science, Philosophy, and Faith," in *Science, Philosophy and Religion: A Symposium* (New York: Conference on Science, Philosophy, and Religion in their Relation to the Democratic Way of Life, 1941), 162–83. In a representative passage, Maritain discusses both "analogical ampleness" and lower and higher forms of knowledge (174). Frank's "lower level" and "higher level" are not quotes but concepts he puts forward to explain Maritain's thinking.
47. Ernst Mach, "On the Principle of Comparison in Physics," in *Popular Scientific Lectures*, trans. Thomas J. McCormack, 3rd ed. (Chicago: Open Court, 1898), 236–58. The lecture was delivered in Vienna before the General Session of the German Association of Naturalists and Physicians on September 24, 1894.
48. Mach, "On the Principle of Comparison," 238.

Chapter 8

1. *The Divine Comedy of Dante Alighieri*, trans. Charles Eliot Norton (Chicago: Encyclopedia Britannica, William Benton, 1952).
2. George Santayana, *Three Philosophical Poets: Lucretius, Dante and Goethe* (Cambridge, MA: Harvard University Press, 1910).
3. Santayana, *Three Philosophical Poets*, 77–78.
4. Santayana, 107.
5. See Plato's *Phaedo* in *The Dialogues of Plato*, trans. B. Jowett, 2nd ed., vol. 1 (London: Macmillan, 1875), 429–31.
6. Santayana, *Three Philosophical Poets*, 75.

7. Plato, *Phaedo*, 476–77. The translation in this edition reads: "I imagined that he would tell me first whether the earth is flat or round; [. . .] and would teach me the nature of the best, and show that this was the best; and if he said that the earth was in the centre, he would further explain that this position was the best, and I should be satisfied with the explanation given, and not want other sort of cause. . . . What hopes I had formed and how grievously was I disappointed! As I proceeded, I found my philosopher altogether forsaking mind . . . but having recourse to air, and ether and water and other eccentricities."

8. Frank's note reads, "The great Rabbinical treasures which was reduced to writing between the second and sixth century of the Christian Era."

9. *Nicholas Copernicus on the Revolutions*, 22.

10. *Nicholas Copernicus on the Revolutions*, 17.

11. Topitsch, "Society, Technology, and Philosophical Reasoning," 276.

12. Werner Jaeger's book on Aristotle was published in 1934 as *Aristotle: Fundamentals of the History of his Development*, trans. Richard Robinson (Oxford: Clarendon) (orig. Berlin, 1923). In the quotation Frank presents, Topitsch discusses a different book, Jaeger's *Paideia: The Ideals of Greek Culture*, trans. Gilbert Highet (Oxford: Basil Blackwell, 1946).

13. Topitsch, "Society, Technology, and Philosophical Reasoning," 283.

14. See Topitsch, 295–96: "These procedures generally lead into purely analytic propositions that are either 'eternal truths' (*verités eternelles*) because they are tautologies and disguised definitions or 'eternal problems' because they are self-contradictions. These empty or self-contradictory formulae have, of course, no factual and/or normative content and therefore neither descriptive nor prescriptive meaning."

15. See, for example, John James Wellmuth, *The Nature and Origins of Scientism* (Milwaukee: Marquette University Press, 1944).

16. See, for example, William of Ockham, *Philosophical Writings: A Selection*, trans. Philotheus Boehner (Edinburgh and London: Thomas Nelson and Sons, 1957), chap. 2.

17. Frank uses "natural theology" (here and later) in an unconventional way. Natural theology is traditionally based upon close observation of nature, as well as reasoning.

18. Frank quotes Bacon from Wellmuth, *Nature and Origins of Scientism*, 37. Wellmuth, in turn, quotes from Étienne Gilson's *La philosophie au moyen âge* (Paris, 1930), 210.

19. Wellmuth, *Nature and Origins of Scientism*, 37.

20. David Hume, *An Enquiry Concerning Human Understanding and Selections from a Treatise of Human Nature* (Chicago: Open Court, 1912), 176.

21. Aquinas, *Summa theologica*, Question 1, Article 1, Answer.

22. Aquinas, Question 1, Article 1, Answer.

23. Max Carl Otto, "The Ethical Neutrality of Science," in *Science and the Moral Life: Selected Writings by Max C. Otto* (New York: Mentor, 1949), 74.

24. Otto, "Ethical Neutrality of Science," 74.

25. Hume, *An Enquiry Concerning Human Understanding*, 23.

26. David Hume, *Dialogues Concerning Natural Religion* (Edinburgh and London: William Blackwood and Sons, 1907), 189. Originally published 1779.

27. Hume, *Dialogues Concerning Natural Religion*, 191.

28. Thomas Hobbes, *Leviathan* (Oxford: Clarendon, 1929), chap. 32, 287. Reprinted from the 1651 edition.

29. See Hans Hahn, "Superfluous Entities, or Occam's Razor," in *Empiricism, Logic and Mathematics*, ed. Brian McGuinness (Dordrecht, The Netherlands: Reidel, 1930/1980), 1–19.

30. August Comte, *The Positive Philosophy of August Comte*, trans. Harriet Martineau (London: George Bell and Sons, 1896), 312.

31. Frank's note points out that Voltaire's "*Lettres Anglaises* (1734) give a popular presentation of Newton's doctrines. His *Dictionnaire Philosophique* (1764) discusses in a lucid and popular way philosophical and scientific topics of his period."

32. Fulton J. Sheen, *Philosophy of Science* (Milwaukee: Bruce, 1934), 3–4. Sheen's book reads: "In the Middle Ages there was a distinction between a scientific method and a philosophy of science, but not a separation. . . . The basic rational explanation is to be found in a science from which all other sciences borrow their first principles, namely, metaphysics. . . . The divorce of science and philosophy never actually took place until the time of Kant." Frank added here the following footnote: "Kant, Immanuel (1724–1804), one of the greatest modern philosophers, born in Germany of Scottish descent. In his native country his philosophy has, in some periods, become almost a national religion. We shall illustrate his doctrines by presenting his conception of geometry (Part II, chap. 7) and Mechanics (Part III, chap. 2, §1)." Despite his cross-references, Frank did not present these topics in his manuscript.

33. Immanuel Kant, *Critique of Pure Reason*, trans. Norman Kemp Smith (London: Macmillan, 1929), 630, A797/B825. This edition reads: "Reason is impelled by a tendency of its nature to go out beyond the field of its empirical employment, and to venture . . . by means of ideas alone, to the utmost limits of all knowledge, and not to be satisfied save through the completion of its course in [the apprehension of] a self-subsistent systematic whole."

34. Kant, *Critique of Pure Reason*, 631, A798/B826.

35. Mortimer J. Adler, *St. Thomas and the Gentiles*, The Aquinas Lecture (Milwaukee: Marquette University Press, 1938), 26.

36. Sheen, *Philosophy of Science*, 190.

37. C. E. M. Joad, *Decadence: A Philosophical Inquiry* (London: Faber and Faber, 1948), 39. In a footnote here, Frank mentions Joad's 1932 book, *Philosophical Aspects of Modern Science* (London: Allen and Unwin). On Joad's book

and Frank's ideas about the metaphysical interpretations of science, see Adam Tamas Tuboly, "Knowledge Missemination: L. Susan Stebbing, Philipp Frank and C. E. M. Joad on the Philosophy of the Physicists," *Perspectives on Science* 28, no. 1 (2020): 1–34.

38. We inserted this transitional paragraph at the conclusion of Frank's manuscript chapter.

Chapter 9

1. Frank's note reads: "*The Encyclopedia Britannica* was first published in 1768. By 1771 it consisted of three moderately-sized volumes. The most recent edition consists of twenty-four large volumes. It is the result of the cordial cooperation of more than 4,400 authorities from all countries of the world. We are using the 1952 edition here."

2. Frank's note reads: "The publication of the *Large Soviet Encyclopedia* started in 1926 by an order of the Executive Committee of the USSR. Its purpose was not only to present all possible information but also to present it from a definite philosophical and political angle. We read in the introduction by the editorial board: 'Previous dictionaries presented a variety of worldviews which were sometimes hostile to each other. To a Soviet Encyclopedia, on the contrary, an accurate worldview is indispensable and, particularly, a strictly materialistic worldview. Our worldview is Dialectical Materialism.' '

3. Frank refers here to "Sir William Cecil Dampier, Vol. XX, page 115 ff." Frank does not point out (and perhaps did not know) that the *Encyclopedia Britannica* was historically connected to Neo-Thomism in Chicago via University of Chicago President Robert Maynard Hutchins, Mortimer Adler (mentioned at the end of the previous chapter), and William Benton, publisher of both the encyclopedia and the *Great Books of the Western World*, edited by Adler and Hutchins. Under the tutelage of Adler and Hutchins, Benton, originally an advertising executive, became broadly sympathetic to Neo-Thomism and supportive of Hutchins and Adler's crusade to promote it. Apropos of Chicago's iconic Palmolive Building (then headquarters of the Colgate-Palmolive-Peet corporation), Hutchins is said to have remarked admiringly that his collaborators Benton and Adler were "made for each other. The two men were consummate promoters of hot products—Palmolive soap and Thomism." See Milton Mayer, *Robert Maynard Hutchins: A Memoir* (Los Angeles: University of California Press, 1993), 303, as well as Mary Ann Dzuback, *Robert M. Hutchins: Portrait of an Educator* (Chicago: University of Chicago Press, 1991, 219).

4. *The Encyclopedia Britannica*, vol. 20 (1952), 122.

5. *Encyclopedia Britannica*, 123.

6. Frank refers to "Andrew Seth Pringle-Pattison, Vol. XVII, page 759 (Philosophy and Natural Science)."

7. Frank's note reads: "Lenin, founder and guiding spirit of the Soviet Republics. His book, *Materialism and Empirocriticism* (in today's language rather *Materialism and Positivism*) (1909) discussed elaborately the contemporary philosophy of science and its sociological background." This remark explains why Frank refers later to Lenin's book as *Materialism and Positivism* in Part II, chap. 7, §6.

8. Lenin, *Materialism and Empirio-Criticism*, 355. The translation reads: "*Not a single one* of these professors, who are capable of making valuable contributions in the special fields of chemistry, history, or physics, *can be trusted one iota* when it comes to philosophy." Frank cites this translation in his "Philosophical Uses of Science" (128) and elsewhere in this book.

9. *Large Soviet Encyclopedia*, vol. 57 (1936), 496.

10. *Large Soviet Encyclopedia*, 496. Frank quoted this earlier (Pt. I. Ch. 4. Sect. 10), where the translation is correctly quoted: "Philosophy was declared a scholastic survival and the thesis of bourgeois positivism was upheld that science by itself is philosophy."

11. Frank's note reads: "In his book, *Creative Evolution*, (1907, English translation 1911), Bergson pointed out that by accepting the current presentation of science, one accepts implicitly also a definite philosophical interpretation."

12. *Large Soviet Encyclopedia*, 496. Frank noted here: "*Natural Science* in the Soviet Encyclopedia (vol. 8, pp. 559–87) by O. Schmidt."

13. *Large Soviet Encyclopedia*, 496.

14. Frank's note reads: "New-Thomism (or, in a more general way, Neo-Scholasticism) has attempted to modify scholastic philosophy in such a way that it becomes compatible with modern science. Aristotle's physics appears, e.g., in Neo-Scholasticism not as a system of physics (that would disagree with modern physics) but as a philosophical interpretation of physics."

15. Duhem quotes Simplicius in *The Aim and Structure of the Physical Theory*, trans. Philip P. Wiener (Princeton, NJ: Princeton University Press, 1991), 40. (The edition was published originally in 1954.) Frank notes, "Simplicius, in his commentary on Aristotle's book *On the Heavens* (Chapter II, footnote 7). He was one of the Greek philosophers who emigrated about 500 B.C. to Persia, when the Christian emperor Justinian liquidated the teaching of 'pagan' philosophy in Athens."

16. Frank's note reads: "Maritain's views are presented in a very readable form in his book: *An Introduction to Philosophy*. In a strictly Thomistic and fairly technical way, Maritain's book, *Philosophy of Nature* (New York 1951), presents the conceptual structure into which science and philosophy can be fitted."

17. Jacques Maritain, *Éléments de philosophie I. Introduction générale à la philosophie* (Paris: Pierre Téqui, 1921), 64. Translated as *An Introduction to Philosophy*, trans. E. I. Watkin (New York: Sheed and Ward, 1937), 102.

18. Frank's note reads: "The Holy Alliance was signed on September 26, 1815, by the Emperors of Russia and Austria and the King of Prussia."

19. August Comte, *The Fundamental Principles of the Positive Philosophy (Being the First Two Chapters of the 'Cours de Philosophie Positive' of August Comte)*, trans. Paul Descours and H. Gordon Jones (London: Watts, 1905).

20. George Boas, *French Philosophies of the Romantic Period* (Baltimore: Johns Hopkins University Press, 1925), 254.

21. Comte, *Fundamental Principles of the Positive Philosophy*, 12. This edition reads: "I use the word *philosophy* in the sense in which it was employed by the ancients, and especially by Aristotle, as comprising the general system of human conceptions . . . [B]y adding the word *positive* I wish to denote that I am considering that particular manner of philosophizing which holds that the purpose of theories, in any class of ideas, is to co-ordinate facts."

22. Comte, *Fundamental Principles of the Positive Philosophy*, 12. This edition reads: "I regret that I have been obliged to employ, for want of another, a term like *philosophy*, which has been so improperly used in a multitude of different meanings."

23. Frank's note reads: "Littré was a French lexicographer and philosopher. He accepted the doctrines of Auguste Comte, except the mystical ideas of Comte's later years. He published *August Comte et la Philosophie Positive*, (1863)."

24. Émile Littré, "Preface d'un Disciple," in August Comte, *Principes de philosophie positive* (Paris: J. B. Bailliére et Fils, 1868), 5–75, 6.

25. Comte, *Fundamental Principles of Positive Philosophy*, 22–23. This edition reads: "The Theological system arrived at its highest form of perfection, when it substituted the providential action of a single being, for the varied play of the numerous independent gods which had been imagined by the primitive mind. In the same way, the last stage of the Metaphysical system consisted in replacing the different special entities by the idea of a single great general entity—Nature—looked upon as the sole source of all phenomena. Similarly, the ideal of the Positive system, towards which it constantly tends, although in all probability it will never attain such a stage, would be reached if we could look upon all the different phenomena observable as so many particular cases of a single general fact, such as that of Gravitation, for example."

26. Littré, "Preface d'un Disciple," 10.

27. Frank's note reads: "The word 'sociology' was invented and introduced by Auguste Comte. *La Grande Encyclopedie* states: 'Although the word was formed from a Latin root and a Greek suffix, and the purists have refused to recognize it, it has been accepted as a citizen in all European languages.'"

28. Frank's note reads: "It would be erroneous to believe that the Catholic Church was more opposed to the Copernican system than other churches. Martin Luther himself called Copernicus a 'Sarmathen fool' and the Lutheran Church was for a long time extremely hostile."

29. At first, Comte called this "social physics" and later "science of man." Frank explains at this point in his manuscript why psychology did not appear

as a special science in Comte's framework: "This science [sociology] was to also treat phenomena which have most frequently been treated in 'psychology.' Comte did not admit psychology to his system of special sciences. He divided the topics of psychology between biology and sociology. He did not consider 'self-observation' as a legitimate scientific method. He retained instead only the concept of a science of human behavior, which led to the proposal to add 'sociology' to the traditional sciences."

30. Émile Littré, *De La Philosophie Positive* (Paris: Librairie Philosophique de Ladrange, 1845), 97–98.

31. Littré, *De La Philosophie Positive*, 98–99.

32. August Comte, *A General View of Positivism*, 2nd. ed., trans. J. H. Bridges (London: Trübner and Co., 1865), 1–2.

33. Frank noted here, "According to the teaching of twentieth-century Positivism, there is no derivation of a general law from sense-observation which would not make use of 'imagination' or 'intuition,'" and then referred the reader to Part II, chap. 5. on logical positivism and to Section 1 of his concluding chapter, "Einstein's Philosophy of Science."

34. Comte, *A General View of Positivism*, 5.

35. Frank's note reads: "The term 'skepticism' is used in two different senses. The first would be that every knowledge is regarded as impossible, while, in the second sense, only knowledge of ultimate reality is denied. In this last sense, modern science leads to skepticism; while in the first sense, skepticism is rejected by the scientist since "scientific knowledge" is declared as valuable.

36. Frank's note reads: "The meaning of "agnosticism" (admission of not knowing) has to be analyzed in a similar way."

37. John Milton, *Paradise Lost, A Poem in Twelve Books*, bk. 8 (Boston: Timothy Bedlington, 1820), ll. 72–73, p. 181.

38. Bergson, "On the Pragmatism of William James," 270. The recent translation reads: "For the ancient philosophies there was, above space and time, a world in which were located from all eternity all possible worlds: the truth of human affirmations was measured by the degree of faithfulness with which they copied these eternal truths. . . . All the work of science consists, so to speak, in piercing the resisting envelope of the facts inside which the truth is lodged, like a nut in its shell."

39. Spencer's *A System of Synthetic Philosophy* was published in nine volumes between 1862 and 1893. It contained his systematic ideas about biology, sociology, ethics, and politics. Frank refers later to the first volume, titled *First Principles*, originally published in 1862.

40. John Fiske, *Outlines of Cosmic Philosophy, Based on the Doctrine of Evolution, with Criticism on the Positive Philosophy*, vol. 1 (London: Macmillan, 1874). Frank noted that Fiske "regarded Spencer as the greatest philosopher of the age."

41. See Part 1 of Spencer's *First Principles*, 6th ed. (London: Watts, 1946).

42. Spencer, *First Principles*, 81.

43. See Spencer's *First Principles*, Part 1, "The Unknowable," especially chap. 5.

44. Fiske was a university lecturer on philosophy at Harvard between 1869 and 1871. We note that "positive philosophy" in Frank's text sometimes refers ambiguously to Comte's famous book and to Comte's positivism.

45. Frank's note reads: "The term 'School Philosophy' for the doctrines that are common to all antipositivistic philosophies refers, in a rather perfunctory way, to 'scholastic philosophy' as the root of the doctrines. The term was first used in my paper, 'Physical Theories of the Twentieth Century and School Philosophy' (in *Modern Science and its Philosophy*, Harvard University Press, 1949, [pp. 90–121]). The paper is an English version of a lecture published in German in 1930. In von Mises's book [referred to in the next footnote], the term 'School Philosophy' is consistently applied."

46. Richard von Mises, *Positivism, An Essay in Human Understanding*, trans. Jeremy Bernstein and Roger Newton (Cambridge, MA: Harvard University Press, 1951), 2. Originally published in German as *Kleines Lehrbuch des Positivismus*, 1939.

47. In his book *The Law of Causality and its Limits* from 1932, Frank said that "If we consider the total attitude toward science, we must rather say that 'school' philosophy is sceptical, but the scientific world-conception is optimistic and full of hope. The assumption of a 'true' world has led people to suppose that there are eternally insoluble problems, riddles of the world, limits of science that can never be transgressed." See page 260 of the 1998 translation.

48. Frank's note reads: "Neurath coined the name 'Vienna Circle' for the group from which the School of Logical Positivism emerged. See Part II, Ch. 11, section 4, 5."

49. See Frank, "Introduction," in *Modern Science and its Philosophy* (Cambridge, MA: Harvard University Press, 1949 §16.

50. Comte, *Fundamental Principles of the Positive Philosophy*, 30. This edition reads: "Is it a course on Positive Philosophy, and not in the Positive Sciences, that I propose to give. We shall only have to consider here each fundamental science in its relations with the whole positive system, and as to the spirit characterizing it; that is to say, under the two-fold aspect of its essential methods and its principal doctrines.'

51. Frank's use of this metaphor may be connected to the intellectual culture at Harvard where he attended the "Science of Science Discussion Group" along with W. V. O. Quine, who made famous a similar metaphor about the "web of beliefs." See Gary L. Hardcastle, "Debabelizing Science: The Harvard Science of Science Discussion Group, 1940–41," in *Logical Empiricism in North America*, ed. Gary L. Hardcastle and Alan Richardson (Minneapolis: University of Minnesota Press, 2003), 170–96; and Joel Isaac, *Working Knowledge: Making the Human Sciences from Parsons to Kuhn* (Cambridge, MA: Harvard University Press, 2012).

52. Frank's note reads: "'Natural selection' is a process by which Darwin explained the evolution of species. In every species there are random variations; by the interaction with the environment the fittest individual survives, and produces in the next generation a shift of the average properties."

53. We might point out here that DNA's structure was discovered roughly about the time that Frank wrote his manuscript; he may, or may not, have known about this development, but it illustrates what he had in mind.

54. John Dewey, "Introduction," in *Reconstruction in Philosophy*, enlarged ed. (Boston: Beacon, 1948), xviii. Originally published in 1920.

55. Frank's note reads: "Democritus was a Greek philosopher, early Greek exponent of the atomic theory and a mechanistic explanation of observable phenomena. Epicurus was a Greek philosopher, too. He judged all knowledge according to its practical usefulness. As a reaction against the role which had been ascribed by Plato and Aristotle to 'idealism' and 'purposiveness' in nature, Epicurus and his school embraced the mechanistic worldview of Democritus."

56. Frank's note reads: "These trends are presented elaborately in Part I, chaps. 4, 5, 6 of the present book."

Chapter 10

1. Henry Thomas Buckle, *History of Civilization in England*, vol. 1, Part 2 (New York: Appleton, 1885), chap. 14, 658–63.

2. Buckle, *History of Civilization in England*, 660.

3. Buckle, 660.

4. Buckle, 658.

5. Buckle, 662.

6. Hermann von Helmholtz, "The Aim and Progress of Physical Science," in *Popular Lectures on Scientific Subjects*, trans. E. Atkinson (New York: Appleton, 1873), 363–97, 375. This edition reads: "The ultimate aim of physical science must be to determine the movements which are the real causes of all other phenomena and discover the motive powers on which they depend; in other words, to merge itself into mechanics."

7. Emil Du Bois-Raymond, "The Limits of our Knowledge of Nature," *Popular Science Monthly* 5 (1874): 17–32, 17 (originally published in German in 1872). The English translation reads: "Natural Science is the resolution of natural processes into the mechanics of atoms. It is a fact of psychological experience that, where such a resolution is practicable, our desire of tracing things back to their causes is provisionally satisfied."

8. John B. Stallo, *The Concepts and Theories of Modern Physics* (New York: Appleton, 1888), 150 (originally published in 1881).

9. Wilhelm Ostwald, "The Modern Theory of Energetics," *Monist* 17, no. 4 (1907): 481–515.

10. William John Macquorn Rankine, "Outlines of the Science of Energetics." *Proceedings of the Royal Philosophical Society of Glasgow*, vol. 3, Royal Philosophical Society of Glasgow, 1855, 121–41. Frank notes that Rankine was a "Scottish engineer and physicist. He was one of the founders of Thermodynamics. He proposed to build up this science as a system of logical deductions from principles which are essentially statements about energy."
    11. *The Catholic Encyclopedia*, vol. 10, 45.
    12. Stallo, *Concepts and Theories of Modern Physics*, 137.
    13. Stallo, 137.
    14. Stallo, 148.
    15. Ernst Mach, "The Economical Nature of Physical Inquiry," in *Popular Scientific Lectures*, trans. Thomas J. McCormack, 3rd. ed. (Chicago: Open Court, 1898), 186–213, 188.
    16. Mach, 'Economical Nature of Physical Inquiry," 189.
    17. Mach, 191.
    18. Mach, 193.
    19. Mach, 193.
    20. Mach, "On the Principle of Comparison in Physics," 240.
    21. Mach, 240–41.
    22. Ernst Mach, *Knowledge and Error: Sketches on the Psychology of Enquiry*, trans. Thomas I. McCormack. Dordrecht, The Netherlands: D. Reidel, 1976 (originally published in 1905).
    23. Ernst Mach, *Space and Geometry in the Light of Physiological, Psychological and Physical Inquiry*, trans. Thomas J. McCormack (London: Regan Paul, Trench, Trübner, 1906), 103–4. The book contains three essays, originally written for *The Monist* between 1901 and 1903. The 1906 edition says "employment with . . . symbolic representations must [e.g., multidimensional manifolds], as the history of science shows us, by no means regarded as entirely unfruitful. . . . Think only of the negative, fractional, and variable exponents of algebra."
    24. Karl Pearson, *The Grammar of Science* (London: Walter Scott, 1892).
    25. Frank noted that Gibbs worked in "American theoretical physics. He advanced highly general and abstract principles in the field of Thermodynamics, in particular rules for the 'Equilibrium of Heterogenous Substances.' His name has become familiar by what is commonly called 'Gibbs's Rule of Phases.'"
    26. Henry Adams, *The Education of Henry Adams* (Boston: Riverside, 1918), 450.
    27. See, for example, Henri Poincaré, "Des Fondements de la Géométrie: A propos d'un Livre de M. Russell," *Revue de Métaphysique et de Morale* 7 (1899): 251–79.
    28. Frank refers his readers here to Édouard Le Roy, "Science et Philosophie," *Revue de Métaphysique et de Moral* 7, no. 5 (1999): 550.

29. See Henri Poincaré, *The Value of Science*, trans. George Bruce Halsted (New York: Science Press, 1907). Originally published in French in 1905.

30. Abel Rey, *La théorie de la physique chez les physiciens contemporains* (Paris: Félix Alcan, 1907). Frank reviewed the German translation of this book: Frank, "Abel Rey, *Die Theorie der Physik bei den modernen Physikern.*" *Monatshefte für Mathematik und Physik* 25 (1910): 43–45. About Frank and Rey, see Matthias Neuber, "Philosophie der Modernen Physik: Philipp Frank und Abel Rey," *Grazer Philosophische Studien* 80 (2010): 131–49.

31. Frank's manuscript contained only "as the description," but since he refers to "metaphysical truth" in the second part of the sentence we added "true" to meet his intentions.

32. Rey, *La théorie de la physique*, 17.

33. Rey, 17.

34. Rey, 19.

35. Frank's note refers readers to Part I, chap. 3, §6, but Part I, chap. 2, §10–11 seems more appropriate in our edition.

36. Rey, *La théorie de la physique*, 20.

37. We were unable to identify the source of this quotation.

38. Pierre Duhem, *Ziel und Struktur der physikalischen Theorien* (Leipzig: Johann Ambrosius Barth, 1908).

39. Armand Lowinger, *The Methodology of Pierre Duhem* (New York: Columbia University Press, 1941), 18–19. Instead of "problems on their metaphysical side," the phrasing in the book actually reads, "problems on their methodological side."

40. Duhem, "The Value of a Physical Theory," in *The Aim and Structure of Physical Theory*, trans. Philip P. Wiener (Princeton, NJ: Princeton University Press, 1954), 312–36. Both quotations are from p. 335. It was originally published in French in 1908 as a separate article entitled, "La valeur de la théorie physique, à propos d'un livre recent," *Review générale des Sciences pures et appliquées* 19, no. 1 (1908): 7–19.

41. Duhem, "Physics of a Believer," in *The Aim and Structure of Physical Theory*, 273–311. Originally published in French in 1905 as a separate article entitled, "Physique de croyant," *Annales de Philosophie Chrétienne* 155 (1905): 44–67, 133–59.

42. Louis de Broglie, "Foreword," in Duhem, *Aim and Structure of Physical Theory*: v–xiii, ix.

43. Duhem, "Physics of a Believer," 283.

44. Duhem, 285.

Chapter 11

1. Frank refers here to the journal *Erkenntnis*, edited by Rudolf Carnap and Hans Reichenbach between 1930 and 1938.

2. Moritz Schlick, 'The Turning-Point in Philosophy," in *Philosophical Papers*, vol. 2: *1925–1936*, ed. Henk L. Mulder and Barbara F. B. Van de Velde-Schlick (Dordrecht, The Netherlands: D. Reidel, 1979), 154–60, 155 (originally published in 1930). The published translation is: "For I am persuaded that we are at present in the midst of an altogether final change in philosophy, and are justly entitled to consider the fruitless conflict of systems at an end. The present age, I maintain, is already in possession of the means to make all such conflict essentially unnecessary; it is only a matter of resolutely using them."

3. Albert E. Blumberg and Herbert Feigl, "Logical Positivism." *Journal of Philosophy* 28, no. 11 (1931): 281–96, 282.

4. Moritz Schlick, *Space and Time in Contemporary Physics*, trans. Henry L. Brose (New York: Oxford University Press, 1920), 83.

5. Schlick, *Space and Time*, 83–84.
6. Schlick, 84–85.
7. Schlick, 84.
8. Schlick, 85.
9. Schlick, 85–86. Schlick's original term was "Zuordnung," which was translated and used in the literature as "coordination."
10. Schlick, 85–86.
11. Schlick, "The Turning-Point in Philosophy," 157. The published translation reads: "The totality of the sciences . . . is *the* system of knowledge; there is no additional domain of 'philosophical' truths, for philosophy is not a system of propositions, and not a science. . . . Philosophy is a system of *acts*; philosophy, in fact, is that activity whereby the *meaning* of statements is established or discovered. Philosophy elucidates propositions, science verifies them."

12. Hans Hahn, "The Significance of the Scientific World View, Especially for Mathematics and Physics," in *Empiricism, Logic and Mathematics*, ed. Brian McGuinness (Dordrecht, The Netherlands: D. Reidel, 1980), 20–30, esp. 20. Originally published in 1930–31. The published translation reads: "The name 'scientific world view' is intended both as a confession and as a delimitation of a subject . . . anything that can be said sensibly at all is a proposition of science, and doing philosophy only means examining critically the propositions of the sciences to see if they are not pseudo-propositions, whether they really have the clarity and significance ascribed to them by the practitioners of the science in question and it means, further exposing as pseudo-propositions those propositions that pretend to a different, higher significance than the propositions of the special sciences."

13. Schlick, "The Turning Point in Philosophy," 159. The published translation reads: "We may add here that the wholly decisive, epoch-making advances in science are of this sort, that they represent a clarification of the meaning of fundamental principles, and hence are accomplished only as those with a talent for philosophizing; the great scientist is thus always a philosopher as well."

14. Otto Neurath, "Unified Science as Encyclopedic Integration," in *Encyclopedia and Unified Science*. vol. 1, no. 1 of *International Encyclopedia of Unified Science* (Chicago: University of Chicago Press, 1938), 17.

15. Neurath, "Unified Science as Encyclopedic Integration," 18.

16. Otto Neurath, "Sociology in the Framework of Physicalism," in *Philosophical Papers 1913–1946*, ed. and trans. Robert S. Cohen and Marie Neurath (Dordrecht, The Netherlands: D. Reidel, 1983), 58–90, esp. 59. Originally published in 1931. The published translation reads: "It is impossible to separate the 'clarification of concepts' from the 'pursuit of science' to which it belongs. Both are inseparably bound up together."

17. Neurath, "Sociology in the Framework of Physicalism," 59. The published translation reads: "Certainly different kinds of laws can be distinguished from each other: for example, chemical, biological or sociological laws; however, it can *not be said of a prediction of a concrete individual process that it depend on one definite kind of law only*. For example, whether a forest will burn down at a certain location on earth depends as much on the weather as on whether human intervention takes place or not. This intervention, however, can only be predicted if one knows the laws of human behaviour. *That is, under certain circumstances, it must be possible to connect all kinds of laws with each other*. Therefore all laws, whether chemical, climatological or sociological, must be conceived as *parts of a system*, namely of *unified science*."

18. Neurath, 66. Instead of "harmonized," Frank used "agreement."

19. Neurath, 66.

20. Neurath's ideas about protocol sentences and their embedded form are much more complex than Frank indicates here. For a detailed elaboration, see Thomas Uebel, *Empiricism at the Crossroads: The Vienna Circle's Protocol-Sentence Debate* (Chicago: Open Court, 2007).

21. Rudolf Carnap, *Logical Syntax of Language*, trans. Amethe Smeaton, Countess von Zeppelin (London: Kegan Paul Trench, Trubner, 1937), §82. Originally published in 1934 as *Logische Sysntax der Sprache*.

22. Rudolf Carnap, *Foundations of Logic and Mathematics*, vol. 1, no. 3 of *International Encyclopedia of Unified Science* (Chicago: University of Chicago Press, 1939), §24.

23. Rudolf Carnap, "The Elimination of Metaphysics through Logical Analysis of Language," trans. Arthur Pap, in *Logical Positivism*, ed. Alfred J. Ayer (New York: Free Press, 1959), 60–81, esp. 72. Originally published in German in 1932.

24. Carnap, "Elimination of Metaphysics," 78–79. The published translation reads: "The (pseudo) statements of metaphysics do not serve for the description of states of affairs. . . . They serve for the expression of the general attitude of a person towards life ("Lebenseinstellung, Lebensgefühl"). . . . But in the case of metaphysics we find this situation: through the form of its works it pretends to be something that it is not. The form in question is that of a

system of statements which are apparently related as premises and conclusions, that is, the form of a theory. In this way the fiction of theoretical content is generated, whereas, as we have seen, there is no such content." Frank often read Carnap's work in special ways: on his efforts to integrate Carnap and Austrian philosophy, see Thomas Mormann, "Philipp Frank's Austro-American Logical Empiricism," *HOPOS* 7, no. 1 (2017): 55–87; on his efforts to integrate Carnap and pragmatism, see Adam Tamas Tuboly, "Philipp Frank's Decline and the Crisis of Logical Empiricism," *Studies in East European Thought* 69, no. 3 (2017): 257–76.

25. Carl G. Hempel, "The Concept of Cognitive Significance: A Reconsideration," *Proceedings of the American Academy of Arts and Sciences* 80, no. 1 (1951): 61–77, esp. 70. The paper appeared in a special issue entitled "Contributions to the Analysis and Synthesis of Knowledge," edited by Frank.

26. Hempel, "Concept of Cognitive Significance," 70.

27. Hempel, 74.

28. Hempel, 74. On Frank's developing views about cognitive significance, see Thomas Uebel, "Beyond the Formalist Criterion of Cognitive Significance: Philipp Frank's Later Antimetaphysics," *HOPOS* 1, no. 1 (2011): 47–72.

29. Willard van Orman Quine, "Semantics and Abstract Objects," *Proceedings of the American Academy of Arts and Sciences* 80, no. 1 (1952): 90–96, esp. 92. (This paper appeared in the special issue, edited by Frank, entitled "Contributions to the Analysis and Synthesis of Knowledge.") The original paper reads, "As an empiricist I consider that the cognitive synonymy of statements should consist in . . ."

Chapter 12

1. Charles Sanders Peirce, "How to Make Our Ideas Clear?," 31.
2. Peirce, 29.
3. Peirce, 29–30.
4. Peirce, 35.
5. Peirce, 29.
6. The quotation is from Peirce's *Harvard Lectures on Pragmatism*, delivered in 1903. See *Collected Papers of Charles Sanders Peirce*, vol. 5: *Pragmatism and Pragmaticism*, ed. Charles Hartshorne and Paul Weiss (Cambridge, MA: Harvard University Press, 1934), 5.198.
7. Peirce, "Pearson's Grammar of Science. Annotations on the First Three Chapters," *Popular Science Monthly* 58, no. 1 (1901): 296–306, 305.
8. Peirce, "Review of English trans. of Mach's *The Science of Mechanics*," *Nation* 57, no. 1475 (1893): 252.
9. The quotation is from Peirce's "What Pragmatism Is?" in *Collected Papers of Charles Sanders Peirce*, vol. 5: *Pragmatism and Pragmaticism*, ed. Charles

Hartshorne and Paul Weiss (Cambridge, MA: Harvard University Press, 1934), 5.423. The paper was originally published in *Monist* 15, no. 2 (1905): 161–81.

10. Mach, *Knowledge and Error*, 176. The published translation reads: "The essential function of a hypothesis is that it leads to new observations and experiments, which confirm, refute or modify our surmise and so widen experience."

11. In the manuscript, Frank added here "as we described in our treatment of inductive inference." His most detailed discussion of induction is in Frank's *Philosophy of Science*, chap. 13 and 14, §1.

12. William James, *A Pluralistic Universe*, Hibbert Lectures at Manchester College on the Present Situation in Philosophy (New York: Longmans, Green, 1909), 8 and 10.

13. William James, "The Sentiment of Rationality," reprinted from *The Will to Believe, and Other Essays in Popular Philosophy* (London and Bombay: Longmans, Green, 1905), 63–110, 64–65.

14. Dewey, *Reconstruction in Philosophy*, 25.

15. Dewey, 26.

16. Dewey, 55–56.

17. Dewey, 59.

18. Dewey, 59.

19. Dewey, 59.

20. Dewey, 65.

21. Dewey, 65–66. After this quotation, Frank's manuscript continues, "This new orientation in political philosophy was certainly rapidly," at which point the manuscript page ends. We have not located any continuation of this sentence on other manuscript pages. Frank's handwritten page numbers indicate that despite this sentence fragment, he intended this chapter to continue as we present it here.

22. Charles Morris, "Scientific Empiricism," in *Encyclopedia and Unified Science*, vol. 1: *International Encyclopedia of Unified Science* (Chicago: University of Chicago Press, 1955), 63–75. Originally published in 1938.

23. Ernest Nagel, *Principles of the Theory of Probability*, vol. 1: *International Encyclopedia of Unified Science* (Chicago: University of Chicago Press, 1955), 342–422. Originally published in 1939.

24. Morris, "Scientific Empiricism," 72.

25. Morris, 72.

26. Morris, 66.

27. Percy W. Bridgman, *The Logic of Modern Physics* (New York: Macmillan, 1927), vii–viii.

28. Percy W. Bridgman, *The Nature of Physical Theory* (New York: Dover, 1936), 4. Bridgman's book does not contain the last part of the sentence ("because in absorbing the new physical theory they absorbed also implicitly the new philosophy of science"), and its source is unidentified.

29. Percy W. Bridgman, "The Nature of Some of Our Physical Concepts. I," *British Journal for the Philosophy of Science* 1, no. 4 (1951): 257–72, 257.

30. Frank refers here to James P. Joule's (1818–1889) famous mid-nineteenth-century experiments to determine the relationship between the work spent to liberate heat and the quantity of the heat liberated in a mechanical system. Through his experiments, Joule was able to establish constant quantifiable measures to prove the equivalence of heat and work.

31. We were unable to identify which of Bridgman's papers Frank has in mind here.

32. We were unable to identify which of Bridgman's papers Frank has in mind here.

33. Bridgman, "Nature of Some of Our Physical Concepts, I," 259.

34. Bridgman, 259.

35. Bridgman, 259.

36. Bridgman, 268.

37. Bridgman, 270–71.

38. Bridgman, 260.

39. Percy W. Bridgman, "Philosophical Implications of Physics," *Bulletin of the American Academy of Arts and Sciences* 3, no. 5 (1950): 2–6, esp. 2.

40. Bridgman, "Philosophical Implications of Physics," 3.

41. Bridgman, 4.

42. Bridgman, 5.

43. Bridgman, 5.

44. Bridgman, 5.

45. Bridgman, 5.

46. Ernest Nagel, "Philosophy and the American Temper," in *Sovereign Reason and Other Studies in the Philosophy of Science* (Glencoe, IL: Free Press, 1954), 50–57. Originally published in 1947.

47. Nagel, "Philosophy and the American Temper," 52.

48. Nagel, 55.

49. Nagel, 53.

50. Nagel, 53–54.

51. Nagel, 54.

52. Ernest Nagel, "Introduction," in *Sovereign Reason and Other Studies in the Philosophy of Science* (Glencoe, IL: Free Press, 1954), 9–16, esp. 16.

Chapter 13

1. See *De Rerum Natura*, 3 volumes with commentary, ed. Cyril Bailey (Oxford, 1947).

2. Julien Offray de La Mettrie, *Man a Machine* (Chicago: Open Court, 1912), 86. Originally published in French as L'Homme Machine in 1748.

3. La Mettrie, *Man a Machine*, 88.
4. La Mettrie, 89.
5. La Mettrie,148.
6. La Mettrie, 148–49.
7. La Mettrie, 132 and 140.
8. La Mettrie, 149.
9. Voltaire, A *Philosophical Dictionary*, vol. 3 (London: John and Henry L. Hunt, 1824). Originally published in 1764.
10. Paul-Henri Thiry, Baron d'Holbach, *System of Nature, or Laws of the Moral and Physical World* (Boston: J. P. Mendum, 1889). Originally published in 1770.
11. Voltaire, A *Philosophical Dictionary*, 341.
12. Voltaire, 342.
13. Frank probably has in mind Newton's arguments in his letters to the theologian Richard Bentley. See *Four Letters from Sir Isaac Newton to Doctor Bentley Containing Some Arguments in Proof of a Deity* (London: R. and J. Dodsley, 1761), 5–10.
14. La Mettrie, *Man a Machine*, 125.
15. George Gaylord Simpson, *The Meaning of Evolution* (New York: New American Library, 1956), 13. Originally published in 1949.
16. Simpson, *Meaning of Evolution*, 13.
17. John Dewey, Sidney Hook, and Ernest Nagel, "Are Naturalists Materialists?," *Journal of Philosophy* 42, no. 19 (1945): 515–30, 516.
18. Dewey, Hook, and Nagel, "Are Naturalists Materialists?," 518. Italics in the original. The article reads: "Materialism of the type now under consideration may then be taken to maintain that every psychological term is *synonymous with*, or *has the same meaning* as, some expression or combination of expressions belonging to the class of physical terms. . . . The word 'red' has the same meaning as the phrase 'electromagnetic vibration having a wave-length of approximately 7100 Angstroms.'"
19. See Rudolf Carnap, "Pseudoproblems in Philosophy," in *The Logical Structure of the World and Pseudoproblems in Philosophy* (Chicago and La Salle: Open Court, 2003), 301–43. Originally published in 1928.
20. Mitin, *Dialehticheskii i istoricheskii materialism*, 36.
21. Mitin, 36–37. A more precise translation reads: "Idealism originated as a product of the limited and ignorant ideas of the aboriginal savages. It seems that the development of scientific knowledge, determined by all the previous development of the means of production of the society, should lead to the complete triumph of materialism and the elimination of all idealistic conceptions."
22. Mitin, 37.
23. Frank refers here to Lenin's *Materialism and Empirio-Criticism*. We note that Frank himself was singled out by Lenin for criticism in this book as a Kantian, conventionalist, and idealist (*Materialism and Empirio-Criticism*, 165–66.).

24. We were unable to locate this quotation in *Materialism and Empirio-Criticism*; Frank is quoting here perhaps from memory.

25. Ralph Winston Fox, *Lenin: A Biography* (London: Victor Gollanz, 1933), 158.

26. We located a similar phrase in *Materialism and Empirio-Criticism*, which reads, "Electricity is proclaimed a collaborator of idealism, because it has destroyed the old theory of the structure of matter, shattered the atom and discovered new forms of material motion, so unlike the old, so totally uninvestigated and unstudied, so unusual and 'miraculous,' that it permits nature to be presented as non-material (spiritual, mental, psychical) motion." Lenin, *Materialism and Empirio-Criticism*, 291.

27. Lenin, *Materialism and Empirio-Criticism*, 145.

28. Mark Moisevich Rosenthal and Pavel Fyodorovich Yudin, eds., *Kratkij Filosofskij Slovar* (Moscow: Gos. Izd. Politiceskoj lit., 1955), 240–41.

29. Jean Druan, "Masse et énergie," *La Pensée: Revue du Rationalisme Moderne* 53 (1954): 29–38, 33.

30. The quotation is from Frank's *The Foundations of Physics*, vol. 1., no. 7., *International Encyclopedia of Unified Science* (Chicago: University of Chicago Press, 1946), chap. 5, §31, 35–36.

31. Druan, "Masse et énergie," 30.

32. Druan, 30.

33. Frank may have in mind the philosopher and physicist I. V. Kuznatsov, head of the Department of Philosophy at the Russian Academy of Sciences between 1947 and 1970. We were unable to locate a published source of Frank's quotation.

Chapter 14

1. The Institute of Red Professorship, also known as the "Institute of Red Professors" was a graduate-level educational institute in Moscow. After its foundation in 1921, some graduates pursued academic careers, but most worked in government. The institute was abolished in 1938.

2. Ernst Kolman, "Stalin i nauka" ["Stalin and Science"], *Pod Znamenem Marksizma* (Under the banner of Marxism) 12 (1939) 172–86. We were unable to identify the page number of the quotation.

3. *Large Soviet Encyclopedia*, vol. 57 (1936), 496.

4. Frank may refer here to Marx and Engels's *The German Ideology* (New York: International, 1939).

5. Mitin, *Dialehticneskii i istoricheskii materialism*, 50.

6. Vavilov, "New Physics and Dialectical Materialism," 151.

7. Lenin *Materialism and Empirio-Criticism*, 355. The published translation reads: "*Not a single one* of these professors, who are capable of making

valuable contributions in the special fields of chemistry, history, or physics, *can be trusted one iota* when it comes to philosophy." Frank cites this translation in his "Philosophical Uses of Science," 128.

8. Lenin, 13. This edition reads: "Scarcely a single contemporary professor of philosophy . . . can be found who is not directly or indirectly engaged in refuting materialism. . . . [They are pretending] that they are refuting materialism from the standpoint of 'recent' and 'modern' positivism, natural science, and so forth." We removed from Frank's quotation a second sentence ("When the physicist, even the bourgeois physicist, works in actual scientific research, he is a common-sense materialist, he records the properties of and changes in 'matter' in the common-sense meaning of this word.") that we do not find in Lenin's text. If, as we believe, the sentence was written by Frank, it supports his criticism here of Lenin and Soviet philosophy for maintaining that scientific practice requires guidance from an independent philosophical system.

9. Abraham F. Ioffe, "Razvitie atomisticheskikh vozzrenii v XX veke," *Pod Znamenem Marksizma* 4 (1934), 52–68. We were unable to identify the page number for this quotation.

10. Mitin, *Dialehticheskii i istoricheskii materialism*, 117. A more precise translation reads: "According to this doctrine, our intuitions and conceptions are not only *produced* by objective things, but they *reflect* them. Intuitions and conceptions are not produced by self-evolution of the subject (as idealists maintain); they are not hieroglyphs (as agnostics think); but they are *reflections, images, copies of things*. . . . Our cognition, in its advancement, reflects the material world more and more precisely, versatile and profoundly. *There are no limits set to our ability to know the world, but there have been in every case, historically determined limitations to our approximations to absolute truth*."

11. We were unable to identify the source of this quotation.

12. We were unable to identify the source of this quotation.

13. Walter Terence Stace, *The Philosophy of Hegel—A Systematic Exposition* (New York: Dover, 1955). Originally published in 1924.

14. Stace, *Philosophy of Hegel*, 73.

15. Hegel develops his position on measurement in his book on logic. See G. W. F. Hegel, *The Science of Logic*, trans. and ed. George Di Giovanni (Cambridge: Cambridge University Press, 2010). See especially Bk. 1, §3.

16. G. W. F. Hegel, *The Logic of Hegel*, translated from his *Encyclopedia of the Philosophical Sciences* (Oxford: Clarendon, 1874), §108, 174. Originally published in 1817.

17. Although Frank mentioned in the previous section three dialectical laws, he discusses in detail only one: the law of the transition from quantity to quality.

18. Mitin, *Dialehticheskii i istoricheskii materialism*, 160. A more precise translation reads: "We have a tempo in the evolution of production that has

never existed before, because the USSR represents a *new quality* as a type of productive relations."

19. Mitin, 158–59.

20. Mitin, 163.

21. Stalin, *History of the Communist Party of the Soviet Union* (New York: International, 1939), 107. The published version reads: "Dialectics does not regard the process of development as a simple process of growth, where quantitative changes do not lead to qualitative changes, but as a development which passes from insignificant and imperceptible quantitative changes to open, fundamental changes, to qualitative changes."

22. B. E. Bykhovsky, "Mechanists, the Servants of Imperialism," *ETC: A Review of General Semantics* 11, no. 3 (1954): 186–92, 187. Translated by Anatol Rapoport, this article was published originally in 1953, in *Nauka i Zhizn'* (Science and life) 20, no. 2.

23. The manuscript reads "physical trends," which we take to be a mistake and rendered "philosophical trends."

Conclusion

1. Albert Einstein, "Ernst Mach," in *The Collected Papers of Albert Einstein*, vol. 6: *The Berlin Years: Writings, 1914–1917*, trans. Alfred Engel (Princeton, NJ: Princeton University Press, 1997), 141–45. Originally published in 1916.

2. Einstein, "Ernst Mach," 142. See, for instance, Philipp Frank, "Einstein, Mach, and Logical Positivism," in *Albert Einstein, Philosopher-Scientist*, ed. Paul A. Schilpp (New York: Harper & Row, 1949), 271–86, 271. The recent English translation reads: "Science is, according to Mach, nothing else but the comparing and orderly arrangement of factually given contents of our consciousness, in accord with certain gradually acquired points of view and methods. . . . As a result of this activity of orderly arrangement, one obtains the abstract concepts and the laws (rules) of their connection . . . concepts make sense only insofar as they can be pointed out in things; also the points of view according to which concepts are associated with things (analysis of concepts)."

3. Albert Einstein, *The Meaning of Relativity, Four Lectures Delivered at Princeton University, May, 1921* (Princeton, NJ: Princeton University Press, 1923), 1–2.

4. Frank's note refers the reader to Part II, chap. 4, §1–4.

5. Einstein discusses "empty ideas" in "Remarks on Bertrand Russell's Theory of Knowledge," in *The Philosophy of Bertrand Russell*, ed. P. A. Schilpp (New York: Tudor, 1944), 277–91, 289.

6. See Carnap's "Elimination of Metaphysics."

7. Albert Einstein, "Remarks on Bertrand Russell's Theory of Knowledge," 287 and 288.

8. Albert Einstein, "On the Method of Theoretical Physics," *Philosophy of Science* 1 no. 2 (1934): 163–69, page 166. The article reads: ". . . were for the most part convinced that the basic concepts and laws of physics were not in a logical sense free inventions of the human mind, but rather that they were derivable by abstraction, i.e., by a logical process, from experiments. It was the general Theory of Relativity which showed in a convincing manner the incorrectness of this view . . ."

9. In a paper from 1936, Einstein said, "The fact that [the world of our sense experiences is comprehensible] is a miracle." See his "Physics and Reality," *Journal of the Franklin Institute* 221, no. 3 (1936): 349–82, 351. Frank quotes this passage exactly as he does in this book in his essay "Einstein, Mach, and Logical Positivism."

10. Einstein, "On the Method of Theoretic Physics," 167. The article reads: "It is my conviction that pure mathematical construction enables us to discover the concepts and the laws connecting them which give us the key to the understanding of the phenomena of Nature."

11. Einstein, 167. The article reads: "Experience, of course, remains the sole criterion of the serviceability of a mathematical construction for physics, but the truly creative principle resides in mathematics."

12. Einstein, 167.

13. Jeans expressed similar concerns in his book, *The Mysterious Universe* (Harmondsworth: Penguin, 1930).

14. Bridgman, *Logic of Modern Physics*, 201.

15. Bridgman, 201.

16. Bridgman, 203. The published version reads: ". . . predisposition to simplicity as formulated in terms of our concepts. . . . There is this observation to be made about all the simple laws of nature that have hitherto been formulated; they apply only over a certain range. . . . It does not seem so very surprising that over a limited domain, in which the most important phenomena are of a restricted type, the conduct of nature should follow comparatively simple rules . . ."

17. Einstein, "Science and Religion," in *Ideas and Opinions*, 41–49, 45.

18. Einstein, *Living Philosophies*, ed. Albert Einstein (New York: Simon and Schuster, 1931), 6.

19. Einstein, "Science and Religion," 46.

Bibliography

(This bibliography contains sources cited or mentioned by Frank in his manuscript as well as sources we mention or cite in our editorial notes—the editors.)

Adams, Henry. *The Education of Henry Adams*. Boston: Riverside, 1918.
Adler, Mortimer J. *St. Thomas and the Gentiles*, The Aquinas Lecture. Milwaukee: Marquette University Press, 1938.
Aeschylus. *Prometheus Bound*, trans. Marion Clyde Wier. New York: Century Co., 1916.
Alberts, Gerard, Luc Bergmans, and Fred Muller, eds. *Significs and the Vienna Circle: Intersections*. Dordrecht, The Netherlands: Springer, forthcoming.
Aquinas, Thomas. *The Summa Contra Gentiles of Saint Thomas Aquinas. First Book*. Translated by English Dominican Fathers. London: Burns Oates and Washbourne, 1924.
Aquinas, Thomas. *Summa theological*. Translated by Fathers of the English Dominican Province. New York: Benziger Bros., 1948.
Bacon, Francis. *Advancement of Learning*, ed. Joseph Devey. New York: Collier, 1901.
Bacon, Francis. *A Description of the Intellectual Globe (Descriptio Globi Intellectualis)*. In *The Philosophical Works of Francis Bacon*, edited by J. H. Robertson, 677–702. London: Routledge & Sons, 1905.
Bacon, Francis. *The Philosophical Works of Francis Bacon*, edited by John M. Robertson. New York: E. P. Dutton & Co., 1905.
Bacon, Francis. *Theory of the Earth (Thema Coeli)*. In *The Philosophical Works of Francis Bacon*, edited by John M. Robertson, London: Routledge and Sons; New York: E. P. Dutton & Co., 1905, 703–9.
Becker, Carl L. *The Heavenly City of the Eighteenth-Century Philosophers*. New Haven, CT: Yale University Press, 1932.
Berdyaev, Nicolas. *Solitude and Society*. Translated by George Reavey. London: G. Bles, The Centenary Press, 1938.

Bergson, Henri. *An Introduction to Metaphysics.* Translated by T. E. Hulme. New York and London: G. P. Putnam's Sons, 1912.

Bergson, Henri. "On the Pragmatism of William James: Truth and Reality." In *Henri Bergson—Key Writings*, ed. Keith Ansell Pearson and John Mullarkey. London: Continuum, 2002, 267–73.

Blumberg, Albert E., and Herbert Feigl. "Logical Positivism." *Journal of Philosophy* 28, no 11 (1931): 281–96.

Boas, George. *French Philosophies of the Romantic Period.* Baltimore: Johns Hopkins University Press, 1925.

Bridgman, Percy W. *The Logic of Modern Physics*, New York: Macmillan, 1927.

Bridgman, Percy W. *The Nature of Physical Theory.* New York: Dover, 1936.

Bridgman, Percy W. "The Nature of Some of Our Physical Concepts. I." *British Journal for the Philosophy of Science* 1, no. 4 (1951): 257–72.

Bridgman, Percy W. "Philosophical Implications of Physics." *Bulletin of the American Academy of Arts and Sciences* 3, no. 5 (1950): 2–6.

Buckle, Henry Thomas. *History of Civilization in England*, vol. 1, Part II, New York: Appleton, 1885.

Bykhovsky, B. E. "Mechanists, the Servants of Imperialism," *ETC: A Review of General Semantics* 11, no. 3 (1954): 186–92.

Carnap, Rudolf. "The Elimination of Metaphysics through Logical Analysis of Language." trans. Arthur Pap, in *Logical Positivism*, edited by Alfred J. Ayer, 60–81. New York: Free Press, 1959.

Carnap, Rudolf. *Foundations of Logic and Mathematics*, vol. 1, no. 3 of *International Encyclopedia of Unified Science.* Chicago: University of Chicago Press, 1939.

Carnap, Rudolf. *Logical Syntax of Language*, trans. Amethe Smeaton, Countess von Zeppelin. London: Kegan Paul Trench, Trubner & Co., 1937.

Carnap, Rudolf. "Pseudoproblems in Philosophy." In *The Logical Structure of the World and Pseudoproblems in Philosophy*, 301–43. Chicago and La Salle: Open Court, 2003.

The Catholic Encyclopedia. Vol. 1. New York: Encyclopedia Press, 1907.

Christian Science Monitor "Letter to the Editor." February 4, 1956.

Clark, Rev. Joseph T. "Toward an Acceptable Philosophy of Science," *Bulletin of the Association of Jesuit Scientists* 28, no. 3 (1951): 74–86.

Colligan, John J. *Cosmology: A Text-Book for Colleges.* New York: Fordham University Press, 1936.

Comte, Auguste. *A General View of Positivism*, 2nd. ed. Translated by J. H. Bridges. London: Trübner and Co., 1865.

Comte, Auguste. *The Fundamental Principles of the Positive Philosophy (Being the First Two Chapters of the 'Cours de Philosophie Positive' of August Comte)*, trans. Paul Descours and H. Gordon Jones. London: Watts, 1905.

Comte, Auguste. *The Positive Philosophy of August Comte.* Translated and edited by Harriet Martineau. London: George Bell & Sons, 1896.

Conant, James B. *On Understanding Science, An Historical Approach.* New Haven, CT: Yale University Press, 1947.
Copernicus, Nicolaus. *On the Revolutions of the Heavenly Spheres.* Translated by Charles Glenn Wallis. In *Great Books of the Western World* 16. Chicago: William Benton, 497–838.
d'Holbach, Paul-Henri Thiry. *System of Nature, or Laws of the Moral and Physical World*, Boston: J. P. Mendum, 1889.
Danilevskii, Nikolai Iakovlevich. *Russia and Europe. The Slavic World's Political and Cultural Relations with the Germanic-Roman West.* Translated by Stephen M. Woodburn. Bloomington. IN: Slavica, 2013.
de Broglie, Louis. Foreword to Pierre Duhem's *The Aim and Structure of the Physical Theory*. Translated by Philip P. Wiener. Princeton, NJ: Princeton University Press, 1991, v–xiii.
Dewey, John. Introduction to *Reconstruction in Philosophy*. Enlarged ed. Boston: Beacon, 1948.
Dewey, John. *Reconstruction in Philosophy*. New York: Henry Holt, 1920.
Dewey, John, Sidney Hook, and Ernest Nagel. "Are Naturalists Materialists?" *Journal of Philosophy* 42, no. 19 (1945): 515–30.
Dickens, Charles. *The Pickwick Papers*, Garden City, NY: International Collector's Library, 1944.
Dingle, Herbert. "Copernicus." *Observatory* 65 (1943): 38–57.
The Divine Comedy of Dante Alighieri. Translated by Charles Eliot Norton. London: Encyclopedia Britannica, 1952.
Druan, Jean. "Masse et énergie," *La Pensée: Revue du Rationalisme Moderne* 53 (1954): 29–38.
Du Bois-Raymond, Emil. "The Limits of Our Knowledge of Nature." Translated by J. Fitzgerald. *Popular Science Monthly* 5 (1874): 17–32.
Ducasse, Curt John. "Philosophy and Natural Science." *Philosophical Review* 49, no. 2 (1940): 121–41.
Ducasse, Curt John. *The Method of Knowledge in Philosophy* (The Howison Lecture for 1944), University of California Publications in Philosophy, vol. 16, no 7. Berkeley and Los Angeles: University of California Press, 1945. Reprinted in *American Philosophers at Work: The Philosophic Scene in the United States*, edited by Sidney Hook, 207–224. New York: Criterion, 1956.
Ducasse, Curt John. *Philosophy as a Science*. New York: Oskar Piest, 1941.
Duhem, Pierre. *The Aim and Structure of the Physical Theory*. Translated by Philip P. Wiener. Princeton, NJ: Princeton University Press, 1991.
Duhem, Pierre. "Physics of a Believer." In *The Aim and Structure of the Physical Theory*. Translated by Philip P. Wiener. Princeton, NJ: Princeton University Press, 1991, 273–311.
Duhem, Pierre. "Physique de croyant." *Annales de Philosophie Chrétienne* 155 (1905): 44–67, 133–59.

Duhem, Pierre. "La valeur de la théorie physique, à propos d'un livre recent." *Review générale des Sciences pures et appliquées* 19, no. 1 (1908): 7–19.

Duhem, Pierre. "The Value of a Physical Theory." In *The Aim and Structure of the Physical Theory*. Translated by Philip P. Wiener, 312–36. Princeton, NJ: Princeton University Press, 1991.

Duhem, Pierre. *Ziel und Struktur der physikalischen Theorien*, Leipzig: Johann Ambrosius Barth, 1908.

Dzuback, Mary Ann. *Robert M. Hutchins: Portrait of an Educator*. Chicago: University of Chicago Press, 1991.

Einstein, Albert. "Ernst Mach." In *The Collected Papers of Albert Einstein*. Vol. 6: *The Berlin Years: Writings, 1914–1917*. Translated by Alfred Engel. Edited by E. L. Schucking, 141–45. Princeton, NJ: Princeton University Press, 1997.

Einstein, Albert. *Geometrie und Erfahrung*. Berlin: Julius Springer, 1921.

Einstein, Albert. "Geometry and Experience." In *The Collected Papers of Albert Einstein*, Vol. 7: *The Berlin Years: Writings, 1918–1921*. Translated by Alfred Engel. Edited by E. L. Schucking, 208–222. Princeton, NJ: Princeton University Press, 2002.

Einstein, Albert. *Living Philosophies*, edited by Albert Einstein. New York: Simon and Schuster, 1931.

Einstein, Albert. *The Meaning of Relativity, Four Lectures Delivered at Princeton University, May 1921*. Princeton, NJ: Princeton University Press, 1923.

Einstein, Albert. "On the Method of Theoretical Physics." *Philosophy of Science* 1 no. 2 (1934): 163–69.

Einstein, Albert. "Physics and Reality." *Journal of the Franklin Institute* 221, no. 3 (1936): 349–82.

Einstein, Albert. "Remarks on Bertrand Russell's Theory of Knowledge." In *The Philosophy of Bertrand Russell*, edited by P. A. Schilpp, 277–91. New York: Tudor, 1944.

Einstein, Albert. "Science and Religion." In *Ideas and Opinions.*, 41–49. New York: Crown, 1954.

Emerson, Ralph Waldo. *Conduct of Life*. Cambridge, MA: Riverside, 1860.

Emerson, Ralph Waldo. *Nature*. Boston: James Munroe and Company, 1836.

Emerson, Ralph Waldo. *Representative Man*. Boston: Houghton, Mifflin, 1884.

Engels, Friedrich. *Development of Socialism from Utopia to Science*. Translated by D. de Leon. New York: National Executive Committee of the Socialist Labor Party, 1900.

Fiske, John. *Outlines of Cosmic Philosophy, Based on the Doctrine of Evolution, with Criticism on the Positive Philosophy*. Vol. 1. London: Macmillan and Co., 1874.

Four Letters from Sir Isaac Newton to Doctor Bentley Containing Some Arguments in Proof of a Diety. London: R. and J. Dodsley, 1761.

Fox, Ralph Winston. *Lenin: A Biography*. London: Victor Gollanz, 1933.

Frank, Hans. "Der Nationalsozialismus und die Wissenschaft der Wirtschaftslehre." *Schmollers Jahrbuch* 58. no. 2 (1934): 641–50.
Frank, Jerome. *Fate and Freedom*, New York: Simon and Schuster, 1945.
Frank, Philipp, ed. *The Validation of Scientific Theories*. Boston: Beacon, 1957.
Frank, Philipp. "Abel Rey, Die Theorie der Physik bei den modernen Physikern." *Monatshefte für Mathematik und Physik* 25 (1910): 43–45.
Frank, Philipp. "Zeigt sich in der modernen Physik ein Zug zu einer spiritualistischen Auffassung?" *Erkenntnis* 5 (1935): 65–80.
Frank, Philipp. *Einstein, His Life and Times*, New York: Alfred A. Knopf, 1947.
Frank, Philipp. "Einstein, Mach, and Logical Positivism." In *Albert Einstein, Philosopher-Scientist*, edited by Paul A. Schilpp, 271–86. New York: Harper & Row, 1949.
Frank, Philipp. *Foundations of Physics, International Encyclopedia of Unified Science*, vol. 1. no. 7. Chicago: Chicago University Press, 1946.
Frank, Philipp. *The Law of Causality and Its Limits*. Translated by Marie Neurath and Robert S. Cohen. Dordrecht, The Netherlands: Kluwer, 1998.
Frank, Philipp. "'Newton Didn't Talk about Why and How': Essay in Physics by Herbert Samuel." *New York Times*, February 17, 1952.
Frank, Philipp. "The Philosophic Meaning of the Copernican Revolution." In *Modern Science and Its Philosophy*, 219–20. Cambridge, MA: Harvard University Press, 1949.
Frank, Philipp. *Philosophy of Science: The Link Between Philosophy and Science*. Englewood Cliffs, NJ: Prentice-Hall, 1957.
Frank, Philipp. "Philosophical Uses of Science,' *Bulletin of the Atomic Scientists*, April 1957, 125–30.
Frank, Philipp. "Physical Theories of the Twentieth Century and School Philosophy." In *Modern Science and its Philosophy*, 90–121. Cambridge, MA: Harvard University Press, 1949.
Frank, Philipp. *Relativity, a Richer Truth*. Boston: Beacon, 1950.
Frank, Philipp. "The Role of Authority in the Interpretation of Science." *Synthese* 10 (1956–1958): 335–38.
Frank, Philipp. "Science Teaching and the Humanities.' *ETC* 4, no. 1 (1946): 1–24.
Garrigou-Lagrange, Réginald. *God, His Existence and His Nature: A Thomistic Solution of Certain Agnostic Antinomies*. Vol. 2. St. Louis and London: Herder, 1949.
Gill, Henry Vincent. *Fact and Fiction in Modern Science*. Dublin: Gill, 1943.
Gilson, Étienne. *La philosophie au moyen âge*. Paris, 1930.
Gilson, Étienne. *The Christian Philosophy of St. Thomas Aquinas*. Translated by Laurence K. Shook. New York: Random House, 1956.
Goblot, Edmond. *Le Systeme des Sciences*. Paris: A Colin, 1922.
Goebbels, Joseph. *Der Angriff*. Munich: Zentralverlag der NSDAP, 1940.

Hahn, Hans. "The Significance of the Scientific World View, Especially for Mathematics and Physics." In *Empiricism, Logic and Mathematics*, edited by Brian McGuinness, 20–30. Dordrecht, The Netherlands: D. Reidel, 1980.

Hahn, Hans. "Superfluous Entities, or Occam's Razor." In *Empiricism, Logic and Mathematics*, edited by Brian McGuinness, 1–19. Dordrecht, The Netherlands: Reidel, 1980.

Haldane, J. B. S. *The Marxist Philosophy and the Sciences*. New York: Random House, 1938.

Hardcastle, Gary L. "Debabelizing Science: The Harvard Science of Science Discussion Group, 1940–41." In *Logical Empiricism in North America*, edited by Gary L. Hardcastle and Alan Richardson, 170–96. Minneapolis: University of Minnesota Press, 2003.

Hegel, G. W. F. *The Logic of Hegel*. Translated from his *Encyclopedia of the Philosophical Sciences*. Oxford: Clarendon, 1874.

Hegel, G. W. F. *The Science of Logic*. Translated and edited by George Di Giovanni. Cambridge: Cambridge University Press, 2010.

Heisenberg, Werner. *The Physicist's Conception of Nature*. London: Hutchinson, Scientific and Technical, 1958.

Hempel, Carl G. "The Concept of Cognitive Significance: A Reconsideration." *Proceedings of the American Academy of Arts and Sciences* 80, no. 1 (1951): 61–77.

Hiscock, W. G. *David Gregory, Isaac Newton and Their Circle: Extracts from David Gregory's Memoranda 1677–1708*. Oxford, 1937.

Hobbes, Thomas. *Hobbes's Leviathan*. Reprinted from the edition of 1651. Oxford: Clarendon, 1929.

Hume, David. *Dialogues Concerning Natural Religion*. Edinburgh and London: William Blackwood and Sons, 1907.

Hume, David. *An Enquiry Concerning Human Understanding and Selections from a Treatise of Human Nature*, Chicago: Open Court, 1912.

Huxley, Aldous. *Science, Liberty and Peace*. London: Chatto & Windus, 1950.

Ioffe, Abraham F. "Razvitie atomisticheskikh vozzrenii v XX veke." *Pod Znamenem Marksizma* 4 (1934): 52–68.

Isaac, Joel. *Working Knowledge: Making the Human Sciences from Parsons to Kuhn*. Cambridge, MA: Harvard University Press, 2012.

Jaeger, Werner. *Aristotle: Fundamentals of the History of His Development*. Translated by Richard Robinson. Oxford: Clarendon, 1934.

Jaeger, Werner. *Paideia: The Ideals of Greek Culture*. Translated by Gilbert Highet. Oxford: Basil Blackwell, 1946.

James, William. *A Pluralistic Universe* (*Hibbert Lectures at Manchester College on the Present Situation in Philosophy*). New York: Longmans, Green, 1909.

James, William. "The Sentiment of Rationality." Reprinted from *The Will to Believe, and Other Essays in Popular Philosophy*, 63–110. London: Longmans, Green and Co. 1905.

Jeans, James. *The Mysterious Universe*. Harmondsworth, UK: Penguin, 1930.
Joad, C. E. M. *Decadence: A Philosophical Inquiry*, London: Faber and Faber, 1948.
Joad, C. E. M. *Philosophical Aspects of Modern Science*. London, Allen and Unwin, 1932.
Kant, Immanuel. *Critique of Pure Reason*. Translated by Norman Kemp Smith. London: Macmillan, 1929.
Koestler, Arthur. *The Sleepwalkers: A History of Man's Changing Vision of the Universe*, New York: Macmillan, 1959.
Kolman, Ernst. "Stalin i nauka" (Stalin and science), *Pod Znamenem Marksizma* (Under the banner of Marxism) 12 (1939): 172–86.
Korzybski, Alfred. *Science and Sanity: An Introduction to Non-Aristotelian Systems and General Semantics*. Lancaster, PA: Science Press, 1933.
Kunstmann, Heinrich K. *Rede zur Feier der Immatrikulation, gehalten in der Aula der Neuen Universität Heidelberg, 23. Nov. 1936*, Heidelberg: Carl Winter, 1937.
La Mettrie, Julien Offray de. *Man a Machine*. Chicago: Open Court, 1912.
Large Soviet Encyclopedia. 1st ed. 65 vols. Moscow: OGIZ, 1926–1947.
Le Roy, Édouard. "Science et Philosophie." *Revue de Métaphysique et de Morale* 7, nos. 4–6 (1899): 375–425, 503–62, 708–31.
Lenin, V. I. *Materialism and Empirio-Criticism: Critical Comments on a Reactionary Philosophy*. Moscow: Foreign Languages Publishing, 1947.
Limbeck-Lilineau, Christoph, and Friedrich Stadler. *Der Wiener Kreis: Texte und Bilder zum Logischen Empirismus*. Vienna: LIT, 2015.
Littré, Émile. *De La Philosophie Positive*. Paris: Librairie Philosophique de Ladrange, 1845.
Littré, Émile. Preface to *Principes de philosophie positive* by Auguste Comte. Paris: J. B. Baillière et Fils, 1868, 5–75.
Lowinger, Armand. *The Methodology of Pierre Duhem*. New York: Columbia University Press, 1941.
Lucretius. *De Rerum Natura*. Edited by Cyril Bailey. 3 vols. Oxford, 1947.
Lukács, Georg. *History and Class Consciousness: Studies in Marxist Dialectics*. Translated by Rodney Livingstone. Cambridge, MA: MIT Press, 1971.
Mach, Ernst. "On the Principle of Comparison in Physics." In *Popular Scientific Lectures*, 236–58. Translated by Thomas J. McCormack. 3rd ed. Chicago: Open Court, 1898.
Mach, Ernst. "The Economical Nature of Physical Inquiry." In *Popular Scientific Lectures*, 186–213. Translated by Thomas J. McCormack. 3rd. ed. Chicago: Open Court, 1898.
Mach, Ernst. *Knowledge and Error. Sketches on the Psychology of Enquiry*. Translated by Thomas I. McCormack. Dordrecht, The Netherlands: D. Reidel, 1976.
Mach, Ernst. *Space and Geometry in the Light of Physiological, Psychological and Physical Inquiry*. Translated by Thomas J. McCormack. London: Regan Paul, Trench, Trübner, 1906.

MacLeish, Archibald. "Why Do We Teach Poetry?" *American Monthly* 197, no. 3 (March 1956), 48–53.

Malik, Charles. *War and Peace: A Statement Made before the Political Committee of the General Assembly, November 23, 1949.* Stamford, CT: Overlook, 1950.

Mannheim, Karl. *Ideology and Utopia.* Translated by Louis Wirth and Edward Shils. New York: Harcourt, Brace, 1954.

Margenau, Henry. *The Nature of Physical Reality. A Philosophy of Modern Physics.* New York: McGraw-Hill, 1950.

Maritain, Jacques. *Distinguish to Unite or the Degrees of Knowledge.* London: G. Bles, 1937.

Maritain, Jacques. *Éléments de philosophie I. Introduction générale à la philosophie.* Paris: Pierre Téqui, 1921.

Maritain, Jacques. *An Introduction to Philosophy.* Translated by E. I. Watkin. New York: Sheed and Ward, 1937.

Maritain, Jacques. *A Preface to Metaphysics: Seven Lectures on Being.* New York: Sheed & Ward, 1939.

Maritain, Jacques. *The Range of Reason.* New York: Charles Scribner's Sons, 1952.

Maritain, Jacques. "Science, Philosophy, and Faith." In *Science, Philosophy and Religion: A Symposium*, 162–83. New York: Conference on Science, Philosophy, and Religion in their Relation to the Democratic Way of Life, 1941.

Marx, Karl. *A Contribution to the Critique of Political Economy.* Translated by N. I. Stone. Chicago: Charles H. Kerr & Co., 1904.

Marx, Karl. *Selected Works*, edited by V. Adoratsky. London: Lawrence and Wishart, 1945.

Marx, Karl, and Friedrich Engels. *The German Ideology.* New York: International, 1939.

Mayer, Milton. *Robert Maynard Hutchins: A Memoir.* Los Angeles: University of California Press, 1993.

McLaughlin, P. J. "Review of O'Rahilly's Electromagnetics." *Studies: An Irish Quarterly Review* 27, no. 108 (1938): 656–66.

McWilliams, James A. *Cosmology: A Text for Colleges.* Rev. ed. New York: Macmillan, 1937.

McWilliams, James A. *Physics and Philosophy. A Study of Saint Thomas' Commentary on the Eight Book of Aristotle's Physics.* Washington, DC: Catholic University of America Press, 1946.

Merton, Robert K. "Karl Mannheim and the Sociology of Knowledge." In *Social Theory and Social Structure*, 489–508. Rev. ed. New York: Free Press, 1957.

Merton, Robert K. "Sociology of Knowledge." In *Twentieth-Century Sociology*, edited by Georges Gurvitch and Wilbert E. Moore, 366–405. New York: Philosophical Library, 1945.

Milton, John. *Paradise Lost, A Poem in Twelve Books.* Bk. VIII. Boston: Timothy Bedlington, 1820.

Mitin, Mark Borisovich. *Dialehticheskii i istoricheskii materialism*, Moscow: Philosophical Institute of the Communist Academy, 1933.
Mitin, Mark Borsovich. "Twenty-Five Years of Philosophy in the U.S.S.R." *Philosophy (The Journal of the British Institute of Philosophy)* 19, no. 72 (1944): 76–84.
Mormann, Thomas. "Philipp Frank's Austro-American Logical Empiricism." *HOPOS* 7 no. 1 (2017): 56–87.
Morris, Charles. *Foundations of the Theory of Signs*, vol. 1, no. 2, International Encyclopedia of Unified Science. Chicago: University of Chicago Press, 1938.
Morris, Charles. "Scientific Empiricism." In *Encyclopedia and Unified Science*, vol. 1, no. 1, International Encyclopedia of Unified Science, 63–75. Chicago: University of Chicago Press, 1955.
Nagel, Ernest. Introduction to *Sovereign Reason and Other Studies in the Philosophy of Science*, 9–16. Glencoe, IL: Free Press, 1954.
Nagel, Ernest. "Philosophy and the American Temper." In *Sovereign Reason and Other Studies in the Philosophy of Science*, 50–57. Glencoe, IL: Free Press, 1954.
Nagel, Ernest. *Principles of the Theory of Probability*, vol. 1, no. 6, International Encyclopedia of Unified Science, 342–422. Chicago: University of Chicago Press, 1955.
Neuber, Matthias. "Philosophie der Modernen Physik: Philipp Frank und Abel Rey." *Grazer Philosophische Studien* 80 (2010): 131–49.
Neurath, Otto. 'Sociology in the Framework of Physicalism." In *Philosophical Papers 1913–1946*. Edited and translated by Robert S. Cohen and Marie Neurath, 58–90. Dordrecht, The Netherlands: D. Reidel, 1983.
Neurath, Otto. "Unified Science and its Encyclopedia." In *Otto Neurath· Philosophical Papers, 1913–1946*, edited by Robert S. Cohen and Marie Neurath, 172–82. Dordrecht, The Netherlands: Reidel, 1983.
Neurath, Otto. "Unified Science as Encyclopedic Integration." In *Encyclopedia and Unified Science*, vol. 1, no. 1 of *International Encyclopedia of Unified Science*. Chicago: University of Chicago Press, 1938.
Neurath, Otto. "Unified Science as Encyclopedic Integration." In *Foundations of the Unity of Science*, edited by Otto Neurath, Rudolf Carnap, and Charles Morris, 1–27. Vol. 1. Chicago: University of Chicago Press, 1971.
Nietzsche, Friedrich. *The Birth of Tragedy*, in *The Complete Works of Friedrich Nietzsche*, edited by Oscar Levy. London: George Allen & Unwin, 1923.
Nitschmann, Leo. "Einstein entsinnlichte den Kosmos. Die Relativitätstheorie als kulturgeschichtliches Ereignis," *Die Zeit* 50, December 16, 1954.
Northrop, Filmer S. C. *The Logic of the Sciences and the Humanities*. New York: Macmillan, 1948.
O'Connell, William Cardinal. *Recollections of Seventy Years*. Boston and New York: Houghton Mifflin, 1934.

O'Rahilly, Alfred. *Electromagnetic Theory: A Discussion of Fundamentals*. London and New York: Longmans, Green, 1938.

Ockham, William. *Philosophical Writings: A Selection*. Translated by Philotheus Boehner. Edinburgh and London: Thomas Nelson and Sons, 1957.

Osiander, Andreas. "Preface to *On the Revolutions*." In *Nicholas Copernicus on the Revolutions*. Translated and commentary by Edward Rosen. London and Basingstoke, UK: Macmillan, 1978.

Ostwald, Wilhelm. "The Modern Theory of Energetics." *Monist* 17, no. 4 (1907): 481–515.

Otto, Max Carl. "The Ethical Neutrality of Science." In *Science and the Moral Life: Selected Writings by Max C. Otto*. New York: Mentor, 1949.

Peirce, Charles S. *Collected Papers of Charles Sanders Peirce*. Vol. 5: *Pragmatism and Pragmaticism*, edited by Charles Hartshorne and Paul Weiss. Cambridge, MA: Harvard University Press, 1934.

Pearson, Karl. *The Grammar of Science*. London: Walter Scott, 1892.

Peirce, Charles S. "How to Make Our Ideas Clear?" In *Philosophical Writings of Peirce*, edited by Justus Buchler, 23–41. New York: Dover, 1955.

Peirce, Charles S. "Pearson's Grammar of Science. Annotations on the First Three Chapters," *Popular Science Monthly* 58, no. 1 (1901): 296–306.

Peirce, Charles S. "Philosophy and the Sciences: A Classification." In *Philosophical Writings of Peirce*, edited by Justus Buchler, 60–73. New York: Dover, 1955.

Peirce, Charles S. "Review of English trans. of Mach's *The Science of Mechanics*." *Nation* 57, no. 1475 (1893): 252.

Phelan, Gerald B. *Saint Thomas and Analogy (Aquinas Lecture, 1941)*, Milwaukee: Marquette University Press, 1941.

Philipps, Richard P. *Modern Thomistic Philosophy: An Explanation for Students*. Vol. 1: *The Philosophy of Nature*. Westminster, MD: Newman Bookshop, 1934.

Planck, Max. *Positivismus und reale Aussenwelt*. Leipzig: Akademische Verlagsgesellschaft, 1913.

Plato. *The Republic*. Vol. 2, Bks. 6–10 (Loeb Classical Library no. 276). Translated by Paul Shorey. Cambridge, MA: Harvard University Press, 1942.

Plato. *Timaeus and Critias*. Translated by A. E. Taylor. London: Methuen, 1929.

Plato. *Phaedo*. In *The Dialogues of Plato*. Translated by B. Jowett. 2nd ed. Vol. 1. London: Macmillan, 1875.

Poincaré, Henri. "Des Fondements de la Géométrie: A propos d'un Livre de M. Russell." *Revue de Métaphysique et de Morale* 7 (1899): 251–79.

Poincaré, Henri. *The Value of Science*. Translated and introduced by George Bruce Halsted. New York: Science Press, 1907.

Polanyi, Michael. *Science, Faith and Society*. London: Oxford University Press, 1946.

Putnam, Hilary. "Review of Philipp Frank's *Philosophy of Science*." *Science* 127, no. 3301 (1958): 750–51.

Quine, Willard van Orman. "Semantics and Abstract Objects." *Proceedings of the American Academy of Arts and Sciences* 80, no. 1 (1952): 90–96.

Rabi, I. I. "To Preserve the Scientific Spirit." *New York Times Magazine*, February 12, 1956.

Rankine, William John Macquorn. "Outlines of the Science of Energetics." *Proceedings of the Royal Philosophical Society of Glasgow*, vol. 3, Royal Philosophical Society of Glasgow, 1855, 121–41.

Reinhardt, Kurt F. *The Existential Revolt: The Main Themes and Phases of Existentialism*. Milwaukee: Burce, 1952.

Reisch, George. "Pragmatic Engagements: Philipp Frank and James Bryant Conant on Science, Education, and Democracy." *Studies in East European Thought* 69, no. 3 (2016): 227–44.

Rey, Abel. *La théorie de la physique chez les physiciens contemporains*. Paris: Félix Alcan, 1907.

Rosenthal, Mark Moisevich and Pavel Fyodorovich Yudin. *Kratkij Filosofskij Slovar*. Moscow: Gos. Izd. Političeskoj lit., 1955.

Russell, Bertrand. *A History of Western Philosophy*, 2nd ed., London: George Allen and Unwin Ltd., 1947.

Rust, Bernhard. "Nationalsozialismus und Wissenschaft." In *Das nationalsozialistische Deutschland und die Wissenschaft. Heidelberger Reden von Reichs-minister Rust und Prof. Ernst Krieck*, 9–22. Hamburg, 1936.

Rust, Bernhard. *Freiheit und Ordnung. Rede bei der Kundgebung der Studenten auf dem Ehrenhof der Universität am 28. Juni 1937. Wissenschaft und Glaube*. Berlin: Oldenburg i.O., 1938.

Samuel, Herbert L. *Essay in Physics*. New York: Harcourt, Brace, 1952.

Santayana, George. *Three Philosophical Poets: Lucretius, Dante and Goethe*. Cambridge, MA: Harvard University Press, 1910.

Schlick, Moritz. *Space and Time in Contemporary Physics*. Translated by Henry L. Brose. New York: Oxford University Press, 1920.

Schlick, Moritz. "The Turning-Point in Philosophy." In *Philosophical Papers*. Vol. 2: *1925–1936*, edited by Henk L. Mulder and Barbara F. B. Van de Velde-Schlick, 154–60. Dordrecht, The Netherlands: D. Reidel, 1979.

Sheen, Fulton J. *Philosophy of Science*. Milwaukee: Bruce, 1934.

Simpson, George Gaylord. *The Meaning of Evolution*. New York: New American Library, 1956.

Snow, Charles Percy. "The Two Cultures." *New Statesman*, October 6, 1956.

Sorokin, Pitirim. *The Crisis of Our Age*. New York: E. P. Dutton, 1942.

Spencer, Herbert. *First Principles*. 6th ed. London: Watts, 1946.

Stace, Walter Terence. *The Philosophy of Hegel—A Systematic Exposition*. New York: Dover, 1955.

Stalin, Joseph. *History of the Communist Party of the Soviet Union*. New York: International, 1939.

Stallo, John B. *The Concepts and Theories of Modern Physics*. New York: Appleton, 1888.
Stevenson, Adlai. "A Purpose for Modern Woman." *Women's Home Companion* (September 1955): 29–31.
Topitsch, Ernst. "Society, Technology, and Philosophical Reasoning." *Philosophy of Science* 21 no. 4 (1954): 275–96.
Topitsch, Ernst. "Kosmos und Herrschaft. Ursprünge der 'politischen Theologie.'" *Wort und Wahrheit: Monatsschrift für Religion und Kultur* 10, no. 1 (1955): 19–30.
Tuboly, Adam Tamas. "Philipp Frank's Decline and the Crisis of Logical Empiricism." *Studies in East European Thought* 69, no. 3 (2017): 257–76.
Tuboly, Adam Tamas. "Knowledge Missemination: L. Susan Stebbing, C. E. M. Joad and Philipp Frank on the Philosophy of the Physicists." *Perspectives on Science* 28, no. 1 (2020): 1–34.
Uebel, Thomas. "Logical Empiricism and the Sociology of Knowledge: The Case of Neurath and Frank." *Philosophy of Science* 67 (2000): 138–50.
Uebel, Thomas. *Empiricism at the Crossroads: The Vienna Circle's Protocol-Sentence Debate*. Chicago: Open Court, 2007.
Uebel, Thomas. "Beyond the Formalist Criterion of Cognitive Significance: Philipp Frank's Later Antimetaphysics." *HOPOS* 1, no. 1 (2011): 47–72.
Vavilov, Sergei I. "The New Physics and Dialectical Materialism." *Modern Quarterly* 2, no. 2 (1939): 146–54.
Voltaire. *A Philosophical Dictionary*. Vol. 3. London: John and Henry L. Hunt, 1824.
von Helmholtz, Hermann. "The Aim and Progress of Physical Science." In *Popular Lectures on Scientific Subjects*, 363–97. Translated by E. Atkinson. New York: Appleton, 1873.
von Mises, Richard. *Positivism, An Essay in Human Understanding*. Translated by Jeremy Bernstein and Roger Newton. Cambridge, MA: Harvard University Press, 1951.
Weber, Max. *The Protestant Ethic and the Spirit of Capitalism*. Translated by Talcott Parsons. New York: Charles Scribner's Sons, 1950.
Wellmuth, John James. *The Nature and Origins of Scientism*. Milwaukee: Marquette University Press, 1944.
Werkmeister, William H. *The Basis and Structure of Knowledge*. New York: Harper and Brothers, 1948.
Whewell, William. *History of the Inductive Sciences—From the Earliest to the Present Times*. Vol. 1. London: John W. Parker, 1837.
Whitehead, Alfred N. *Adventures of Ideas*. Cambridge: Cambridge University Press, 1961.
Whitehead, Alfred North. *The Principle of Relativity with Applications to Physical Science*. Cambridge: Cambridge University Press, 1922.

Whitehead, Alfred North. *Science and the Modern World*. New York: Felican Mentor, 1948.
Wuest, Amy. "Simplicity and Scientific Progress in the Philosophy of Philipp Frank." *Studies in East European Thought* 69, no. 3 (2016): 245–55

Index

acceptance of theories. *See* sociology of knowledge
Adler, Kathia, 8
Adler, Mortimer, 20, 22–24, 32, 43, 224, 337n10, 343n3; at CSPR, 29–30
alchemy, 160
American Academy of Arts and Sciences, 32, 37; Frank being member of, 27
American Physical Society, Frank being member of, 27
analogies, 107–108, 109, 124–26, 129, 132, 150, 193, 204–207, 216, 220, 259, 321; as philosophical interpretations of science, 109, 111–12, 183; between science and society 184; and physical laws, 201–204
angels, 199–201, 216
Aristotle (Aristotelian), 85, 98, 102, 109, 112, 143, 152, 158, 180–83, 189, 197, 198, 199, 200, 209, 213, 215, 222, 230, 234, 236, 238, 249, 252, 257, 262, 264, 266, 280, 289, 290, 308, 309, 344n14–15; on humanizing science, 105, 108, on matter, 195
astrology, 82, 160, 213

atomic bomb, 43n59, 72, 75, 84, 89, 160, 175, 197, 336n3
atoms, 97, 117, 194, 197, 207, 210, 245, 277, 293; debate over, 4
auxiliary concepts, 148, 222, 264
Avenarius, Richard, 299
axioms, 79, 122, 125, 132, 134, 135–37, 168, 255, 256, 262, 268, 272, 282, 287, 298, 319

Bacon, Francis, 89, 108, 144, 238, 326n21; on Ptolemy and Copernicus, 101–105, 117
Bacon, Roger, 217, 259
Becker, Carl, 189–90
Bellarmine, Cardinal, 140, 173, 331n20
Benjamin, A. Cornelius, 47n65
Bentley, Richard, 356n13
Benton, William, 23, 343n3
Berdyaev, Nicolas, 98, 101
Bergmann, Peter G., 9
Bergson, Henri, 49, 76, 110–12, 229, 239, 323n10, 344n11
Bernstein, Jeremy, 3
Berwald, Ludwig, 9, 15n24
Bible, 49, 127, 173, 185, 211–12, 234
biology, 15, 82, 83, 86, 87, 91, 115, 171, 195, 196, 202, 227, 234, 235,

biology *(continued)*
  243, 261, 265, 269, 279, 280, 294, 305 307, 310–13
Black, Max 282
Blumberg, A. E., 14n19, 263, 282
Boas, George, 232
Bohr, Niels, 36n51, 41n56, 107, 115, 162, 194, 265, 319
Boltzmann, Ludwig, 3, 4–5, 54
Bolzano, Bernard, 13
Born, Max, 14
Boston Colloquium for the Philosophy of Science, 36, 40
Bragg, Raymond, 47
Bridgman, Percy, 25, 31, 32, 34, 42, 78, 283, 320–21, 354n28; on theory of meaning, 284–86
Bronk, Detlev, 34
Buckle, Henry Thomas, 247–48
Bulletin of the Atomic Scientists, 33
Butts, Robert, 52

Caldin, Edward Francis, 193
Calvinism, 157
Carnap, Rudolf, 2, 14–18, 20, 22, 23, 25, 26, 27, 32, 41, 44, 45, 48, 49, 51, 84, 269, 273, 297, 329n2, 350n1; criticism of metaphysics 46–47, 270–72, 318
Catholic Encyclopedia, 204, 250
Christ, 216–17
Church, 85, 89, 140–41, 148, 152, 173, 191, 215, 259, 260, 292, 293, 345n28
City College of New York, 27
Clarke, Joseph T., 193–94, 197
Cohen, I. Bernard, 8, 9
Cohen, Robert S., 36, 39, 329n2
Colligan, John J., 196
combustion, 196
commonsense: analogies, 129, 131, 134, 139, 141, 150, 262, 321; experience 46, 103–104, 107–109, 112, 116, 117–20, 123, 124–27, 132, 147, 190, 194, 205, 209, 311; language 87, 103–104, 113, 121, 124, 127, 129, 144, 163, 168, 174–75, 183, 190, 205–207, 287
Communist Party, 86, 136, 165, 179, 191
complementarity: principle of, 194
Comte, August 44, 77, 90–91, 162, 222, 232–34, 240–43, 246, 251, 252, 254, 258, 261, 266, 272, 276–77, 279, 280, 290, 318, 335n25, on facts 168, on sociology 235–36, 345n27, 345n29
Conant, James Bryant, 25–26, 43n59, 50, 162, 324n18, 329n2
conceptual scheme, 50, 81, 82, 162, 324n18
Conference on the Scientific Spirit and Democratic Faith, 30
Conferences on Science, Philosophy, and Religion, 28–29, 43, 337n15, 340n46
Congress for the Unity of Science, 15, 25
conventionalism, 172, 255–57, 267, 356n23
Copernicus: system, 85, 89, 91, 108, 140–41, 143–44, 148, 152–53, 173, 212, 229, 345n28; vs. Ptolemaic system, 79, 100, 102, 104, 117, 172–73, 230, 234, 237, 239, 252; system as fiction, 104, 238
cosmic state, 184
cosmology, 171, 211
Coulomb's law, 147

d'Alembert, Jean le Rond, 222
d'Autrecourt Nicholas, 217, 220
Daedalus, 33
Danilevskii, Nikolai I., 149–50, 332n35
Dante, Alighieri, 210

Index 377

Darwin, Charles, 229, natural selection, 244, 348n52
de Broglie, Louis, 259
de La Mettrie, Julian Offray, 44, 290–92, 294, 295
democracy, 28, 30, 32, 43, 86, 139, 178–83, 191, 247–48, 280, 281
Democritus, 181, 245, 348n55
Descartes, René, 117, 205, 222
desired and undesired way of life, 134–35, 139, 143, 213
determinism, 87, 90, 171, 172, 179, 213
Deutsche Physikalische Gesellschaft, 12
Dewey, John, 20, 22, 24, 28, 30, 42, 44, 46, 48, 49, 168, 184, 245, 271, 279, 280–82, 287, 295, 329n2, 354n21
dialectical materialism, 23, 43, 49, 54, 74, 76, 86, 88, 137, 173, 179, 181, 191, 213, 214, 296, 298, 299, 301, 302, 305, 307, 308, 310, 312, 313, 314, 318
Dickens, Charles, 202
Dingle, Herbert, 331n18
Druan, Jean, 301–302
Du Bois-Raymond, Emil, 249–50
Ducasse, John C., 146
Duhem, Pierre, 46, 77, 84, 142–43, 162, 230, 241–42, 249, 258–60, 261, 275, 277–79, 344n15

economy, 23, 74–75, 82, 83, 157, 161, 165, 166, 171, 313, 324n22; of science, 4, 131, 153, 253, 256, 259, 277, 278, 279
Ehrenfest, Paul, 3, 7
Ehrenhaft, Felix, 3
Einstein, Albert, 1, 3, 7–9, 28, 30, 39, 54, 85, 86, 89, 115, 141, 147, 152, 162, 171, 220, 244–45, 263, 265, 286, 301, 322; and positivism, 317–19; comprehensibility of the world, 320–21; cosmic religion 220; $E = mc^2$, 301–302; Herbert Spencer Lecture, 319; member of Russian Academy, 136; supporting Frank in New York, 27n40; theory of gravitation, 97–98, 147
Emerson, Ralph Waldo, 94–95, 101, 105, 325n2
Encyclopedia Britannica, 227–28, 343n1, 343n3
energetics, 145, 249–50
Engel, Walter, 15n24
Engels, Friedrich, 80, 155, 156, 161, 306, 312–13, 334n3
Epicurus, 89, 117, 245, 289, 291, 293, 348n55
Erkenntnis, 350n1
Erkenntnislehre der exakten Wissenschaften, 12–14
Ernst Mach Society, 12
ether, 244
Exner, Franz, 3

facts: and creative imagination, 162; and observation, 76, 79, 83, 87, 93, 105, 122, 125, 126, 134, 140, 144, 145, 146, 151, 190, 216, 223, 230, 233, 234, 236–39, 240, 243–46, 251, 252, 255–57, 265, 269, 272, 286, 290, 312, 319, 322; and theory, 168; and values, 71–72, 79; experimental, 102, 119, 285; gathering and finding, 32, 80, 95, 231, 236, 277–78; historical, 102; intelligible, 189; knowledge of, 73; mental, 137, 151; mere and pure, 77, 161, 162, 168; moral interpretation of, 86; role of human mind, 77; social 159, 181, 184, 241
Fanta, Berta, 9
Faraday, Michael, 147
FBI investigations, 48n67

Feigl, Hebert, 16, 32, 48n66, 263, 282
Feyerabend, Paul 2, 36, 54
Finkelstein, Louis, 28, 30
Fiske, John, 240–42, 346n40, 347n44
formalism, 100, 132, 151
Fortner, Paul, 15n24
Foscarini, Paolo Antonia, 173, 331n20
Fox, Ralph, 299
Frank, Hania, 13, 36, 39n55
Frank, Hans, 176
Frank, Jerome, 90, 179, 213
Frank, Jozef, 3
Frank, Philipp: on Mach-Boltzmann, 4–5; on relativity, 6; biography of Einstein, 8–9; lectures in Prague, 10n14; Frank-Mises (see von Mises, Richard); Prague Circle, 15–16; sociology research group, 51–52; teaching at MIT, 35; Festschrift, 36–37, 40
Fraydas, Stan, 39n55
free will, 90, 108, 109, 143, 258
French Revolution, 232, 247
Freud, Sigmund, 157
Führer of German University Teachers, 138–39, 330n13
Fürth, Reinhold, 9, 16

Galileo, Galilei, 85, 86, 90, 115, 140, 173, 198, 234, 249
Garrigon-Lagrange, Reginald, 142
Genesis, 185
geometry, 39, 79, 107, 122, 136, 137, 157, 255, 268, 269, 271, 272; Euclidean, 77, 262, 319; non-Euclidean, 77, 118, 119, 245
Gibbs, Josiah Willard, 254, 349n25
Gicklhorn, Joseph, 15n24
Gill, Henry Vincent, 120–21
Gilson, Étienne, 198

Goblot, Edmond, 49, 118–20
Gödel, Kurt, 16
Goebbels, Joseph, 149, 332n34
Goering, Hermann, 176
Goudsmit, Samuel, 175, 336n4
Gregory, David, 106

Hahn, Hans, 6, 16, 49, 265
Haldane, J. B. S., 88
Hasenöhrl, Friedrich, 3
Hegel, Georg Wilhelm Friedrich, 180, 181, 183, 191, 222, 310–12, 314, 358n15
Heisenberg, Werner, 115, 128–29, 265, 308
heliocentric vs. geocentric, 144, 153, 173, 212, 230, 239; choice between 103, 184, 234, 237–38, agreement with facts 236
Helmholtz, Hermann, 248, 249
Hempel, Carl G., 36, 45, 272–73
Hessen, Boris, 333n38
Hessen, Sergius, 15n24
Hilbert, David, 3
Hitler, Adolf, 18, 29, 149, 174, 336n3
Hobbes, Thomas, 221
d'Holbach, Baron, 292
Holton, Gerald, 27
Holy Office, 143, 144
Hook, Sidney, 14, 20, 24, 28, 30, 34, 48, 295
humanism, 34, 43, 45, 47, interview with Frank on, 48, 54n74
Hume, David, 127, 202, 218, 220–21, 252, 318
Hutchins, Robert M., 20–24, 25, 29, 32, 43, 48, 343n3
Huxley, Aldous, 73, 81–82

idealism, 14, 76, 78, 85, 100, 136, 137, 138, 158, 160, 165, 179, 183,

191, 248, 250–51, 252, 262, 263, 270–71, 283, 287, 296, 299, 300, 301, 310–11, 312, 315, 318; and formalism, 151, animistic, 297, Menshevik, 305–307
ideology, 89, 167, 183; as distortion of truth, 156–57, 159, 165–66, general concept of, 158–59, 167
indeterminism. See determinism
inertia: law of, 97, 106–107, 120, 125, 142–43, 157–58, 198–99
inner eye, 108–10, 113–14, 117, 119, 134, 152, 155
Institute for the Unity of Science, 27, 32–33
Institute of International Education, 18, 24, 25
Institute of Red Professorship, 306, 357n1
intelligentsia: socially unattached, 167
intelligible principles, 98, 105–108, 109, 189–90, 194, 222, 233, 236, 243, 251, 257, 289, 290
Inter-Science Discussion Group, 27
International Encyclopedia of Unified Science, 83–84, 301
intuition, 114, 120, 123, 126, 155, 264, 346n33
Ioffe, Abraham, 49, 138, 308
iron curtain, 183

Jaeger, Werner, 213, 341n12
James, William, 20, 77, 168, 239, 278–79, 323n10
Jeans, James, 222, 320, 360n13
Jerusalem, Wilhelm, 335n23
Jewish Theological Seminary, 28
Joad, C. E. M., 224, 342n37
Joergensen, Joergen, 281
Joule, James P., 284, 355n30

Journal for General Science (Zeitschrift für die Gesamte Naturwissenschaft), 150

Kafka, Franz, 9
Kallen, Horace, 20, 24, 28, 30
Kant, Immanuel, 149, 191, 222, 342n32, on metaphysics 223–24, 272
Katkov, Georg, 15n24
Kaufmann, Felix, 16
Klein, Felix, 3
knowledge: meanings of, 169, pragmatic theory of, 172
Koestler, Arthur, 331n21, 336n1
Kohl, Emil, 7
Kolman, Ernst, 306
Korzybski, Alfred, 79, 324n12
Kotarbiński, Tadeusz, 281
Kuhn, Thomas, 3–4, 27, 36, 50–52
Kunstmann, Heinrich, 139, 330n15
Kuznatsov, I. V., 302, 357n33

Lampa, Anton, 7
Langer, Suzanne, 34
Large Soviet Encyclopedia, 151–53, 227–29, 306, 343n2
laws of nature. See physical laws
laws: dialectical, 310, 312, 358n17; of causality, 126, 185, 211–12, 255, 262; of motion, 90, 94, 106, 145, 189, 194, 200, 255, 281, 290, 312
Le Roy, Édouard 49, 114, 255–56
Leibniz, G. W. 222
Lenin, V. I. 48n67, 139, 176, 228, 299–300, 302, 307–309, 329n8, 344n7, 356n23, 358n8
Littré, Émile 49, 91, 233–35, 345n23
Locke, John 182–83
Lowinger, Armand 258
Löwner, Karl 9
Lucretius 181, 289

Lukács, George, 49, 156, 161, 164, 166, 333n2, 335n23, 335n17

Mach, Ernst 3, 4–5, 7, 8, 13, 44, 54, 77, 136, 152–53, 162, 229, 248, 249, 251, 254, 255, 257, 258, 260, 261, 262, 265, 269, 275, 276, 277–78, 279, 283, 317–18, 340n47, 349n23; and Max Planck, 222; as idealist, 318; as reactionary, 299; on descriptions, 253, 270; on natural laws, 252; on physical laws, 252, on science and metaphysics, 206–207
MacLeish, Archibald, 53, 72–73
Mainx, Felix, 15n24
Malik, Charles, 179–83
Mannheim, Karl, 158–60, 164, 166–68, 172, 335n23, 336n27, on social sciences, 162–63
Margenau, Henry, 332n32
Maritain, Jacques, 21, 22, 53, 118, 125, 230, 231, 242, 340n46, 344n16, on metaphysics and intelligibility, 110, 121–23, 124–25, 206 124
Marx, Karl, 80, 155–57, 161, 164–66, 182, 306, 312–14
Marxism, 14, 21, 49, 74, 88, 156–58, 161–65, 166–68, 182, 214, 305
Masaryk, Thomas G., 15
Maxwell, James Clerk, 147, 269
McLaughlin, P. J., 141
McWilliams, James A., 193
Mead, George H., 20, 182
measurement, 133, 206, 218, 231, 264, 311, 358n15
Mehring, Franz, 334n3
Mendel's genetics, 89, 229
Menger, Karl, 16

Merton Robert K., 51, 157–59, 164, 168
metaphysics, 20, 22, 29, 32, 41n58, 46–47, 51, 110–11, 113, 198, 205, 223, 224, 228, 231–35, 237, 256–59, 261, 276–79, 282, 287, 288, 292, 317–18; and common sense, 118–20; as confusion, 78, 206; and positivism, 215, 222, 261–62, 270–73; fallacies, 250–51; idealistic, 301, 307, 315; theory of knowledge, 160, 167–68
Mill, John Stuart, 77
Milton, John, 89, 239
Mitin, Mark Borisovich, 49, 99–100, 101, 137, 165, 174, 297, 300, 306, 308–309, 314, 333n44
Moore, G. E., 78, 282
Morris, Charles, 20, 22, 23, 29, 32, 42, 45, 50, 84, 282–83, 287
Moses, 216
Mumford, Lewis, 34

Næess, Arne, 281
Nagel, Ernest, 20, 30, 32, 48, 51, 282, 283, 287–88, 295
National Science Foundation, 37
natural theology, 341n17
naturalism, 20, 47–48, 287–88
Neurath, Otto, 6, 16, 21, 23, 27, 33, 43, 46–47, 49, 242, 266–69, 271, 272, 273, 284, 324n22, 347n48
Newski (Nevski, Vladimir Ivanovich), 136, 329n8
Newton, Isaac, 86, 90, 106–107, 112, 115, 120–21, 143, 151, 169, 182, 183, 189, 194, 198–99, 202, 206, 222, 252, 250, 275, 281, 287, 289, 293, 295, 298, 300, 305, 312, 356n13; mechanics, 77, 86, 109,

132, 136, 145, 157, 248–49, 252, 262, 290, 294, 319
Nietzsche, Friedrich, 74, 323n7
Nitschmann, Leo 336n6
nominalism, 148–49, 221–22, 259, 309
Northrop, F. S. C., 114, 181–83

O'Connell, William Henry, 192
O'Rahilly, Alfred 141–42
Ockham, William, 216, 217, 220, 221, 259
operational definition, 79, 87, 119, 122, 125, 131, 132, 135, 136, 145, 150–51, 157, 206–207, 216, 255, 256, 264, 268, 272, 284, 301–302; meaning 31, 85, 87, 134, 141, 149, 157, 196
operationalism, 78, 87, 164, 169, 302–303
Osiander, Andreas, 140
Ostwald, Wilhelm, 249–50
Otto, Max, 219

Pearson, Karl, 254, 276, 277
Peirce, C. S., 42 44, 46, 49, 77, 83, 117, 119, 168, 275–78, 282, 284, 287, 324n11; triadic scheme, 84; classifications of the sciences, 115–16
Phelan, Gerald B., 124, 204–205, 339n20
Philipps, Richard Percival, 195
Philosophy of Science Association: Frank as president, 27
philosophy of science, 25 34, 39, 42–45, 46, 50 53, 132, 150, 171, 181, 184, 191 263, 267, 273, 278, 282, 283 287–88, 309, 314; critical, 134, 145; history and, 26; history of, 2, 4, 52; integrated

40; mechanistic, 254; obsolete 82; relation to political philosophy, 183–84; Russian, 14; and political philosophy, 182–83
philosophy: analytical, 78, 281, 287; chance, 133–35, 143, 145; isolationist attitude, 224, 227, mechanistic, 205–206, 298, 310; proper, 222–23, 228–30, 26; relation to science, 87; schools and politics, 86
physical laws. 252; analogy with social law, 125–28, 201–204, 294, 320
physical universe: relation to human behavior and morality, 209–10, 212–15
physics: purging of philosophy, 132, real and reality in, 144–48, 150, 153, 263
Pick, Georg, 7, 9
Planck, Max. 115, 145–46, 222, 264
Plato, 85, 89–91, 158, 171, 180, 195, 209–11, 222, 230, 327n25
Poincaré, Henri, 77, 84, 136, 172, 255–58, 260, 261, 262, 263, 275, 276, 279, 318, on creative imagination, 162, 168, on arbitrariness, 256
Polanyi, Michael, 190–91, 337n10
Pope Leo XIII, 192, 338n9
Popper, Karl. 18n27, 282
positivism, 191, new 77–78, 162, 164, 168, 254, 256, 318, 335; old, 162, old vs. new, 126–27, 282
pragmatics, 85, 91, 325n25
pragmatism, 20, 25, 26, 42–45, 48, 53, 77, 86, 164, 169, 180, 191, 215
prediction, 82–83, 269, 324n22
prime mover, 198–99, 203

Prometheus, 75
psychoanalysis, 157
Putnam, Hilary, 329n2

quality and quantity, 214, 311–13
quantum theory (physics, mechanics), 12, 14, 41n56, 86, 109, 115, 118, 119, 120, 128, 132, 158, 174, 270, 275, 286, 295, 318
Quine, W. V. O., 32, 45–46, 273, 347n51, 353n29

Rabi, I. I., 80
racial segregation, 281
Radcliffe College, 27
Rankine, William J. M., 249, 349n10
reality 131, 141, 169, 173, 180, 194, 240–41, 252, 257–60, 264, 271, 277–78, 280, 288, 308, 312; in the social sciences, 149–50; true picture of, 72–73, 76, 79, 268, 275, 308
Reichenbach, Hans, 14, 17, 26, 27, 32, 44, 281, 329n2, 350n1
Reinhardt, Kurt, 71
relativity. *See* theory of relativity
revelation, 216–18, 220–21, 258
Rey, Abel, 49, 77, 162, 256–57, 259, 350n30
Rhine, Joseph Banks, 295
Rockefeller Foundation, 27, 33
Rothe, Hermann, 6
Rougier, Louis, 281
Rousseau, Jean-Jacques, 280
Russell, Bertrand, 78, 100, 114, 281
Rust, Bernhard, 176, 330n13, 336n6
Rutherford, Ernest, 210
Ryle, Gilbert, 282

Samuel, Herbert, 38, 96–98, 101
Santayana, George, 210

Saturday Review, 33
saving the appearances, 140, 143, 230–31, 233–34
Scheler, Max, 335n23
Schlick, Moritz, 4, 6, 14, 16, 46n63, 49, 262–65, 266, 270, 284
Schmidl-Waehner, Trude, 15n24
school philosophy 41n58, 87–88, 191, 242, 243, 245, 325n28, 347n45
Schrödinger, Erwin 3
Schultze, Walter 330n13
Science of Science Discussion Group 27, 347n51
science: and common sense, 103–104, 108, 118; as instrument, 79–80, 89; as picture of reality, 78–79; bourgeois, 161; collecting facts, 230–31; conscience, 176–77; dehumanization of, 99–100, 121; hierarchy and pyramid, 90–91; humanizing, 105–107, 109, 121, 122; idealist interpretation, 248; metaphysical interpretations of, 87, 88, 155, 157, 161, 178, 180, 222, 229, 241, 260; method, 178; philosophical interpretations of, 109, 179–81, 280, 296–98, 300, 319; purging of philosophy, 132, 134, 142; rationality, of 189–90; rehumanization of, 99; religious interpretation of, 321–22; separation from philosophy, 100, 222–23, 231; specialization, 82; unified, 83, 232, 267–68; vs. humanities, 75, 80–81
scientific empiricism, 283
Sellars, Roy Wood, 47
Semantics, 84–85, 325n25, general, 79, rules, 206
sense data, 139, 145
Shapley, Harlow, 25, 28, 33

Sheen, Fulton, 53, 223, 224
Short Philosophical Dictionary, 300
Significs, 79, 324n12
simplicity, 82, 131, 144, 147, 237–38, 329n1
Simplicius 230, 344n15
Simpson, George Gaylord 294
Sixth Congress of Russian Physicists 14
skepticism 220, 238, 239, 346n35
Snow, C. P., 33, 323n9
social sciences, 166, 167, 181, 237–38, 280; and choice of theory, 178, 235; and social status, 162; objective, 164
socio-cosmic universe, 185, 213, 215
sociology of knowledge, 159–61, 167–69, acceptance of theories, 90, 99, 104, 171, 175, 237
sociology of science: strong program, 52n73
Socrates, 210–11
Söderman, Max, 10
Sorokin, Pitirim, 95–96, 101, 105
Soviet Union: official philosophy of, 99
Spencer, Herbert, 77, 240–42, 346n39
Stace, Walter Terence, 311
Stalin, Joseph, 53, 306, 314
Stallo, John Bernard, 248, 250–51, 254, 255, 257, 318
statistical laws, 90, 107, 209, 293
Stens, Jānis, 333n44
Stevens, S. S., 27
Stevenson, Adlei, 38, 72

Talmud, 212
Tarski, Alfred, 27, 45
technology, 135, 171, 175, 197; and ideology, 89, 160–61; vs. social institutions, 80

theory of relativity, 85, 86, 147, 173, 183, 192, 263, 295, 318, debunking, 157; Soviet objection to, 136; reducing to formal system, 141, general, 147, 245, 270, 318, German compromise on, 175
Third Reich, 99, 139, 146
Thirring, Hans 3
Thomas Aquinas, 109, 119, 182, 193, 198–200, 209, 210, 215–16, 217, 218, 222, 230, 233, 242, 249, 289
Thomism, 22–24, 28–30, 32, 43–44, 49, 53, 54, 76, 86, 109, 110, 120, 121, 125, 152, 158, 182–83, 209, 217, 220, 222, 224, 230, 236, 238, 241–42, 275, 287, 295, 307, 309, 313, 318, 338n9, 343n3, 344n14, 344n16; interpretation of science, 88, 108; laws of nature, 252, 320; matter and form, 249–51, 290; metaphysics, 121, 124, 142, 315; on truth, 257, 264, 308; proof of God, 216, 218; proof of objects, 293
Topitsch, Ernst, 185, 213, 215
Trotsky, Leon, 24
truth, 236, social influences on, 155–56, distortion of, 159; double, 219–21, relativity of 238–39

Uebel, Thomas, 31
uncertainty principle, 128–29
unified field theory, 243–44
unified science. *See* science
Unity of Science Movement, 20, 23, 84
unknowable, 240–41

validity: criterion of, 155, 168, 172, 264, 279; metaphysical, 168; of hypothesis, 277; of statements,

validity (continued)
  160, 167, 216; of theories, 155, 169, 171–72, 175, 178, 264, 320
values, 23, 30–32, 44, 48, 181, 280, 295. See also facts
Vavilov, Sergei, 49, 138, 307, 330n11
Vienna Circle, 16–17, 78, 215, 281–82, 287, 347n48; First, 6
Voltaire, 222, 292, 293, 342n31
von Mises, Richard, 11–12, 14, 18n27, 27, 242
von Neumann, John, 27

Waismann, Friedrich, 16, 282
Walsh, Michael P., 178, 337n14
Wartofsky, Marx W., 39, 329n2
wave function, 147, 285
Weber, Alfred, 167
Weber, Max, 157

Wellmuth, John James, 217
Weltsch, Felix, 15n24
Werkmeister, William H., 332n32
Whewell, William, 326n21
Whitehead, Alfred North, 49, 122–23, 145, 179, 180, 201, 222, 229, on laws of nature, 126–27, 201, on chance philosophy, 133–34, on scientific rationality, 189
Wilson, Edwin H., 48
Wittgenstein, Ludwig, 78, 282
worldview, 43, 73, 221, 266, 280; antiscientific, 129; Communist, 76; evolutionary, 47; mechanistic, 254, 348n55; naturalistic, 47–48; positivistic, 81

Zetkin, Kostja, 15n24
Zilsel, Edgar, 17